RADIATION PROTECTION AND DOSIMETRY

Michael G. Stabin

Radiation Protection and Dosimetry

An Introduction to Health Physics

 Springer

Michael G. Stabin
Department of Radiology and Radiological Sciences
Vanderbilt University
1161 21st Avenue South
Nashville, TN 37232-2675
Email, personal: michael.g.stabin@vanderbilt.edu

ISBN 978-1-4419-2391-2   e-ISBN 978-0-387-49983-3

Printed on acid-free paper.

9 8 7 6 5 4 3 2 1

springer.com

To the four strongest and most beautiful people that I know: Laura, Daniel, Mark, and Julia Stabin.

To Dr. Charles E. Roessler, who taught me not only health physics, but the love of health physics, which I will carry with me always.

Table of Contents

Preface

This text is meant to serve as the basis for a two-course series in the study of radiation protection (a.k.a. "health physics"). The first course would be an introduction to and fast-paced overview of the subject. For some, this is the only course in radiation protection that they will take, and thus all material must be covered in a fairly superficial and rapid fashion. The second course is a more in-depth and applied study of radiation protection, bringing in current materials from the literature, a detailed study of regulations, practice with real-world dose and shielding calculations, and perhaps application in a semester-long student project assigned by the instructor. Several chapters include an additional section of suggested readings and other resources that can be used by the instructor to build such detailed investigations in a second course of this nature. In the first course, the chapter may be basically studied, with reference to the idea that a much richer literature base exists than can be covered in a broad overview of radiation protection. Through exploration of this literature base, and other similar materials that the instructor may be aware of that are not specifically cited, this second, more in-depth course may be developed. A routine part of any good health physics program is a complete course in radiation detection and measurement. My brief overview chapter here cannot provide the depth needed for this subject. Most instructors are familiar with the excellent book by Glenn Knoll on the subject, which I heartily recommend as the basis for such a full course.

Most chapters have some exercises for the student. I have provided the problem and the answer and worked-out solution in all cases. It is not advisable that the students be provided with the worked-out solutions, as this will certainly encourage laziness of approach with some students. These are for the instructor's use; students should be given the problems and correct answers, to guide their work, and then shown individual solutions as needed to clarify difficult areas. All of the problem questions and solutions are available from the RADAR web site (www.doseinfo-radar.com); it is hoped that they will be of value to the instructor. Any errors found in the textbook or problem solutions should naturally be forwarded to the author.

Acknowledgments

I note the input and inspiration of other professors and teachers whose excellence gave me models to follow in my work and teaching. There are, of course, many to whom credit is due. Besides Dr. Charles E. Roessler, to whom the book is dedicated, I single out for notable mention, Dr. Roger A. Gater and Dr. Genevieve S. Roessler of the University of Florida and Dr. Paul W. Frame of the Oak Ridge Institute for Science and Education.

Special thanks to Ms. Mary Ann Emmons, who convinced me to organize my many scattered thoughts about health physics in this volume and make them available to others.

1

Introduction to Health Physics

1.1 Definition of Health Physics

"Health physics" may be defined as the protection of human beings and the environment from the harmful effects of ionizing radiation while permitting its beneficial applications. Health physicists can also be called "specialists in radiation protection." The origin of the term "health physics" is not entirely clear. Dr. Paul Frame of the Oak Ridge Institute for Science and Education suggests:[1]

Health Physics refers to the field of radiation protection. How appropriate the name is has been a matter of some debate (Taylor 1982[2]).

The term Health Physics originated in the Metallurgical Laboratory at the University of Chicago in 1942, but it is not known exactly why, or by whom, the term was chosen. Most likely, the term was coined by Robert Stone or Arthur Compton. Stone was the head of the Health Division, of which Health Physics was one of four sections. Arthur Compton was the head of the Metallurgical Laboratory.

Because the first task of the Health Physics Section was to design shielding for the reactor (CP-1) Fermi was constructing, the original HPs were mostly physicists who were trying to solve health-related problems. This is the crux of the following explanation given by Robert Stone (1946):[3] The term "Health Physics" has been used on the Plutonium Project to define that field in which physical methods are used to determine the existence of hazards to the health of personnel.

A slight variation on this explanation was given by Raymond Finkle, an early Health Division employee (Hacker 1987[4]): The coinage at first merely denoted the physics section of the Health Division. . . . The name also served security: "Radiation protection" might arouse unwelcome interest; "Health physics" conveyed nothing.

1.2 Overview of the Role of Health Physics

The basic function of any health physicist is in the measurement and/or calculation of radiation-related quantities, forming an assessment of the associated risks, and acting in accordance with these findings, either in taking direct action or in communication of appropriate actions to other authorities (if any action is deemed necessary). The health physicist may, in the extreme, be immediately responsible for saving human lives in situations involving very high potential levels of exposure to radiation. Mostly, health physicists monitor and record real or potential exposures, document their findings, and possibly recommend or implement modifications to existing practices to eliminate unacceptable exposures or optimize exposures (limit the exposures to the lowest level possible, with consideration of practical, social, and other factors). They may suggest engineering or behavioral solutions to potential problems involving radiation exposure (i.e., introduction of shielding material or other controls, or changing procedures in a given location). They may be responsible for the tracking of the release of radioactive material from a location and following its movement through the environment, most notably to estimate the ultimate delivery of radiation dose to human beings, but also perhaps to evaluate direct impacts on the environment itself (contamination of the environment, radiation exposure to living species).

1.3   Employment of Health Physicists

Health physicists work in many venues, wherever exposure to radiation may occur. Sometimes the first application that comes to mind is the protection of workers and the public in applications involving nuclear power, but this represents only one form of employment of health physicists. Health physicists also:

- Practice in medical environments
- Work for government agencies
- Are widely employed in many applications involving the monitoring and possible remediation of environmental releases of radioactive materials
- Are instructors in universities and other sites
- Work in the military
- Are employed on university campuses to monitor a wide variety of uses of radiation and radioactive materials
- Work in the space program
- Work for consulting groups to provide any number of services to different clients

1.4   Educational Background

Practicing health physicists generally will have a college education, typically at the baccalaureate or master's level. Many as well are interested in obtaining a doctorate degree, particularly if one's goal includes becoming a university instructor or national laboratory researcher. Many others practice with training provided on the job, or as provided in various professional short courses of a few days' to a few weeks' duration, perhaps without having attained a college degree. The location and strength of university-based training programs in health physics tends to vary. These programs are typically small, often headed by one or two individuals with a modest student base (compared to other disciplines), and as these individuals retire or move, so does the focus of these training programs. Only recently has a formal accreditation process for university health physics programs been established, by the Accreditation Board for Engineering and Technology (ABET, Inc.). Other training programs are offered from time to time by private training firms; typically one-week to several-week intensive short courses are offered to train health physics professionals or technologists in general, or more commonly, very specific topics.

A group at Oak Ridge Institute for Science and Education has performed periodic salary surveys for health physicists and nuclear engineers. Some results from the most recent salary survey for health physicists are shown in a document provided by the Oak Ridge Institute for Science and Education (http://hps.org/documents/). From these data, one notes that education level, years of experience, and health physics certification are all factors that determine the salary level of a health physicist in the current market.

The American Board of Health Physics (ABHP) annually administers an examination to candidates who wish to be considered for health physics certification. The test format has varied over the years. Currently Part I of the exam consists of 150 multiple-choice questions, and can be taken soon after graduation from a recognized health physics program. Part II can be taken with a number of years of experience, and consists of multiple choice, short answer, and mathematical problem solving. ABHP certification is generally not a requirement for employment at most sites, but it can be helpful in finding employment and gaining promotion.

1.5 Interaction of Health Physicists with Other Disciplines

The term "health physics," as noted above, reflects the "marriage," if you will, between the disciplines of biology and physics, although the field is quite interdisciplinary, involving contributions from many fields of science and engineering. Health physicists are most often attracted to the field from either engineering or biology backgrounds. The health physicist must have a strong understanding of basic physics, particularly of primary particles and their interactions with matter. The health physicist must be comfortable with performing relatively complex calculations and correctly expressing the answer in appropriate units. The complexity of the calculations is not carried so much in theoretical aspects; health physics is a relatively straightforward science compared to, for example, nuclear, chemical, or aerospace engineering. The health physicist is expected to have mastered calculus and be able to apply it practically; however, the day-to-day use of such higher mathematics is infrequent. The complexity of most calculations is in the understanding of the physical principles in play, and in the careful manipulation of the appropriate numerical quantities, often with extensive application of fundamental quantities (e.g., the amount of charge on an electron) and extensive conversion of units. This requires patience and attention to detail; small mistakes can lead to huge errors in the final result, because very large conversions are often involved in moving from the atomic world to the world of experiential reality.

Health physicists in daily practice work more like engineers than physicists, but with a strong understanding of the biological consequences of measured or calculated radiation doses. Radiation biology is briefly treated in this text, but is the subject of an entire course of study in any established health physics training program. The human body is marvelously complex, and the interactions of these invisible particles and rays with the cells and tissues of the body create a variety of reactions that we are still struggling to understand completely. At high enough doses of radiation, where effects may be observed in short periods of time (days to weeks), our understanding is well established. At lower levels, it is known that later effects (primarily cancer) may occur, but all relationships of dose and effect are not well understood. Research in this area is ongoing and intensive, and will certainly ultimately lead to a clear understanding of effects on cells and organisms at lower levels of dose and dose rate. For the present, the health physicist must work with these uncertainties remaining unresolved; in the face of this uncertainty, conservative assumptions are usually applied to ensure reasonable protection of workers and the public while allowing technologies involving the use of radiation to continue and develop.

1.6   This Text and Its Relation to a Training Program

I hope that you find the study of the material in this book as enjoyable as I do. I was an engineering student, not really sure what I wanted to do with my career, when I took this strange elective called "Introduction to Radiological Protection" (health physics). As did my instructor (Dr. Charles Roessler, then at the University of Florida at Gainesville), I developed a passion for the study of this material that has not decreased to this day (in fact it continues to increase!). The applications continue to be fascinating and challenging, from use in the medical arts to environmental sciences to protection of society from terrorist activities.

The book is designed to provide a complete academic introduction of all information needed to understand the discipline; however, a few chapters are designed as well to be a guide to the working professional (particularly the chapters on internal and external dosimetry). In these cases, somewhat more technical detail is provided than may be appropriate for use in an introductory semester course on health physics, and many citations are given to other professional resources (books, published articles, electronic data compendia) that should be helpful. It is also suggested that these chapters may be used with these expanded resources to design a second semester course (advanced health physics, health physics design, etc.). Student projects could be developed, related possibly to real facilities (power plants, laboratories, etc.) to implement some of these more advanced ideas. Similar to radiation biology, there is a chapter on health physics instrumentation, but a complete course should be devoted to this study, and the excellent text by Glenn Knoll[5] is suggested as a companion to this work, to expand on this subject.

Endnotes

1. http://www.orau.org/ptp/articlesstories/names.htm#healthphysics.
2. L. Taylor, Who is the father of health physics? *Health Physics*, **42**, 91–92 (1982).
3. R. Stone, Health protection activities of the plutonium project, *Proc. Am. Phil. Soc.* **90**(1), (1946).
4. B. Hacker, *The Dragon's Tail* (University of California, 1987).
5. G. Knoll, *Radiation Detection and Measurement*, 3rd ed. (Wiley and Sons, Hoboken, NJ, 1999).

2

Scientific Fundamentals

2.1 Quantities and Units in Science and Engineering

Science and engineering depend on the quantification of variables, by measurement, calculation, or both. When we have a quantity that expresses some measure of the variable of interest, it is quite useless without the correct units. What is the acceleration on the earth due to gravity? People will often quickly answer "32" or "9.8", but that is a meaningless answer without the appropriate units (ft/s^2, m/s^2). Perhaps I would like the answer (just to be absurd) in furlongs/fortnight^2; if that's what I asked for on a test, and the answer was given as 9.8, it would clearly be wrong (the correct answer, by the way, is 7.1×10^{10} furlongs/fortnight^2).

The quantities that we have to deal with are rooted ultimately in measures of length, mass, and time. *Quantities* are entities as basic as these (length is a quantity) or combinations thereof (velocity has units of length per unit time, acceleration is the rate of change of velocity, and has units of length per unit time to the second power). To become specific, we have to apply numerical values and specific units. A quantity is a phenomenon which may be expressed as a numerical value and a specific unit.

A number of unit systems have been proposed and accepted over the years. The British system of units was accepted for many years in Europe and the United States. Many of the units in this system were tied to commonly encountered, but somewhat arbitrary measures (such as the distance from the thumb of the extended arm to the nose of a king, or the height of the shoulder of a horse). The Conférence Général des Poids et Mesures (CGPM) developed a new system of units in 1960, called the International System of units (SI) (from the French Système International d'unités).

A unit is a selected reference sample of a quantity. There are seven base units in the SI system: the kilogram (kg), meter (m), second (s), ampere (A), kelvin (K), mole (mol), and candela (cd). Combinations of base units form "derived units" (Table 2.1). For example, a Newton is the unit of force required to accelerate a mass of one kilogram one meter per second per second, $1\,\text{N} = 1\,\text{kg m/s}^2$. Derived units may have special names. However some of the special names are restricted to certain quantities; for example, Hz (s^{-1}) is the unit for frequency, but becquerel (s^{-1}) is the unit of activity.

Table 2.1 Common quantities, units, and symbols for the physical sciences.

Quantity	Unit	Symbol
Length	Meter	m
Velocity	Meters per second	m/s
Area	Square meters	m^2
Force	Newton	N (kg m/s^{-2})
Energy (force × distance)	Joule	J (kg m^2/s^{-2})
Absorbed dose	Gray	Gy (1 J/kg ≡ 1 m^2/s^{-2})
Absorbed dose	Rad	rad (100 erg/g = 0.01 Gy)

Derived units can be obtained from the base units using prefixes that multiply the base unit by some power of 10 (Table 2.2).

Examples

1 kilometer = 1 km = 1000 m = 10^3 m.
1 micrometer = 1 μm = 0.000001 m = 10^{-6} m.

2.2 Background Information

Health physics in practice appears much like an engineering science applied at the atomic/molecular level, but with understanding of many aspects of biological and environmental sciences. Students should be familiar with basic principles of mechanics (work, energy, basic dynamics), electricity (nature of electric charges, basic aspects of electrical circuits: this becomes most relevant when understanding radiation detectors), quantum mechanics, statistics, basic chemistry and biochemistry, and biology. That sounds like a tall order to fill. No one should feel that she cannot undertake a study of health physics without formal coursework in all of the above areas. However, all of these areas are touched on in the study of health physics, and at least a basic undergraduate familiarity with all of these subjects will be needed to understand this field.

Table 2.2 Unit prefixes.

Factor	Prefix	Symbol	Factor	Prefix	Symbol
10^{24}	yotta	Y	10^{-1}	deci	d
10^{21}	zetta	Z	10^{-2}	centi	c
10^{18}	exa	E	10^{-3}	milli	m
10^{15}	peta	P	10^{-6}	micro	μ
10^{12}	tera	T	10^{-9}	nano	n
10^9	giga	G	10^{-12}	pico	p
10^6	mega	M	10^{-15}	femto	f
10^3	kilo	k	10^{-18}	atto	a
10^2	hecto	h	10^{-21}	zepto	z
10^1	deca	da	10^{-24}	yocto	y

An introductory study of health physics may be taught at an undergraduate level, although typically in the junior or senior years of study. Training in health physics as a discipline is best taught at the graduate level, with complementary, in-depth courses in radiation detection and instrumentation, radiation biology, and a number of other electives designed to complete a professional understanding of environmental, medical, engineering, or other sciences. Students often enter a study of health physics from either an engineering or biology background. The former students will wish to refresh their memories or undertake some additional reading in biology, chemistry, and related sciences, and students with a biology background may need similar help with the physical sciences, statistics, quantum mechanics, or other mathematical concepts.

2.3 Nature of Matter—Molecules, Atoms, Quarks

The physical world that we see around us is made up of many materials of different qualities. Objects vary in size, color, texture, and behavior. We know that the observable characteristics of matter are ultimately related to the inherent nature of its constituents. Matter is composed of building blocks. The quest for the most fundamental element of these building blocks has been a fascination of mankind since we were able to think. The Greeks coined the term *atomos*, meaning "indivisible" long before we understood even the most basic concepts of atomic structure. Today we know that matter is composed of molecules, which are composed of atoms, but we also know that atoms are not fundamental particles. Atoms are composed of various components (neutrons, protons, electrons) that are themselves composed of other components. Scientists have struggled with many models of the world that exists beyond our ability to see with our eyes (even with the aid of other machines or devices). For many years before we developed our current atomic models, we knew that the components of the atom are electrically charged and that atoms usually exist in an electrically balanced state.

The Greek thinker Empedocle (492–432 BC) thought of matter as divided into four elements: water, earth, air, and fire. The popular music group, Earth, Wind and Fire (formed in 1969 and still performing in the early 2000s) is also known as "The Elements," alluding to this idea. In Empedocle's system, the basic elements are walled in by the forces of love and hate. Other Greeks, specifically Leucippe of Abdere and his student, Democrite of Abdere, suggested that matter was made of particles that had certain qualities:

- Very small and invisible
- Solid
- Eternal
- Surrounded by empty space (to explain material deformations, changes in density, etc.)
- Have an infinite number of shapes (to explain the diversity of material shapes observed in nature)

They thought that the particles were in perpetual motion within observable matter.

Thompson discovered the electron in 1897 and suggested the "plum pudding" model of the atom, in which the atom is a sphere made of some electrically positive substance, with the negative electron embedded in it "like the raisins in a cake."

In 1911, Rutherford proposed the model that we retain today. In this model, all positive charge is concentrated in a small massive nucleus, with the electrons at remote points, orbiting the nucleus. The atom is electrically balanced. In 1913, Geiger and Marsden derived a very ingenious experiment to test this model. They directed alpha particles from a polonium source at a thin gold foil. If Rutherford's model was correct, two phenomena should be observed:

- Most of the alphas should pass through the foil undeflected.
- Those that scattered from a gold nucleus should be observed in a fairly uniform pattern.

Indeed, these expected findings were observed. The majority of alphas passed directly through the foil and were detected in line with the incident beam, but there was also a continuous pattern of scattered αs from 0 to 150°.

The small massive (in the atomic world, anyway) nucleus is made up of positively charged protons and uncharged neutrons. Their mass is approximately 1.7×10^{-27} kg (1 amu) each. The negatively charged electrons, which are tiny by comparison (9.1×10^{-31} kg) revolve around the nucleus in orbits of around 5×10^{-11} m. An empirical relationship that approximately gives the radius of a nucleus of atomic mass number A is:

$$\text{Nuclear radius (m)} = 1.2 \times 10^{-15} A^{1/3}. \tag{2.1}$$

If we drew a proton or neutron with a diameter of 1 cm, the whole atom would have a diameter of more than 10 football fields, with each electron having a diameter of about one human hair.[1] A typical nuclear radius, then, has dimensions of the order of 10^{-15} m. We know that the forces of repulsion between two protons at this distance will be very high. The presence of neutrons, discovered in 1932 by the British physicist Chadwick, accounts for the differences in atomic number and atomic mass, but their presence alone does not explain how these particles can stay together in the nucleus. The answer is found at a lower level of the particle structure. Protons and neutrons (collectively called *nucleons*) are themselves made of particles called quarks. There are six quarks in nature, each with different charges, and six other fundamental particles, called leptons (Table 2.3).

All visible matter in the universe is made of particles in the first generation of matter. The higher generations are unstable, and decay quickly when formed into first-generation particles. Quarks do not exist alone. Composite

Table 2.3 The generation of matter.

	I	II	III
Quarks*	Up ($^2/_3$)	Charm ($^2/_3$)	Top ($^2/_3$)
	Down ($^{-1}/_3$)	Strange ($^{-1}/_3$)	Bottom ($^{-1}/_3$)
Leptons*	Electron (-1)	Muon (-1)	Tau (-1)
	Electron neutrino (0)	Muon neutrino (0)	Tau neutrino (0)

*Charge shown in parentheses.

particles made of quarks are called *hadrons*. There are two types of hadrons: *baryons* and *mesons*. Baryons are composed of three quarks, whereas mesons are composed of one quark and one antiquark. The two important examples of baryons for this course are the proton and neutron. The proton contains two up quarks and a down quark (and thus has a net positive charge of +1); the neutron is made up of one up quark and two down quarks (and thus is electrically neutral). Carrier particles, called *gluons*, are exchanged between quarks to form hadrons. "Strong forces" between quarks in one proton and other protons thus overwhelm the electromagnetic forces of repulsion, and the nucleus stays together.

The classical model of atomic structure also predicted an unstable atom. It is logical to think of electrons rotating around the nucleus, with the attractive force between it and the positively charged nucleus being balanced by the centrifugal force of its circular movement. The problem is that accelerating charged particles constantly radiate energy, so the electrons in this system would be constantly losing energy (because of their radial acceleration), and moving in a decaying orbit and eventually crashing into the nucleus. In 1913, Neils Bohr "colored outside the lines" in order to solve this dilemma. He simply assumed that classical electromagnetic theory was not valid in the defining of atomic structure. He worked from the observation noted in other experiments that excited atoms radiate light only at certain frequencies. Thus, he thought that it would be reasonable to assume that electrons exist only in certain fixed orbits. That thinking led to the idea of an electron's angular momentum being some integer multiple of $h/2\pi$ (h is Planck's constant, 6.6262×10^{-34} J s):

$$mvr = \frac{nh}{2\pi} \tag{2.2}$$

Here m is the mass of the electron, v is its velocity, r is the radius of its orbit, and n is an integer. Now using this assumption, if we return to the idea that in an electron's orbit around the nucleus, the electrostatic forces of attraction are balanced by centrifugal forces, we can now write:

$$\frac{k_0 Z e \times e}{r^2} = \frac{mv^2}{r}. \tag{2.3}$$

where:

k_0 = Coulomb's law constant = $9 \times 10^9 N - m^2/C^2$
Z = atomic number
e = electron/proton charge, $1.6 \times 10^{-19} C$

From Equation (2.2), we can write:

$$v^2 = \frac{n^2 h^2}{4\pi^2 m^2 r^2}. \tag{2.4}$$

and combining these two equations, we can solve for the exact radius as

$$r = \frac{n^2 h^2}{4\pi^2 m e^2 Z k_0}. \tag{2.5}$$

The energy of an electron in any orbit is a sum of its kinetic and potential energy:

$$E = E_k + E_p. \tag{2.6}$$

The kinetic energy (if we assume that relativistic effects are not significant, which is reasonable), is $\frac{1}{2}\, mv^2$. The potential energy can be derived from the relationship:

$$E_p = W = \frac{k_0 Q q}{r} = -\frac{k_0 Z e^2}{r}. \tag{2.7}$$

From Equations (2.3), (2.5), and (2.7), we can derive that

$$E = \frac{-2\pi^2 k_0^2 m Z^2 e^4}{h^2} \frac{1}{n^2}. \tag{2.8}$$

This defines the total energy of the orbit corresponding to the integer value of n. The energy of photons is defined as $E = hv = hc/\lambda$, where c is the speed of light, v is the photon frequency, and λ is the photon wavelength. If an electron falls from a higher to a lower orbit, manipulation of Equation (2.8) shows that the difference in energy levels is released as electromagnetic radiation with frequency:

$$v = \frac{2\pi^2 k_0^2 m Z^2 e^4}{h^3} \left(\frac{1}{n_f^2} - \frac{1}{n_i^2} \right). \tag{2.9}$$

If we wish instead to designate the wavelength, we simply write:

$$\frac{1}{\lambda} = R \left(\frac{1}{n_f^2} - \frac{1}{n_i^2} \right). \tag{2.10}$$

where R is known as Rydberg's constant, and is:

$$R = \frac{2\pi^2 k_0^2 m Z^2 e^4}{c h^3}. \tag{2.11}$$

2.4   Excitation and Ionization

The two processes that are key to understanding almost all of health physics are excitation and ionization. These processes are the basis for our understanding of most radiation interactions of importance, which are relevant to radiation shielding, biological effects, the operation of radiation detection instruments, and many other processes. *Excitation* is the process in which energy absorbed by an electron, raised to an excited state, returns to ground or lower state with emission of electromagnetic (EM) radiation. *Ionization* is the situation in which enough energy is imparted to an electron to remove it from the electric field of the nucleus, creating an "ion pair". After excitation, the excited electron will return to its ground state, and a photon will be released, whose energy represents the difference in the energy levels of the two orbits. This photon may in itself be important in the delivery of energy to the system of interest, but is often of interest as a characteristic emission that may be helpful in identification of the elements in a system.

The *ionization potential* is the amount of energy needed to ionize the least tightly bound electron in an atom of that element. This quantity is generally of the order of a few eV. For example, in Equation (2.10), if you put ∞ for the value of n_2 for a hydrogen atom, you can show easily that the ionization potential for a hydrogen atom is 13.6 eV.

For a photon whose energy exceeds the ionization potential of an atom with which it interacts, it is possible that a "photoelectron" may be formed. This subject is discussed in more detail in a subsequent chapter on photon interactions with matter. In this kind of interaction, all of the photon energy disappears and is gained by the electron, which was formerly in an orbit of the atom. The electron is ejected from the atom, with an energy equal to

$$E = h\nu - IP. \tag{2.12}$$

For example, imagine that an ultraviolet photon of $\lambda = 2500$ Å strikes a potassium atom (IP 4.32 eV). Can we calculate the kinetic energy of the photoelectron? The energy of a photon (as discussed above) is:

$$E = h\nu = h\frac{c}{\lambda}. \tag{2.13}$$

In this example:

$$E = \frac{(6.6262 \times 10^{-34} \, J - s)\left(3 \times 10^8 \frac{m}{s}\right)}{\left(1.6 \times 10^{-19} \frac{J}{eV}\right)(2.5 \times 10^{-7} \, m)} = 4.97 \text{ eV}$$

$$E_{pe} = 4.97 - 4.32 = 0.65 \text{ eV}$$

$$E_{pe} = 4.97 - 4.32 = 0.65 \text{ eV}.$$

2.5 Refinements to the Bohr Atom

As spectroscopes with higher resolving power were developed, we learned that there were more lines visible in the atomic spectra of many atoms than predicted by the simple Bohr model. These additional lines suggest the presence of sublevels of energy within the principal energy levels. Some characteristics of the atom that were added to explain these additional levels include:

- All orbits are not circular, some are elliptical.
- Electrons rotate on their own axes as they orbit the nucleus.
- Electron motion and spin create a magnetic moment.

With these new considerations, three new quantum numbers were defined, in addition to the principal quantum number:

n	Principal quantum number	$1, 2, \ldots$
l	Azimuthal quantum number	0 to $(n - 1)$
m	Magnetic quantum number	$-l$ to $+l$
s	Spin quantum number	$-\frac{1}{2}$ or $+\frac{1}{2}$

The Periodic Table of the Elements (Table 2.4) was derived from consideration of the advanced Bohr model, with the addition of the *Pauli exclusion*

Table 2.4 The Periodic Table of the Elements.

Period	\multicolumn{18}{c}{Group}																
	1	2	3	4	5	6	7	8	9	10	11	12	13	14	15	16	17	18
1	H																	He
2	Li	Be											B	C	N	O	F	Ne
3	Na	Mg											Al	Si	P	S	Cl	Ar
4	K	Ca	Sc	Ti	V	Cr	Mn	Fe	Co	Ni	Cu	Zn	Ga	Ge	As	Se	Br	Kr
5	Rb	Sr	Y	Zr	Nb	Mo	Tc	Ru	Rh	Pd	Ag	Cd	In	Sn	Sb	Te	I	Xe
6	Cs	Ba	La	Hf	Ta	W	Re	Os	Ir	Pt	Au	Hg	Tl	Pb	Bi	Po	At	Rn
7	Fr	Ra	Ac	Rf	Db	Sg	Bh	Hs	Mt	Uun	Uuu	Uub						
	\multicolumn{2}{l}{Lanthanides}			Ce	Pr	Nd	Pm	Sm	Eu	Gd	Tb	Dy	Ho	Er	Tm	Yb	Lu
	\multicolumn{2}{l}{Actinides}			Th	Pa	U	Np	Pu	Am	Cm	Bk	Cf	Es	Fm	Md	No	Lr

principle, which states that no two electrons in an atom may have the same set of four quantum numbers. Consider the first element, hydrogen, which has just one electron. The spin quantum number may be either $+1/2$ or $-1/2$. The principal quantum number must be 1, so the azimuthal and magnetic quantum numbers must be 0. The second element, helium, has two protons in the nucleus, thus there are two electrons. The first electron may have the same set of quantum numbers as hydrogen. The second electron must be different, but the only place for difference can be in the spin, and this uses up all of the possibilities for the principal quantum number $n = 1$. Considering the third element, lithium, which has three electrons, principal quantum numbers $n = 1$ and $n = 2$ are possible. The third electron will have $n = 2$, and have an azimuthal quantum number of $l = 0$ or $l = 1$; that is, the orbit may be circular or elliptical. When $l = 1$, the magnetic quantum number m can be -1, 0, or $+1$. Each of these states may have electrons with spin $+1/2$ or $-1/2$. Therefore, eight different electrons, each with a unique set of quantum numbers, may exist, and these possibilities are exhibited in the elements Li, Be, B, C, N, O, F, and Ne, which include elements with atomic numbers 3–10.

2.6 Characteristic X-Rays

The idea that electrons exist in discrete energy levels in the orbits around the nuclei of atoms leads to the idea that transitions of electrons between those energy levels will result in the release of discrete amounts of energy. This energy is in the form of electromagnetic radiation that is defined as "X-rays" (a formal definition of the different forms of electromagnetic radiation is given later). *Characteristic X-rays* are those discrete forms of radiation that are emitted from atoms when electron transitions occur in the orbitals of electrons. The names of these transition X-rays are defined for identification. For example, when a K shell electron is filled by an L shell electron, the X-ray is called a "K-alpha" (K_α) X-ray, and this is reflected in most tables of decay data for radionuclides. Other examples:

K shell filled by L e^- — K_α
K shell filled by M e^- — K_β
L shell filled by M e^- — L_α
L shell filled by N e^- — L_β

Other examples follow from this straightforward naming system.

Other, more complicated models of the atom, for example, Schrodinger's 1933 wave description of e^- behavior, have been given to more carefully and accurately describe the probability of finding an e^- at a given distance from the nucleus. Understanding of these principles is necessary for complete description of all theoretical considerations, but the Bohr model explains most phenomena well, and this is all that is necessary for an adequate practice of health physics. The science of "logical positivism" may be applied here; this is a form of empiricism that bases all knowledge on perceptual experience (not on intuition or revelation), and allows for the use of the simplest model that explains all of the available data.

Some other definitions that we introduce at this point are those of the *isotope*, which describes different species of the same element made up of nuclei with different numbers of neutrons, of the *isotone*, which describes different atoms with the same number of neutrons, but with different numbers of protons (not as widely quoted), and of the *atomic mass unit*, which is just a definition of mass, with one amu $= 1.6604 \times 10^{-24}$ g, which is 1/12 the mass of a ^{12}C atom.

2.7 Binding Energy

If you could actually weigh an atom, you would find that it weighs less than the sum of its constituent parts. The difference between the sum of the rest masses of the constituent parts and the actual mass observed may be expressed in mass units, or using Einstein's well-known $E = mc^2$ relationship, equally in mass units. The "missing" mass may be thought of as the energy that is "holding" the nucleus together.

For example, for ^{17}O, we can calculate the mass of the individual protons and neutrons in the nucleus:

$$W = 8(1.007825) + (17 - 8)(1.008665)$$

$$W = 17.14059 \text{ amu.}$$

This value would be 17.14102 if we included the orbital electrons, and thus calculated the mass of an atom (rather than a nucleus) of ^{17}O. The actual mass, as found experimentally and reported in tables of the elements is 16.999132 amu. The *Mass Defect*, $\delta = W - M$, is thus 0.1419 amu. Using the $E = mc^2$ relationship, the binding energy multiplied by 931.5 MeV/amu gives the mass defect in energy units, so we can also say that for ^{17}O the Binding Energy (BE) is:

$$BE = 131.7 \text{ MeV.}$$

We might also express the binding energy per nucleon, by dividing the binding energy by (the number of nucleons minus 1). In the ^{17}O nucleus (8 protons and 9 neutrons) 131.7/16 = 8.23 MeV/nucleon. If one plots values of

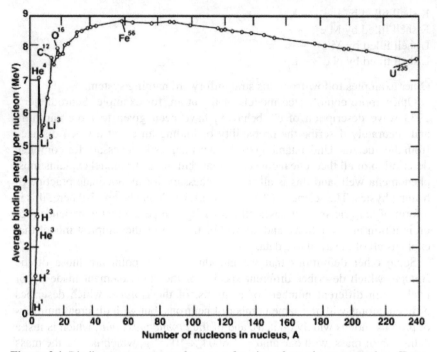

Figure 2.1 Binding energy per nucleon as a function of atomic mass number. (From http://en.wikipedia.org/wiki/Binding_energy.)

the binding energy per nucleon for all of the elements, an interesting picture emerges (Figure 2.1). The binding energy per nucleon reaches a maximum at mass numbers of about 30–40. This explains why large quantities of energy are released in the *fission* of heavy nuclei (e.g., ^{238}U ($Z = 92$), which may split into two elements of mass number each around 80–140, or the *fusion* of light elements (e.g., H) into elements of higher mass number. In each case, if we can capture the energy released, we can in theory provide energy to societies for consumption. This theory has been a reality for fission reactors for many years, but has still only been shown in limited research settings for fusion reactors (this subject is explored in more detail in a later chapter).

2.8   The Chart of the Nuclides

For health physicists, the Chart of the Nuclides is the companion to the Periodic Table of the Elements. This chart shows all known isotopes of the elements and a brief summary of their properties. Figure 2.2 shows an excerpt from the 15th edition of *Nuclides and Isotopes* by Parrington et al.[2] The grey-shaded blocks are the stable isotopes of each element; others are radioactive isotopes. For each element, at the far left, the element name, atomic weight, symbol, and thermal neutron absorption cross-section (defined in a later chapter) is given. For the stable isotopes running across each row, the atomic percent abundance is shown under the name, and for radioactive elements the nuclide's half-life is given. Other data include nuclear reaction cross-sections, decay energies, modes of decay, and other information. The chart serves as a handy reference to obtain important basic information about any radioactive nuclide.

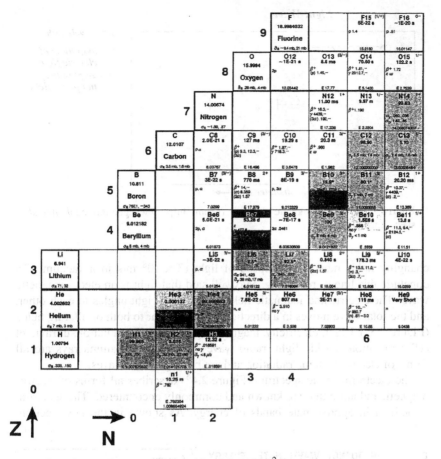

Figure 2.2 Excerpt from *Nuclides and Isotopes*, 1996.[2]

In addition, decay products can easily be traced once the mode of decay is known (as covered in the next chapter) and the properties of the chart are understood. For example, because ^{11}C decays by positron decay, it is easy to see that the species it decays to is (stable) ^{11}B, one square down and to the right. Lithium-9, on the other hand, decays by beta-minus decay to ^{9}Be, one square up and to the left. Sometimes these decay products are themselves radioactive, and so the decay of such radioactive "chains" can be quickly traced, to find the half-lives and basic decay properties of the various species. Complete decay data, however, must be found in more complete paper or electronic compendia (e.g., *ICRP Publication* 38[3], the RADAR decay data compendium[4]).

2.9   Some Elements of Quantum Theory

2.9.1 Electromagnetic Radiation

James Clerk Maxwell first explained in the mid-1800s that as changing magnetic fields create electric fields, and as changing electric fields create magnetic fields, changing electric and magnetic fields influence each other, and these

Light Wave

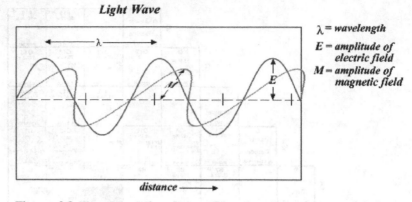

λ = *wavelength*

E = *amplitude of electric field*

M = *amplitude of magnetic field*

distance ⟶

Figure 2.3 Electromagnetic waves. (From http://www.geo.mtu.edu/rs/back/spectrum/).

changing fields move at the speed of light (3×10^8 m/s in a vacuum). To conclude this line of reasoning, Maxwell said that light is an electromagnetic wave. The electric and magnetic fields oscillate at right angles to each other, and the total wave moves in a direction perpendicular to both oscillating fields (Figure 2.3[5]). Light, electricity, magnetism, infrared (IR) radiation, ultraviolet (UV) radiation, visible light, radio waves, X-rays, and gamma rays are all forms of electromagnetic radiation and are propagated as waves.

The electromagnetic spectrum (Figure 2.4[6]) describes all forms of electromagnetic radiation that are known and commonly encountered. The spectrum is defined in approximate bands of energy. As shown in the next section,

CLASS	FREQUENCY	WAVELENGTH	ENERGY
γ	300 EHz	1 pm	1.24 MeV
HX	30 EHz	10 pm	124 keV
	3 EHz	100 pm	12.4 keV
SX	300 PHz	1 nm	1.24 keV
	30 PHz	10 nm	124 eV
EUV	3 PHz	100 nm	12.4 eV
NUV			
NIR	300 THz	1 μm	1.24 eV
MIR	30 THz	10 μm	124 meV
FIR	3 THz	100 μm	12.4 meV
EHF	300 GHz	1 mm	1.24 meV
SHF	30 GHz	1 cm	124 μeV
UHF	3 GHz	1 dm	12.4 μeV
VHF	300 MHz	1 m	1.24 μeV
HF	30 MHz	1 dam	124 neV
MF	3 MHz	1 hm	12.4 neV
LF	300 kHz	1 km	1.24 neV
VLF	30 kHz	10 km	124 peV
VF	3 kHz	100 km	12.4 peV
ELF	300 Hz	1 Mm	1.24 peV
	30 Hz	10 Mm	124 feV

Legend:

γ = Gamma rays
HX = Hard X-rays
SX = Soft X-Rays
EUV = Extreme ultraviolet
NUV = Near ultraviolet
Visible light
NIR = Near infrared
MIR = Moderate infrared
FIR = Far infrared

Radio waves:
EHF = Extremely high frequency (Microwaves)
SHF = Super high frequency (Microwaves)
UHF = Ultra high frequency
VHF = Very high frequency
HF = High frequency
MF = Medium frequency
LF = Low frequency
VLF = Very low frequency
VF = Voice frequency
ELF = Extremely low frequency

Figure 2.4 The electromagnetic spectrum. (From http://en.wikipedia.org/wiki/Electromagnetic_spectrum.)

energy has been related to the frequency and wavelength of the radiation, as is commonly defined for any wave phenomenon.

2.9.2 Wave/Particle Duality of Nature

Near the beginning of the 20th century, Max Planck suggested that energy is emitted or absorbed by atoms only in discrete chunks (which he called *quanta*). We thus defined a "quantum behavior" of energy (and by relation, matter). Albert Einstein suggested that we consider light as "packets" of energy called *photons*. As experiments proliferated based on our expanding nature of the universe, we often found that light sometimes behaves like a wave and at other times like a particle, depending on exactly how the experiment is designed. Other small particles, like the electron, similarly can exhibit either a wavelike or particlelike behavior in different situations. Einstein described the relationship between the energy (E) of a photon and its frequency (v) and wavelength (λ):

$$E = hv = \frac{hc}{\lambda}$$

Here, h is Planck's constant, as in Equation (2.2) above, and c is the velocity of light in a vacuum, cited in the previous section. Photons of high energy have high frequency and small wavelengths, and photons of low energy have low frequency and long wavelength.

2.9.3 The Heisenberg Uncertainty Principle

An important principle of quantum mechanics was defined by Werner Heisenberg. This principle states that it is not possible to simultaneously know exactly both the position and momentum of a particle. The better the position is known, the less well the momentum is known (and vice versa). The principle can be stated mathematically as

$$\Delta_x \Delta_{\rho x} \geq \frac{1}{2}\hbar$$

In this equation, Δ_x is the uncertainty in the position of the particle, $\Delta_{\rho x}$ is the uncertainty in the particle's momentum, and $\hbar = h/2\pi$.

Endnotes

1. Taken from *The Particle Adventure* (http://particleadventure.org/particleadventure/index.html, The Particle Data Group, 2002).
2. J. R. Parrington, H. D. Knox, S. L. Breneman, E. M. Baum, and F. Feiner, *Nuclides and Isotopes*, 15th edition (Lockheed Martin/GE Nuclear Energy, Schenectady, NY, 1996).
3. International Commission on Radiological Protection, *Radionuclide Transformations—Energy and Intensity of Emissions. ICRP Publication* 38 (Pergamon Press, Oxford, 1983).
4. M. G. Stabin and C. Q. P. L. da Luz, New decay data for internal and external dose assessment, *Health Phys.* **83** (4), 471–475 (2002).
5. http://www.geo.mtu.edu/rs/back/spectrum.
6. http://en.wikipedia.org/wiki/Electromagnetic spectrum.

3

Radioactive Atoms—Nature and Behavior

We understood in the previous chapter that nature is made of atoms whose configurations vary. In radiation protection, our principal focus is on a particular subset of atoms that are by nature unstable. Our interest extends as well to protection of persons and the environment from radiation that may not have come from a radioactive atom (e.g., a radiation-producing machine). An important basic distinction in our discussion is the difference between radioactivity and radiation. *Radioactivity* is the process of spontaneous nuclear transformation that results in the formation of new elements. *Radiation* includes the various forms of particles and/or rays that are emitted by these atoms (and perhaps other secondary processes) during this nuclear rearrangement. Some atoms (in nature or artificially derived) are stable, and some are unstable. Unstable atoms rearrange themselves to achieve stability. The principal guiding issues defining which atoms are stable and which are not stable, and how stability is achieved (what transitions occur and what particles or rays may be emitted) are determined by

- The neutron/proton ratio within a nucleus
- The mass–energy relationships of parent, progeny, and emitted particles

Among the smaller nuclei in nature, one finds that a stable arrangement of neutrons and protons in the nucleus involves equal numbers of neutrons and protons in many cases. For example, ^{12}C, perhaps the most fundamental building block of living organisms, contains 6 protons and 6 neutrons. Nitrogen is principally ^{14}N, which has 7 protons and 7 neutrons. As the atomic number increases, however, we find that stable atoms are those that have more neutrons than protons. Figure 3.1 shows a comparison of a line of identity for a plot of the number of protons versus the number of neutrons and the approximate location of the number of neutrons and protons for stable elements (the curved line above the line of identity in Figure 3.1).

Every student who has taken even a high school level chemistry course has become familiar with the Periodic Table of the Elements, which shows groupings of all of the elements according to their chemical properties. The basic principles behind how atoms are organized and electron structures are established (which all leads to the chemical behavior of the elements) was covered in the previous chapter. Another table of the elements exists, which

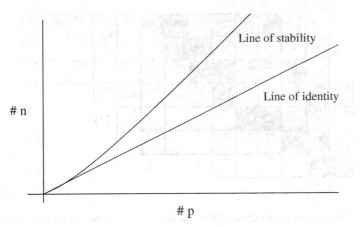

Figure 3.1 Theoretical plot of all nuclides, showing the line of identity and line of stability.

shows all isotopes of all elements, indicating whether they are stable or radioactive, and, if radioactive, some basic information about the particular isotope of the particular element (its half-life, defined later, modes of decay, and other information). This chart is called the Chart of the Nuclides. The chart is arranged opposite to the order of Figure 3.1 in that the number of protons is organized on the vertical axis (ordinate) and the number of neutrons on the horizontal axis (abscissa). A schematic view of the Chart of the Nuclides is shown in Figure 3.2. Now the "line of stability" curves downward from the line of identity.

A portion of the chart, as distributed by Lawrence Berkeley Laboratory[1] is shown in Figure 3.3. In this presentation, the black squares are the naturally abundant species, and none are radioactive. So the "line of stability" is not really a line, but a general region where stable species exist. Note that certain forms of radioactive decay exist above the area of stability and others exist below it. Each box, which represents a particular isotope of a particular element, has some more information about the species, which is also defined in the chart legends. Another representation of the chart is as a three-dimensional

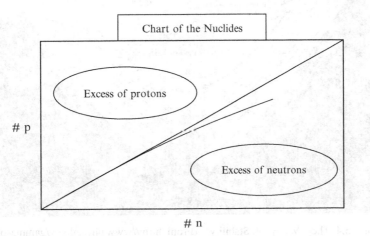

Figure 3.2 Schematic layout of the Chart of the Nuclides.

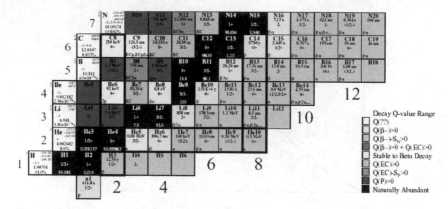

Figure 3.3 Portion of the Chart of the Nuclides, with legend. (From http://ie.lbl.gov/toi/pdf/chart.pdf.)

plot, showing the stable elements at the bottom of the so-called "valley of stability" (Figure 3.4).

We develop some of this information shortly. We now discuss the modes of nuclear rearrangement that atoms use to achieve stability, and the particles or rays that are emitted in the process of rearrangement. As alluded to above, stability is achieved by emitting particles and/or EM radiation. The types of radiation are named after letters of the Greek alphabet (alpha particles, beta particles, gamma rays) or other naming conventions that we define.

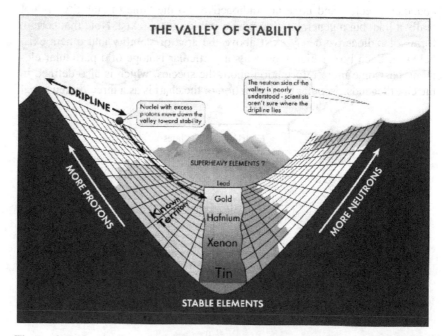

Figure 3.4 The "Valley of Stability." (From http://www.phy.anl.gov/gammasphere/pub/logos_98.html.)

3.1 Alpha Emission

Alpha emission is one mode of nuclear decay that occurs in nuclei that have an excess of protons in their nucleus (i.e., they will be above the line of stability on the Chart of the Nuclides). We can also say that the neutron-to-proton ratio in these nuclides is low. One might also say, for the majority of alpha emitters, that they simply have too many nucleons in general. The heavier alpha emitters can only reach the line of stability through a series of alpha and beta-minus emissions, as we show later. Alpha emission is usually considered first in discussions of modes of decay, but only because of the position of the Greek letter alpha (α) in the Greek alphabet. Alpha emission is not as commonly encountered as other modes of decay, and generally occurs only in heavier nuclei. There are only a few naturally occurring α emitters under $Z = 82$ (e.g., ^{147}Sm, ^{148}Sm ($Z = 62$), ^{152}Gd ($Z = 64$)). An alpha particle is comprised of 2 neutrons and 2 protons, and is really a helium nucleus:

$$2 \, {}_{0}^{1}n + 2 \, {}_{1}^{1}p = {}_{2}^{4}\text{He nucleus}$$

An example of alpha decay is the decay of ^{226}Ra to ^{222}Rn:

$$^{226}_{88}\text{Ra} \rightarrow \, ^{222}_{86}\text{Rn} + \, ^{4}_{2}\alpha$$

Radon-222 itself also undergoes alpha decay (you should be able to easily identify the species that it decays to, by following the example above and accounting for the number of protons and neutrons in the product). Actually, ^{226}Ra and ^{222}Rn are part of a long series of decay products that started with ^{238}U, as discussed later. Every radioactive decay does not lead to a stable product; many times the product itself is radioactive (as in this example) and decays to another product that itself may or may not be stable. As an aside, ^{222}Rn is an extremely important radionuclide in our exposures to natural sources of radiation. Radon is a noble gas, and thus moves through environmental systems (and homes) via diffusion after being formed by decay of ^{226}Ra. We have much more to say about ^{222}Rn in subsequent chapters.

We noted above that modes of decay are related to neutron/proton ratios, and also the mass/energy relationships between the reactants, currently known affectionately as "parent" and "progeny"[2]. If we plot a diagram of the energy levels in and around a nucleus, the potential drops off with distance from the nucleus, where a large reservoir of positive charge exists, and increases dramatically as we approach the nucleus. Within the nucleus, an "energy well" exists that essentially traps positively charged entities within the nucleus unless their kinetic energy exceeds the level at the surface of the nucleus. In heavier nuclei, electrostatic repulsive forces are greater, but the kinetic energy still must be sufficient to overcome the potential barrier at the surface of the nucleus, and this level is not observed in emitted alpha particles. Alpha emission is thus explained as the α particle "tunneling" (Figure 3.5) through the potential barrier (see Evans' classic book, *The Atomic Nucleus*[3]).

The illustration represents an attempt to model the alpha decay characteristics of ^{212}Po, which emits an 8.78 MeV alpha particle with a half-life of 0.3 microseconds. The Coulomb barrier faced by an alpha particle with this energy is about 26 MeV, so by classical physics it cannot escape at all. Quantum mechanical tunneling gives a small probability that the alpha can penetrate the barrier. To evaluate this probability, the alpha

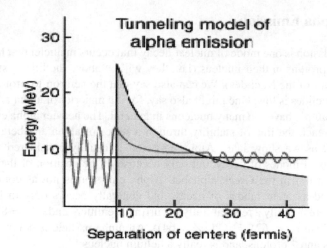

Figure 3.5 Alpha "tunneling" through the nucleus. (From http://hyperphysics.phy-astr.gsu.edu/hbase/nuclear/alptun.html.)

particle inside the nucleus is represented by a free-particle wavefunction subject to the nuclear potential. Inside the barrier, the solution to the Schrodinger equation becomes a decaying exponential. Calculating the ratio of the wavefunction outside the barrier and inside and squaring that ratio gives the probability of alpha emission.

Still, the kinetic energy must exceed 3.8 MeV for these heavy nuclei, according to theoretical calculations. The energy equation describing α decay is:

$$M_p = M_d + M_\alpha + 2M_e + Q$$
$$Q = M_p - M_d - M_\alpha - 2M_e.$$

Here, M_p is the mass of the parent, and M_d is the mass of the progeny, M_α is the mass of the α particle, M_e is the mass of an electron, and Q is the energy released in the reaction. For the ^{226}Ra example above:

$$Q = 226.025 - 222.0176 - 4.0015 - 2(0.00055)$$
$$Q = 0.00523 \text{ amu} = 4.78 \text{ MeV}$$

Radium-226 will decay either with or without an accompanying γ-ray emission. With the γ emission (0.186 MeV, 3.6% of decays), the α particle has an energy of about 4.6 MeV). When there is no γ emission in this case, the α particle has the full energy of 4.78 MeV, and we can look also at the energy of the recoil nucleus from a simple consideration of conservation of energy and momentum:

$$Q = MV^2/2 + mv^2/2$$
$$MV = mv$$
$$V = \frac{mv}{M}$$
$$Q = \frac{Mm^2v^2}{2M^2} + \frac{mv^2}{2}$$

$$E = \frac{mv^2}{2}$$

$$Q = E\left(\frac{m}{M} + 1\right)$$

$$E = \frac{Q}{1 + m/M}$$

$$E = \frac{4.78}{1 + 4/222} = 4.6954 \text{ MeV}$$

$$E_{\text{recoil}} \approx 0.088 \text{ MeV}$$

Note that in the above calculations, a large number of significant figures was used in some cases. This is necessary in this kind of calculation, and values such as the mass of an electron are known to a high degree of precision. In general, caution should be used in the expression of results with a reasonable number of significant figures.

An important consideration in any kind of nuclear transition is whether the emitted particles are monoenergetic or emitted with a spectrum of energies. This has implications for detection and identification of nuclides, for radiation dosimetry, and shielding considerations. Alpha emissions are monoenergetic. We noted in the above example that most ^{226}Ra α particle decays involve 4.78 MeV α particles, and a few percent will be observed as 4.6 MeV α particles. When a 4.78 MeV α particle appears from a ^{226}Ra decay, however, it will always be exactly 4.78 MeV. Small differences in the energy levels of the ^{222}Rn nucleus here caused the α with another energy to appear, with the γ-ray representing the difference between the progeny's excited and ground states. Some nuclides may also emit more than one group of α particles.

We spend entire chapters on radiation interactions and dosimetry, but we can make some general comments here about the different forms of radiation that we study in this chapter. Alphas are very heavy, compared to other forms of radiation. You may not think that 6.64×10^{-27} kg is very heavy, but in the atomic world it is (compare it to the mass of an electron, 8.68×10^{-30} kg, nearly 800 times lighter). Alpha particles also have a 2+ charge, thus they do not penetrate far into matter. Alpha particles are like huge bulldozers plowing into material, whereas electrons behave more like bullets, bouncing around freely while giving up energy. We explore this in more detail in the next chapter. Alpha particles cannot even penetrate the outer protective layer of dead skin, and so are an internal, not an external hazard.

3.2 Positron Emission

To follow the Greek alphabet, we would next cover beta (minus) decay. Instead, however, we consider two other forms of radioactive decay that occur in nuclei which have an excess of protons in their nucleus, which are, again above the line of stability on the Chart of the Nuclides, and have a low neutron-to-proton ratio. The next mode of decay for these nuclides that we study is positron emission. These nuclides attain stability by releasing a *positron*, which is a positively charged electron. It has the same mass as an electron (0.000548 amu) and same charge (1.6×10^{-19} C), except that the charge is positive. An example of positron decay (which is quite important in medical applications) is the decay of ^{18}F to ^{18}O, with the emission of a positron:

$$^{18}_{9}F \rightarrow {}^{18}_{8}O + {}^{0}_{+1}\beta + \nu$$

Positrons have a transitory existence in nature. They are a form of antimatter; they are the antimatter counterpart of the electron. Matter and antimatter annihilate each other when they come together. Matter is of course comprised of atoms, and electrons are everywhere. Positrons thus quickly combine with a free electron in nature, and the two particles are annihilated. All of the mass of the two particles is converted to energy, and two photons whose energies are equal to the masses of the two particles are formed of energy 0.000548 amu $\times$ 931.5 MeV/amu = 0.511 MeV.

To be energetically possible,

$$M_p = M_d + M_{\beta+} + M_e + Q.$$

For ^{18}F,

$$Q = 18.00094 - 17.99916 - 2\,(0.000548)$$

$$Q = 0.0068 \text{ amu} = 0.633\,\text{MeV}.$$

The positron energy is 0.633 MeV (maximum, 0.25 MeV average).

Positrons are emitted from the nucleus. They do not exist in the nucleus, but are postulated to be formed as

$$^{1}_{1}p \rightarrow {}^{1}_{0}n + {}^{0}_{+1}e + \nu$$

Positrons, unlike alpha particles, are emitted with a continuous energy spectrum, ranging from 0 to the maximum theoretical β energy. To satisfy conservation laws, $\beta+$ decay is postulated to be accompanied by the emission of a particle called a *neutrino*. The neutrino (ν) carries the energy that the $\beta+$ does not (i.e., 0.633 MeV minus the $\beta+$ energy). The neutrino has no charge, and a vanishingly small mass. Every distribution of positrons (and beta particles, as we show soon) has its own characteristic shape. The distribution of the ^{18}F positron is shown in Figure 3.6.

Positrons penetrate farther into matter than α particles; this is very dependent on the positron energy. We discuss this in more detail when we discuss beta (minus) emitters later. Positrons may penetrate the skin's dead layer and give an appreciable dose to skin, so they may be an external hazard, and are always a potential internal hazard. In addition to the positron itself, we must always consider also the creation of the 0.511 MeV γ-rays (two per positron emitted). As we show in subsequent chapters, it is not too difficult to stop positrons (or beta minus emitters) with a small amount of material. The 0.511 MeV photons, however, are very penetrating and must be shielded separately. Neutrinos, on the other hand, are absolutely no hazard at all. They have a vanishingly small mass (until only recently it was not known if they had a nonzero mass), and are not charged. Thus, they do not interact much at all with matter. They are therefore very hard to detect (they do not interact with detectors very well), but because they do not interact with human tissue, they are not a hazard with which to be concerned.

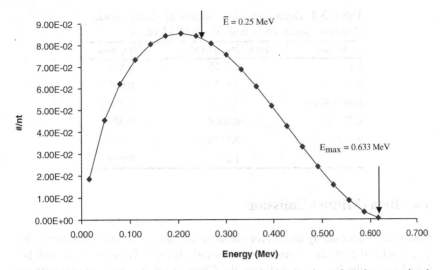

Figure 3.6 F-18 positron spectrum. The ordinate is the number of positrons emitted per nuclear transition.

3.3 Orbital Electron Capture

Another way to resolve a low neutron-to-proton ratio does not involve an emission of a particle at all, but the absorption of a particle by the nucleus. In orbital electron capture, the same nuclear imbalance exists as for alpha and positron emission, but the relationship $M_p > M_d + 2M_e$ is not satisfied. In these cases, an orbital electron may combine with a proton in the nucleus to form a neutron:

$$_1^1 p + \, _{-1}^0 e \; \rightarrow \; _0^1 n + \nu$$

The energy balance equation that applies is now $M_p + M_e - \phi = M_d + Q$, where ϕ is the binding energy of the captured electron.

All these three modes of decay compete with one another within the nuclei of different atoms. Note that the same transformation of nucleons occurs here as occurs for positron decay. Whether one mode or the other is predominant depends on the mass/energy relationships between the parent and progeny atoms. Some nuclides decay by both positron emission and electron capture, in various proportions.

An example reaction of orbital electron capture is ^{22}Na, which decays to ^{22}Ne by either electron capture ($\sim$10%) or positron emission ($\sim$90%). After an electron capture event occurs, an additional $Q = 3.352\,\text{MeV}$ of energy remains. A 1.274 MeV photon is observed with about 100% abundance (in addition to the two 0.511 MeV photons emitted per positron, thus 180% of the time). The remaining energy ($3.352 - 1.274\,\text{MeV} = 2.078\,\text{MeV}$) is carried off by a monoenergetic neutrino. Neutrinos from positron decay carry off various amounts of energy, depending on the energy of the positron.

One other form of emission will occur after electron capture. The electron was captured from a particular electron orbit, usually the K shell. This vacancy will be filled by other outer shell electrons, resulting in the emission of characteristic X-rays. The full decay scheme for ^{22}Na is shown in Table 3.1 (the notation $\gamma +/-$ is used for the 0.511 annihilation photons).

Table 3.1 Decay data for sodium-22 decay mode: electron capture/$\beta+$, half-life: 2.6087 years.

Emission	Mean Energy (MeV)	Frequency
$\beta+$	0.2155	0.8984
$\beta+$	0.8350	0.0006
Auger-K e−	0.0008	0.0920
K X-ray	0.0008	0.0013
$\gamma +/-$	0.5110	1.7979
γ	1.2745	0.9994

3.4   Beta (Minus) Emission

Now we consider decay modes for nuclides that have an excess of neutrons in their nuclei; that is, the neutron-to-proton ratio is high. These nuclides will be found below the line of stability on the Chart of the Nuclides. Theory states that the neutron/proton ratio is enhanced as a neutron spontaneously converts to a proton plus an electron.

$$ {}^{1}_{0}n \rightarrow {}^{1}_{1}p + {}^{0}_{-1}e $$

Theoretical considerations preclude the existence of an electron within the nucleus. Nonetheless, the phenomenon observed is that a nucleus with one more proton and one less electron is observed after the transition, and a particle that is in all aspects equivalent to an orbital electron is emitted at high energies from the nucleus of the atom. To be energetically possible,

$$ M_p > M_d + M_e $$

$$ M_p = M_d + M_e + Q. $$

An important beta emitter is ^{32}P (phosphorous-32):

$$ {}^{32}_{15}P \rightarrow {}^{32}_{16}S + {}^{0}_{-1}\beta + 1.71 \text{ MeV} $$

The Q value for the reaction can be easily obtained by subtracting the mass of the two atoms from each other: $Q = 31.973907 - 31.97207 = 0.001837$ amu = 1.71 MeV.

The recoil nucleus energy is negligible, due to the large difference in mass between the nucleus and the ejected electron, so the energy available to the reaction is 1.71 MeV. Like positrons, β^- particles are emitted with a continuous energy spectrum, ranging from 0 to the maximum theoretical β energy. As with positron emission, β^- emission is accompanied by the emission of a neutrino (actually called an antineutrino) that carries away the energy not borne by the β particle:

$$ {}^{1}_{0}n \rightarrow {}^{1}_{1}p + {}^{0}_{-1}e + \nu $$

A number of important β^- emitters, for example, ^{32}P, ^{3}H, ^{14}C, ^{90}Sr, and ^{90}Y, emit no γ rays, and are called *pure beta* emitters. Another class of beta emitter, *beta–gamma* emitters undergo nuclear rearrangement resulting in the progeny nucleus being in an excited state, with excess energy appearing as a γ ray:

$$ {}^{131}_{53}I \rightarrow {}^{131}_{54}Xe + {}^{0}_{-1}\beta + \nu + 0.968 \text{ MeV} $$

The most prominent gamma ray from ^{131}I has an energy of 0.365 MeV. Some nuclides emit complex β^- ray spectra, including more than one distinct group of β^- particles. An example is K-42:

$$^{42}_{19}K \rightarrow\ ^{42}_{20}Ca +\ ^{0}_{-1}\beta$$

Most of the decay scheme (82%) has a β^- particle with a maximum energy of 3.55 MeV, whereas the other 18% has a β^- particle with a maximum energy of 2.04 MeV, and is accompanied by a gamma emission at 1.53 MeV.

As with positrons, β^- particles penetrate farther into matter than αs: this is very dependent on β energy and varies from perhaps 0.1–2.5 MeV (α energies vary only from about 4–8 MeV). Some β emitters are not external hazards (e.g., ^{3}H, ^{14}C, ^{35}S); others can penetrate the skin's dead layer and give an appreciable dose to the skin. They are always a potential internal hazard. Shielding of βs gives rise to EM radiation called *bremsstrahlung* radiation, discussed later.

3.5 Gamma Ray Emission

Gamma rays (γ) are monochromatic electromagnetic (EM) radiations emitted from the nuclei of excited atoms after radioactive transformations. Several types of EM radiation are of interest to health physics. They are identical in nature to all other forms of EM radiation (visible light, infrared radiation, ultraviolet radiation, etc.) and are distinguished not by their energies, but by their origins:

1. γ ray: originates in the nucleus
2. X-ray: originates in electron orbital shells of atom
3. Annihilation photons: from positron/electron annihilation (always 0.511 MeV)
4. Bremsstrahlung radiation: from deceleration of energetic electrons in matter

These precise distinctions are not held by all working in this field. Some, for example, refer to bremsstrahlung radiation as "gammas" or "X-rays". Indeed, medical X-ray generators produce bremsstrahlung radiation to form images. So it is somewhat of a philosophical question how the terms are used, but many hold the distinctions shown above. Gamma rays are in general of higher energy than characteristic X-rays, but there are many low-energy γ rays that have lower energy than many characteristic X-rays. Gamma-ray emissions leave the atomic and mass numbers of the atom involved unchanged; these transitions are termed *isomeric*, and the original and final nuclides are referred to as *isomers*. Most nuclear de-excitation occurs within about 10^{-10} sec, but in some cases the nucleus is left in an excited state, and these nuclides are called "metastable" and given the designation "m"; for example, ^{99}Tc ($T_{1/2} = 213{,}000$ y) versus ^{99m}Tc ($T_{1/2} = 6.01$ h). This kind of transformation is referred to as *Isomeric Transition* (IT). The Q value is simply the difference in energy states between the parent and progeny.

Gamma rays are highly penetrating. They have no mass and no charge, and interact with matter via special mechanisms that we discuss in the next chapter, primarily with orbital electrons. Although we can place enough material to completely stop all alpha particles and negative or positive electrons, with gamma rays we can only reduce their intensity. They are external as well as internal hazards.

3.6   Internal Conversion Electrons

An alternate method exists for nuclei to release excess energy. The de-excitation energy that may have been released as a γ ray is instead absorbed by a tightly bound electron, which is then ejected from the atom. The electron looks like a β particle, but we do not refer to it as a β particle, again because of its origin. β particles originate in the nuclei of atoms, whereas Internal Conversion Electrons (ICEs) were ejected from electron orbits. One might also imagine that a γ ray, on its way out of a nucleus, incidentally interacted with an electron, transferring its energy to the electron. The kinetic energy of the electron that appears is indeed:

$$KE = E_\gamma - \phi$$

and is monoenergetic, that is, always the same value. The energy of the γ ray is always the same, and the binding energy of the orbit is of course fixed. As with orbital electron capture, the removal of an electron from an electron orbital is always followed by the filling of that gap by an electron from a higher orbital, and thus by the emission of characteristic X-ray emissions.

3.7   Auger Electrons

Characteristic X-rays emitted from an atom may themselves interact with orbital electrons, causing their ejection. The kinetic energy of the emitted electron is:

$$KE = E_{\text{X-ray}} - \phi$$

and is again monoenergetic, for obvious reasons parallel to those developed for ICEs. Auger electrons (pronounced "oh-zhay", not "aw-ger") are generally very low in energy, but are potentially important to radionuclide therapy, as their short range may cause cell killing if the activity can enter the cell nucleus.

3.8   Summary and Examples

Table 3.2 shows fairly complete decay schemes for a number of important radionuclides, chosen to illustrate typical examples of all the decay modes discussed above and to demonstrate some of the energy relationships within atoms. The value beneath the nuclide identity is its physical half-life (defined shortly). Technetium-99m has as its most prominent emission the 140.5 keV gamma ray (89%). Another minor gamma ray exists at about 142 keV. Note the energy of the conversion electrons (E-CE, the following letter denotes the orbit, or shell, from which the electron originated): they are 140.5 keV minus the binding energy of that shell (except for the few that are 142.6 keV minus the binding energy). The release of conversion electrons causes rearrangements in the electron orbitals and emission of Kα and Kβ X-rays at around 18 and 20 keV. Then, Auger electrons (E-AU, again followed by the shell designation) are emitted at the X-ray energies minus the shell binding energies. For ^{18}F, we note the principal emission of the 633 keV (maximum energy)

Table 3.2 Detailed decay data for three important radionuclides.

	Rad. Type	Energy (keV)	Endpoint Energy (keV)	Radiation Intensity (%)
^{99m}Tc	E-CE-M	1.6286		74.6
6.01 h	E-AU-L	2.17		10.2
	E-AU-K	15.5		2.07
	E-CE-K	119.467		8.8
	E-CE-K	121.59		0.55
	E-CE-L	137.4685		1.07
	E-CE-L	139.59		0.172
	E-CE-M	139.967		0.194
	E-CE-N+	140.443		0.0374
	E-CE-M	142.09		0.034
	G-X-L	2.42		0.48
	G-X-Kα2	18.2508		2.1
	G-X-Kα1	18.3671		4.02
	G-X-Kβ	20.6		1.2
	G	140.511		89.06
	G	142.63		0.0187
^{18}F	β^+	249.8	633.5	96.73
110 min	E-AU-K	0.52		3.072
	G-X-K	0.52		0.01795
	G-AN	511		193.46
^{131}I	β^-	69.36	247.9	2.1
8.04 d	β^-	86.94	303.9	0.651
	β^-	96.62	333.8	7.27
	β^-	191.58	606.3	89.9
	β^-	200.22	629.7	0.05
	β^-	283.24	806.9	0.48
	E-AU-L	3.43		5.1
	E-AU-K	24.6		0.604
	E-CE-K	45.6236		3.54
	E-CE-L	74.7322		0.464
	E-CE-M	79.043		0.094
	E-CE-N+	79.977		0.0239
	E-CE-K	142.6526		0.0507
	E-CE-L	171.7612		0.0114
	E-CE-K	249.744		0.252
	E-CE-L	278.852		0.0439
	E-CE-K	329.928		1.55
	E-CE-L	359.036		0.246
	E-CE-M	363.347		0.0507
	E-CE-N+	364.281		0.0123
	E-CE-K	602.428		0.0288
	G-X-L	4.11		0.57
	G-X-Kα2	29.458		1.38
	G-X-Kα1	29.779		2.56
	G-X-Kβ	33.6		0.91
	G	80.185		2.62
	G	177.214		0.27
	G	272.498		0.0578
	G	284.305		6.14
	G	318.088		0.0776
	G	324.65		0.0212
	G	325.789		0.274
	G	358.4		0.016

(Continued)

Table 3.2 (Continued)

Rad. Type	Energy (keV)	Endpoint Energy (keV)	Radiation Intensity (%)
G	364.489		81.7
G	404.814		0.0547
G	503.004		0.36
G	636.989		7.17
G	642.719		0.217
G	722.911		1.77
^{226}Ra			
1600 y			
α	4601		5.55
α	4784.34		94.45
E-AU-L	8.71		0.99
E-AU-K	62.7		0.0193
E-CE-K	87.807		0.693
E-CE-L	168.162		1.32
E-CE-M	181.729		0.351
E-CE-N+	185.114		0.122
G-X-L	11.7		0.88
G-X-Kα2	81.07		0.197
G-X-Kα1	83.78		0.328
G-X-Kβ	94.9		0.149
G	186.211		3.59

positron (96.7%) and the accompanying annihilation photons at 511 keV (at $2 \times 96.7 = 194\%$) and some other minor emissions. The ^{131}I decay scheme has a number of different beta-minus (β^-) emissions (the most prominent being the 606 keV maximum, 192 keV average energy beta), a complex scheme of photon emissions (the most prominent being the 365 keV line) and a number of CE and AU emissions that can be explained as we did above for ^{99m}Tc. Radium-226 is primarily an alpha emitter, but also has a 186 keV photon emission at about 4%, and some accompanying low abundance CE and AU emissions.

3.9   Transformation Kinetics

Unstable atoms rearrange themselves in an attempt to achieve stability. The process of radioactive decay may be represented by a first-order, linear differential equation:

$$\frac{dN}{dt} = -\lambda N$$

In this equation, N is the number of atoms of some radioactive species and λ is a rate constant (with units of time^{-1}) that describes the rate of decay of a particular radionuclide. We can easily solve this equation to give an expression for the number of radioactive atoms present at some time T, given a number of atoms at some established time 0, N_0:

$$\frac{dN}{N} = -\lambda dt$$

$$\int \frac{dN}{N} = \int -\lambda dt$$

$$\ln(N) \Big|_0^T = -\lambda t \Big|_0^T + C$$

$$\ln(N_T) - \ln(N_0) = -\lambda T$$

$$\ln\left(\frac{N_T}{N_0}\right) = -\lambda T$$

$$\boxed{N_T = N_0 e^{-\lambda T}}$$

So we see that the number of atoms at any time decreases exponentially with time. The activity of any sample A is related to the number of atoms present by

$$\boxed{\boxed{A = \lambda N}}$$

Note: Important equations are denoted in the text by being enclosed in a box. This equation is so important to the study and practice of health physics that it is enclosed in a double box. Because λ is a constant, we can also write:

$$A_T = A_0 e^{-\lambda T}$$

Activity is the number of nuclear transformations per unit time occurring in a given sample of radioactive material. Two major units are used to describe activity. The first was the curie (Ci) $= 3.7 \times 10^{10}$ transformations/sec (named for Madame Curie, multiple Nobel Prize recipient for the discoveries of radium and polonium). In the SI unit system, activity is measured in becquerels, where 1 becquerel (Bq) $= 1$ transformation/sec (named for Henri Becquerel, who discovered radioactivity). The term *disintegration* is much more commonly used in the practice of health physics, but some people object to its use (as the atom does not disintegrate and become disorganized, but is transformed into a new state). All activity units need to be used with some multipliers to express quantities encountered in practice. The curie was far too large a quantity, and very small multipliers were needed to express normally used quantities. The becquerel is a very small quantity, and very large multipliers are needed to express normally used quantities. It is an easy quantity to remember; most SI units are based on unit quantities.

$$kBq = 10^3 Bq \qquad mCi = 10^{-3} Ci$$
$$MBq = 10^6 Bq \qquad \mu Ci = 10^{-6} Ci$$
$$GBq = 10^9 Bq \qquad nCi = 10^{-9} Ci$$
$$TBq = 10^{12} Bq \qquad pCi = 10^{-12} Ci$$
$$1 Ci = 3.7 \times 10^{10} Bq$$
$$1 mCi = 37 MBq$$
$$1 \mu Ci = 37 kBq, \ldots .$$

It should be immediately clear that A_0 is the amount of activity present in the sample at time 0. If we plot the activity of a sample as a function of time,

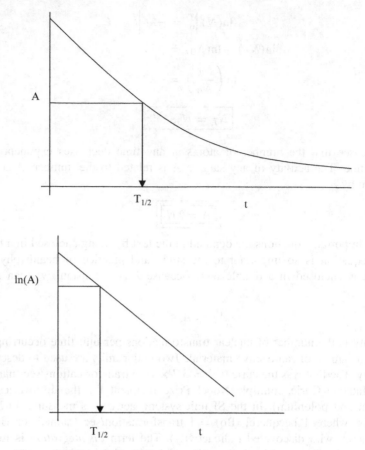

Figure 3.7 Hypothetical plot of activity of a sample as a function of time. Top graph: linear–linear plot, bottom graph: semilogarithmic plot.

we will obtain a graph that looks like those in Figure 3.7. The first figure is a plot using linear scales on both axes; the second figure uses a logarithmic scale for the abscissa (i.e., a semilogarithmic plot).

An important quantity in the study of radioactive nuclides is that of the radionuclide *half-life*, which is the time needed for any radionuclide for activity in a sample to be reduced to one-half of its initial value. Mathematically, we can define the half-life as being the natural logarithm of 2 over the decay constant:

$$\ln \frac{N}{N_0} = -\lambda t$$

$$\frac{N}{N_0} = 0.5 \quad \ln(0.5) = -0.693$$

$$T_{1/2} = \frac{0.693}{\lambda}$$

After each half-life, the fraction of the activity remaining is reduced by one-half. It thus in theory never goes completely to zero.

# Half-Lives	Fraction Remaining
1	0.500000
2	0.250000
3	0.125000
4	0.062500
5	0.031250
6	0.015625
7	0.007813
8	0.003906
9	0.001953
10	0.000977

Eventually, given a finite number of atoms at time zero, they will all decay. Radioactive decay is a stochastic process. It is not possible to say when any given atom will decay. The differential equation that describes radioactive decay describes the average behavior of large populations of atoms.

The half-life of any given nuclide is characteristic, generally well known, and cannot be altered by any known physical, chemical, or other processes. Some limited evidence has been given that may show exceptions to this rule. The decay rate is not always constant for electron capture, as occurs in nuclides such as ^{7}Be, ^{85}Sr, and ^{89}Zr. For this type of decay, the decay rate may be affected by local electron density.[4] Theoretically, as one approaches temperatures near absolute zero, the velocities of electrons could be affected, and the half-life of nuclides undergoing electron capture could be affected. In an article discussing the possibility of cold fusion,[5] an experiment done by Otto Reifenschweiler was discussed. In searching for a compound that could protect Geiger–Mueller tubes from damage when irradiated, Dr. Reifenschweiler reported that tritium emissions fell as the temperature of the system increased from 115°C to 160°C. This is, however, most likely an erroneous result due to changes in instrument response not properly accounted for, as this kind of temperature change could not affect the rate of beta minus decay.

Table 3.3 gives half-lives for a number of important radionuclides. Half-lives of commonly encountered radionuclides vary from microseconds to billions of years. Note that λ represents the instantaneous fractional decrease, or probability of transformation, per unit time. The differential equation above is an expression of the relationship:

$$\lim_{\Delta t \to 0} \frac{\Delta N / N}{\Delta t} = -\lambda$$

Table 3.3 Half-lives for some selected radionuclides.

^{32}P	14.3 d	^{99m}Tc	6.02 h
^{60}Co	5.3 y	^{11}C	20.3 m
^{3}H	12.3 y	^{15}O	122 s
^{137}Cs	30.1 y	^{18}F	1.83 h
^{226}Ra	1600 y	^{238}U	4.48×10^9 y
^{222}Rn	3.8 d	^{235}U	7.04×10^8 y

For example, ^{222}Rn has a half-life of 3.8 days. The decay constant may be given in any time units of interest:

$$\lambda = 0.693/3.8\,d = 0.182d^{-1}$$
$$= 0.0076h^{-1}$$
$$= 2.1 \times 10^{-6}s^{-1}.$$

As the value gets smaller, the number is more representative of the fraction of atoms that are converted per unit time. But it is incorrect to think that 18.2% of ^{222}Rn atoms in a sample will decay in one day; this is simply not true.

3.10   Average Life (Mean Life)

We can calculate the mean life of atoms in a sample by calculating the sum of the lifetime of the individual atoms in a sample and dividing it by the number of atoms originally present.

$$\tau = \frac{1}{N_0} \int_0^\infty t\lambda N\,dt = \frac{1}{N_0} \int_0^\infty t\lambda N_0 e^{-\lambda t}\,dt$$

$$\tau = \int_0^\infty t\lambda e^{-\lambda t}\,dt$$

integrate by parts, $v = -e^{-\lambda t}$, $u = t$

$$\tau = \frac{1}{\lambda} - te^{-\lambda t}\,|_0^\infty$$

$$\boxed{\tau = \frac{1}{\lambda} = 1.443\,T_{1/2}}$$

3.11   Specific Activity

Specific activity is the amount of activity per unit mass of a radioactive substance. Mathematically, it is defined as

$$SA = \frac{A}{m} = \frac{\lambda N}{m} = \frac{\lambda N_A}{A_0}$$

where:

N_A = Avogadro's number (6.025×10^{23} atoms/mol)
A_0 = nuclide mass number (g/mol) (Note: Not atomic number, Z)

for example, for a sample of pure ^{226}Ra, which has a $T_{1/2} = 1600$ y

$$SA = \left(\frac{0.693}{1600\ y}\right)\left(\frac{1\ y}{\pi \times 10^7 s}\right)\left(\frac{6.025 \times 10^{23}\,\text{atoms}}{\text{mol}}\right)$$

$$\times \left(\frac{1\ \text{mol}}{226\ g}\right)\left(\frac{1\ \text{dis}}{\text{atom}}\right)\left(\frac{1\ Bq - s}{\text{dis}}\right)$$

$$SA = 3.66 \times 10^{10}\ Bq/g \approx 1\ Ci/g$$

Thus we see how the specification of the amount of activity in one curie was derived.

3.12 Series Decay

So far we have considered only the situation in which species A decays to species B, which is assumed to be stable. Now we consider the situation, which is not uncommonly encountered, in which species B is itself radioactive. Species B will decay with its own characteristic half-life, and thus rate constant, λ_B. The relationships established for species A will not change:

$$A \xrightarrow{\lambda_A} B$$

$$N = N_0 e^{-\lambda t}$$

$$A = A_0 e^{\lambda_A t}$$

$$\tau_A = \frac{1}{\lambda_A}$$

We can now develop a series of differential equations that describe the kinetics of the transformations of species B (assumed to decay to species C):

$$A \xrightarrow{\lambda_A} B \xrightarrow{\lambda_B} C$$

$$\frac{dN_B}{dt} = \lambda_A N_A - \lambda_B N_B$$

$$\frac{dN_B}{dt} + \lambda_B N_B = \lambda_A N_{A_0} e^{-\lambda_A t}$$

$$\frac{dN_B}{dt} e^{\lambda_B t} + \lambda_B N_B e^{\lambda_B t} = \lambda_A N_{A_0} e^{-\lambda_A t} e^{\lambda_B t}$$

This relationship can be resolved by noting that:

$$\frac{d}{dt}\left(N_B e^{\lambda_B t}\right) = \lambda_A N_{A_0} e^{-\lambda_A t} e^{\lambda_B t}$$

$$\int d\left(N_B e^{\lambda_B t}\right) dt = \int \lambda_A N_{A_0} e^{-\lambda_A t} e^{\lambda_B t} dt$$

Applying the initial condition that there were no atoms of B present at time zero (if there were, they could be treated in the same way as for our original case for species A, above, and added in later) and assuming that the branching ratio for A $\rightarrow$ B is 1.0:

$$N_B e^{\lambda_B t} = \lambda_A N_{A_0} \left(\frac{e^{(\lambda_B - \lambda_A)t}}{\lambda_B - \lambda_A}\right) + C$$

$$N_B = 0 \quad at \quad t = 0:$$

$$C = \frac{-\lambda_A N_{A_0}}{\lambda_B - \lambda_A}$$

$$N_B e^{\lambda_B t} = \lambda_A N_{A_0} \left(\frac{e^{(\lambda_B - \lambda_A)t}}{\lambda_B - \lambda_A}\right) - \frac{\lambda_A N_{A_0}}{\lambda_B - \lambda_A}$$

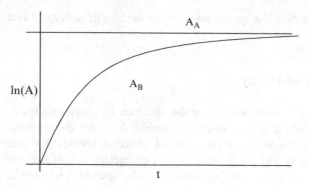

Figure 3.8 Activity of parent and progeny in secular equilibrium.

Dividing both sides by $e^{\lambda_B t}$ and rearranging, we obtain:

$$N_B = \left(\frac{\lambda_A N_{A_0}}{\lambda_B - \lambda_A}\right)(e^{-\lambda_A t} - e^{-\lambda_B t})$$

Then, applying the all-important relationship $A = \lambda N$, we obtain finally:

$$\boxed{A_B = \lambda_B N_B = \left(\frac{\lambda_B A_{A_0}}{\lambda_B - \lambda_A}\right)(e^{-\lambda_A t} - e^{-\lambda_B t})}$$

This is a general solution. We consider now three important special cases.

Case 1: Half-Life of Species A ≫ Half-Life of Species B: "Secular" Equilibrium

Many cases are encountered in this category. If the half-life of species A is much greater than the half-life of species B, then it is also true that $\lambda_A \ll \lambda_B$. In this case, the ratio $\lambda_B/(\lambda_B - \lambda_A) \approx 1$, and $e^{-\lambda_A t} \approx 1$, so we have the simpler equation:

$$\boxed{A_B \approx A_{A_0}(1 - e^{-\lambda_B t})}$$

This situation is called "secular" equilibrium (Figure 3.8). In this situation, the activity of species B asymptotically approaches that of species A (a $(1 - e^{-\lambda t})$ function) and then disappears with the half-life of species A.

In Figure 3.8, the activity of species A is shown to be approximately constant; for many cases of secular equilibrium, the half-life of species A is very long, and does not decrease appreciably over normal periods of observation. For example, ^{226}Ra (half-life 1600 y) decays to ^{222}Rn (half-life 3.8 d). Note that the ingrowth of species B goes as the inverse of radioactive decay. That is, the level of species B reaches 50% of that of species A in one half-life (of species B), it reaches 75% of the level of species A in two half-lives (of species B), and so on. Note also that there is no "equilibrium" until species B reaches nearly the same activity as species A.

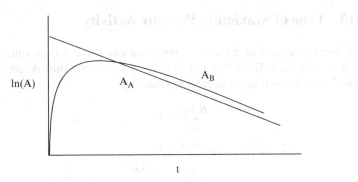

Figure 3.9 Activity of parent and progeny in transient equilibrium.

Case 2: Half-Life of Species A > Half-Life of Species B: "Transient" Equilibrium

In the second case, the half-life of species A is longer than that of species B, but not markedly so. In this case, $\lambda_A < \lambda_B$. This situation is termed "transient" equilibrium (Figure 3.9). No simplification of the general case equation is possible, but a solution will show the following behavior for the two species.

Note that the activity of species B is actually higher than that of species A after the crossover, by the factor:

$$\left(\frac{\lambda_B}{\lambda_B - \lambda_A} \right)$$

A kind of equilibrium is established after this point, with species B again disappearing with the half-life of species A, as with secular equilibrium.

Case 3: Half-Life of Species A < Half-Life of Species B: No Equilibrium

In this case, species A decays rapidly, and then species B builds in to some level and then decays with its own half-life (Figure 3.10).

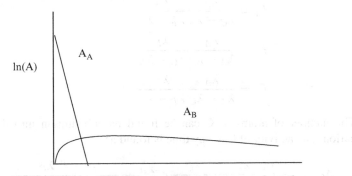

Figure 3.10 Activity of parent and progeny when the progeny half-life is longer than that of the parent (no equilibrium).

3.13   Time of Maximum Progeny Activity

In either the second or third case above, we can solve for the time at which the level of species B reaches a maximim, by setting the time derivative of the activity of species B equal to zero and solving:

$$\frac{dA_B}{dt} = 0$$

$$-\lambda_A e^{-\lambda_A t} + \lambda_B e^{-\lambda_B t} = 0$$

$$\frac{\lambda_B}{\lambda_A} = \frac{e^{-\lambda_A t}}{e^{-\lambda_B t}} = e^{-(\lambda_A - \lambda_B)t}$$

$$t_{MB} = \frac{\ln(\lambda_B/\lambda_A)}{\lambda_B - \lambda_A}$$

$$t_{MB} = \tau_B \left(\frac{T_A}{T_A - T_B}\right) \ln\left(\frac{T_A}{T_B}\right)$$

Now, I know the question burning in your mind is, "What if species C is also radioactive?" This is certainly possible; in fact some of the most important and interesting problems in health physics involve long chains of products, one decaying to the next until a stable species is reached. So now let's solve for the situation:

$$A \xrightarrow{\lambda_A} B \xrightarrow{\lambda_B} C \xrightarrow{\lambda_C} D$$

$$\frac{dN_C}{dt} = \lambda_B N_B - \lambda_C N_C$$

The solutions for A and B do not change; they are not dependent on what happens to other members of the chain. The solution for C will be of the form:

$$N_C = N_{A_0}\left(F_1 e^{-\lambda_A T} + F_2 e^{-\lambda_B T} + F_3 e^{-\lambda_C T}\right)$$

where F1, F2, and F3 are coefficients that will depend on the initial conditions of the problem. If we assume that all branching ratios are 1.0 and that:

$$N_A(0) = N_{A_0}, N_B(0) = 0, N_C(0) = 0 \text{ we find:}$$

$$F_1 = \frac{\lambda_A}{\lambda_C - \lambda_A}\frac{\lambda_B}{\lambda_B - \lambda_A}$$

$$F_2 = \frac{\lambda_A}{\lambda_A - \lambda_B}\frac{\lambda_B}{\lambda_C - \lambda_B}$$

$$F_3 = \frac{\lambda_A}{\lambda_A - \lambda_C}\frac{\lambda_B}{\lambda_B - \lambda_C}$$

The number of atoms of C can be found by substitution into the above equation. The activity of C at any time is found as

$$A_C = A_{A_0}\left(\frac{\lambda_B}{\lambda_B - \lambda_A}\frac{\lambda_C}{\lambda_C - \lambda_A}e^{-\lambda_A t} + \frac{\lambda_B}{\lambda_A - \lambda_B}\frac{\lambda_C}{\lambda_C - \lambda_B}e^{-\lambda_B t}\right.$$

$$\left. + \frac{\lambda_B}{\lambda_B - \lambda_C}\frac{\lambda_C}{\lambda_A - \lambda_C}e^{-\lambda_C t}\right)$$

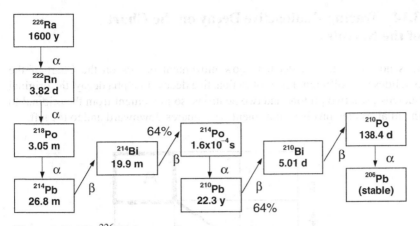

Figure 3.11 The ^{226}Ra decay series.

We can continue on with a species D, E, F, and so on, but the relationships among the species obviously become more complicated and are difficult to categorize. If Species A is very long-lived, however, relative to other members of the chain, after a long time (seven to ten half-lives of the longest-lived progeny species), all members of the chain will be in secular equilibrium and decaying with the half-life of Species A, and all having the same activity as Species A. An important example is the ^{226}Ra decay series (Figure 3.11).

Radium-226 is itself produced by other species that ultimately lead back to ^{238}U; we show this shortly. But considering only ^{226}Ra and its progeny for the moment, we note that ^{226}Ra decays with a very long half-life (about 1600 years) to ^{222}Rn, which has relatively a very short half-life, about 3.8 days. So after about 30–40 days, ^{222}Rn will be in equilibrium with ^{226}Ra. All of the progeny down to ^{210}Pb are even more short-lived, and so will rapidly come into equilibrium with ^{222}Rn, which in turn is in equilibrium with ^{226}Ra, so all of these species will have the same activity as ^{226}Ra, and will demonstrate a 1600 year half-life. If the species are also decaying for around 200 years or so, all of the progeny including and beyond ^{210}Pb will also be in equilibrium.

Now this assumes that all of the species stay in the same place. If we are talking about uranium ores buried deep in the earth and of large size, this may be reasonable. Radon-222 is a noble gas, so it will tend to diffuse through most structures. If a uranium deposit is large, throughout most of the ore, radon diffusion out of a region will be balanced by other radon diffusion in, and all of the species will have about the same activity. If we dig up that ore, however, the digging and excavation process will disturb the radon and it will be dispersed, so the equilibrium will be maintained only from ^{238}U down to ^{226}Ra, and we will need another 30 days or so with the ore in an enclosed space before all of the progeny through ^{214}Po will come back into equilibrium. In phosphate fertilizers from Florida phosphate deposits (which are rich in uranium), chemical separation processes cause the uranium and similar products to be separated from the radium species, so the ^{226}Ra becomes the controlling parent. The uranium portions go with the fertilizer (these are not a significant radiation hazard, but they are present), and the radium remains behind in the leftovers ("tails"). If this material is left on reclaimed lands, and people build houses on it, seepage of ^{222}Rn into the homes can result in significant radiation exposures, from a cancer risk perspective.

3.14   Tracing Radioactive Decay on the Chart of the Nuclides

It is not difficult to understand how movement occurs on the Chart of the Nuclides with different forms of radioactive decay. In alpha decay, the original nucleus loses two protons and two neutrons, so movement from the original to the final species involves movement two squares downward and to the left.

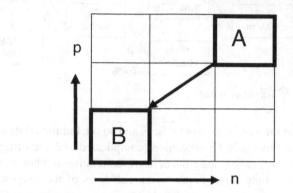

Example: ^{226}Ra (1600 y) $\rightarrow$ ^{222}Rn (3.8 d).

224Ra 3.66 D	225Ra 14.9 D	226Ra 1600 Y
α	β–	α
223Fr 22.00 M	224Fr 3.33 ʌ	225Fr 4.0 M
β–	β–	β–
222Rn 3.8235 D	223Rn 24.3 M	224Rn 107 M
α	β–	β–

In beta minus decay, the original nucleus gains a proton and loses a neutron, so the movement is one square up and to the left.

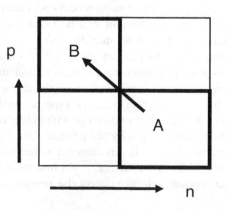

Example: ^{90}Sr (29 y) → ^{90}Y (2.7 d), also ^{90}Y (2.7 d) → ^{90}Zr (stable). ^{90}Sr/^{90}Y secular equilibrium is very important in several applications in health physics.

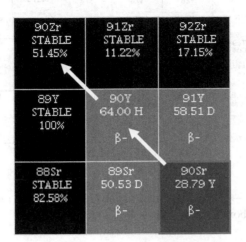

In positron decay (also electron capture), the original nucleus gains a neutron and loses a proton, so the movement is one square down and to the right.

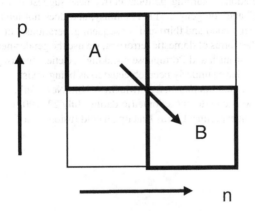

Example: ^{22}Na (2.6 y) → ^{22}Ne (stable)

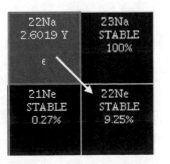

In gamma decay and isomeric transition, no movement occurs.

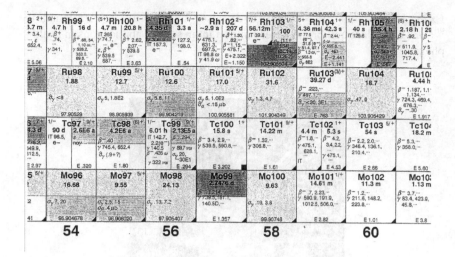

Example: ^{99m}Tc → ^{99}Tc. The Chart of the Nuclides shows metastable states within the same box as the nonmetastable states.

Endnotes

1. http://ie.lbl.gov/toi/pdf/chart.pdf.
2. In health physics, in English-speaking countries, we once spoke of "parent", "daughter", "granddaughter", and so on, to describe members of a decay series. In the late 1900s, considerations regarding political correctness suggested use of the gender-neutral "parent" and "progeny". This is clumsy, and does not lend itself to distinguishing between second and third and subsequent generations, but to avoid endless lawsuits and other forms of domestic terrorism, we use the gender-neutral terms here. Interestingly, in Spanish and Portuguese speaking societies, the usage was "parent" and "son", which have similarly been objected to as being sexist.
3. Robley D. Evans, *The Atomic Nucleus* (McGraw-Hill, New York, 1955).
4. http://www.answers.com/topic/radiometric-dating, July 22 (2005).
5. W. Bown, Ancient experiment turns heat up on cold fusion, *New Scientist*, January 8, p. 16 (1994).

4

Interaction of Radiation with Matter

The particles or rays that are emitted during radioactive decay are emitted with a certain energy (which we learned how to calculate in Chapter 3), and they may or may not have charge. All of them interact with the environment into which they are released, transferring energy to that medium, and eventually dissipating all of their energy. We quantify the transfer of energy to matter in the next chapter. For now, it should be clear that such transfers of energy will have important implications for radiation biology, radiation shielding, radiation detection, and almost every practical application of radiation protection.

The main processes by which radiation interacts with matter are:

- Ionization
- Excitation
- Capture

Ionization and excitation are the most important processes for the majority of radiation types and interaction situations. Capture reactions occur for particular radiation types under some circumstances, and lead to some important outcomes that we discuss. We can think of some interactions of radiations being like "billiard ball" collisions, that is, involving only elastic scattering. If we consider a particle of mass M approaching a stationary particle of mass m:

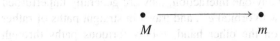

The maximum energy that can be transferred in a single collision may be derived through a conservation of energy and momentum, in an interaction assumed to be a "head-on" collision. Assume that particle M had velocity V before the collision and velocity V_1 after the collision. Particle m had no velocity before the collision (of course if this is an orbital electron or most any particle, it will have been in motion, but from a relative point of view, in terms of this interaction, it can be considered to have been at rest) and had velocity v

after the collision. Then, conserving total energy and momentum, we find that:

$$\frac{MV^2}{2} = \frac{MV_1^2}{2} + \frac{mv^2}{2} \tag{4.1}$$

$$MV = MV_1 + mv \tag{4.2}$$

Solving Equation (4.2) for v, we obtain $v = (MV - MV_1)/m$. If this is substituted into Equation (4.1), we find that:

$$MV^2 = MV_1^2 + \frac{M^2 (V - V_1)^2}{m}$$

$$mMV^2 = mMV_1^2 + M^2 (V - V_1)^2$$

$$m\left(V^2 - V_1^2\right) = M (V - V_1)^2$$

$$m(V - V_1)(V + V_1) = M(V - V_1)(V - V_1)$$

$$V_1(m + M) = V(M - m)$$

$$V_1 = \frac{(M - m)V}{M + m}$$

If we use this expression for V_1, the maximum energy that can be transferred in a single collision is given as

$$Q_{max} = \frac{MV^2}{2} - \frac{MV_1^2}{2} = \frac{4mME}{(M + m)^2}$$

Example
An α interacting with an orbital e^-:

$$E = \frac{4 \times (0.00055) \times 4}{(4 + 0.00055)^2} = 0.00055 \approx 0.055\%$$

A β interacting with an orbital e^-:

$$E = \frac{4 \times (0.00055) \times (0.00055)}{(2 \times 0.00055)^2} = 1.0 = 100\%$$

Now most β particles of interest travel at velocities in which relativistic concerns are involved, and the simple equations above describing the particle energy are not correct. Thus, this simple energy transfer equation cannot be used. The principle, however, that they may give up significant portions of their energy in a single collision still applies. As αs give up only a tiny fraction of their energy in any one interaction, they are generally unperturbed in individual interactions with orbital e^-, and travel in straight paths of rather well-defined length. βs, on the other hand, follow tortuous paths through matter and, although the total path length of a given β or e is well known, the actual distance one will penetrate is quite variable, depending on the shape of the path. This is further complicated by the fact that the betas (plus and minus) are emitted with a spectrum of energies. A device designed by C. T. R. Wilson in 1911 permits the visualization of tracks from radiation particles (Figures 4.1–4.3). A chamber is filled with methyl alcohol vapor in a supersaturated environment. Alcohol vapor will condense around ion trails left by the ionizing particles. Ionization trails are left, which can be seen with the naked eye.

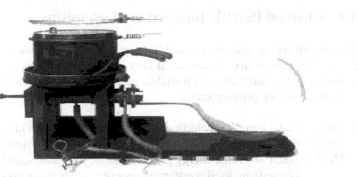

Figure 4.1 Wilson's original cloud chamber from 1911. (From http://www.physics.brown.edu/physics/demopages/Demo/modern/demo/7d3050.htm.)

Figure 4.2 Cloud chamber photograph of α-particle tracks. (From: http://www.physics.hku.hk/academic/courses/.)

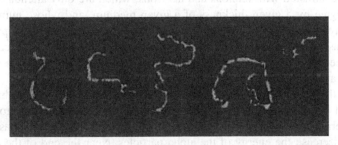

Figure 4.3 Cloud chamber photograph of beta particle tracks. (From: http://newton.hanyang.ac.kr/pds/phywe/5_2_04.pdf.)

4.1 Charged Particle Interaction Mechanisms

Photons (including gamma rays and X-rays) and neutrons have unique inter-action mechanisms that are discussed later in the chapter. Charged particles, including alpha particles, beta particles, electrons, and recoil atoms interact with matter in two primary ways:

- *Ionization*: An outer shell electron is removed from an atom in the medium, and an *ion pair* (the free electron plus the positively charged atom) is formed. The free electron has the energy gained from the interaction, and will travel through the medium, itself transferring energy to other atoms (possibly even causing new ionization events) until it comes to rest. The positively charged atom will attract a free electron from somewhere in the medium and return to its neutral state.
- *Excitation*: An electron within one of the orbits of an atom absorbs energy and is moved into a higher energy state. This electron may eventually fall back to its original state, with the emission of the difference in the energy states being emitted as photon energy, or the excited atom may dissoci-ate. We evaluate this in more detail when we study biological effects in Chapter 6.

These two mechanisms are absolutely vital to understanding all aspects of health physics. Ionization and excitation are the basis for most charged particle interactions (including the charged particles formed by the indirectly ionizing photons and neutrons), and thus are the underlying mechanisms that explain radiation interaction and shielding, radiation detector response to radiation, causation of biological effects, and many other fundamental aspects of health physics. We study other interaction mechanisms, but these two should always be borne in mind in any and every application of health physics.

4.2 Alpha Particle Interactions

Alpha particles are the least penetrating of all (commonly encountered) forms of radiation. Alpha particles will travel only a few cm in air, and will penetrate soft tissue only to a distance of micrometers. Both alpha (α) and beta (β) particles may be characterized as having fixed ranges in a given medium (this is to be contrasted with photons and neutrons, which are only attenuated to a finite degree by a given thickness of a given medium, and whose number is not always reduced completely to zero). Recall from Chapter 3 that α particle emissions are monoenergetic. If we were to place a detector near a source of αs and place different thicknesses of absorber material between them, we would observe a curve like that shown in Figure 4.4.

Note that the relative count rate is very flat as more and more absorber is added, until a point comes where the count drops very suddenly to zero. The intermediate layers of absorber do not completely stop any alphas; they simply decrease the energy of the alpha particles. Near the end of the curve, the count rate does not immediately drop to zero, but drops sharply and goes quickly to zero, in a way that demonstrates the stochastic nature of radiation interactions.

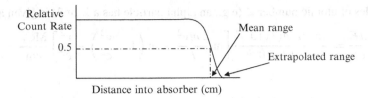

Figure 4.4 Alpha particle count rate versus range curve.

Individual alpha particles passing through the absorber have slightly different histories, depending on exactly where and when individual interactions with atoms of the absorber occurred. Thus when the alpha particles reach the thickness of the absorber that will on average attenuate all of the alphas (the mean range in Figure 4.4), the alphas have a distribution of energies about an average value. This variability in energy values is called *energy straggling*. As every alpha particle will not have exactly the same history and have the same remaining energy at a given distance from the starting point, the alpha particles will also travel slightly different distances in total into the absorber. Thus, the relative count rate in Figure 4.4 does not drop to zero instantly at a point, but *range straggling*, the corollary to energy straggling, is also observed (Figure 4.5, taken from Knoll[1]).

Another term that is defined when dealing with charged particle interactions in matter is *specific ionization*, which is the number of ion pairs formed per unit distance traveled by a charged particle. The quantity can be obtained by dividing the linear rate of energy loss by the energy needed to remove an electron from an atom, that is, to form one ion pair:

$$\frac{\text{linear rate of energy loss (eV/cm)}}{\text{energy needed to form an ion pair (eV/i.p.)}} \equiv \frac{\text{i.p.}}{\text{cm}}$$

The numerator of this expression is called the *collisional stopping power* for the radiation. A formula for the stopping power, *dE/dx*, for heavy charged

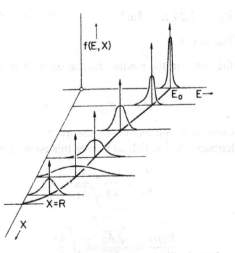

Figure 4.5 Plot of the energy distribution of a beam of initially mononergetic charged particles at various penetration distances. *E* is the particle energy and *X* is the distance along the track.

particles of atomic number Z (e.g., an alpha particle has a Z of 2) is given as

$$\frac{dE}{dx} = \frac{4\pi z^2 e^4 n (3x10^9)^4}{mv^2 1.6x10^{-6}} \left[\ln \frac{2\,mv^2}{I} - \ln \left(1 - \frac{v^2}{c^2} \right) - \frac{v^2}{c^2} \right] \frac{MeV}{cm}$$

where:

z = atomic number of the ionizing particle
e = unit electrical charge (1.6×10^{-19} C)
n = number of electrons per unit volume of the absorber
m = the mass of an electron
v = projectile velocity
c = speed of light
I = effective ionization potential of the absorber atoms (1.38×10^{-10} erg for air)

An important fact of much use in health physics is that the work function for air or tissue is approximately 34 eV/ip. We make use of this fact in a number of applications in dosimetry and detector technology. Alpha particles are very highly ionizing and create between 40,000–80,000 ion pairs/cm. Using an average value of 60,000 ip/cm and this assumed work value of 34 eV/ip, we see that alpha particles will expend about 2 MeV/cm (assuming constant stopping power at all energies, which is not observed, but this is reasonable for this qualitative evaluation). Because alpha particles are emitted with energies of about 4–8 MeV, one can see that alpha particles will travel only a few cm in air. In soft tissue, the number of electrons per unit volume is higher, so the specific ionization is higher, and alphas will travel only a few tens of micrometers. As the body is covered with a depth of 0.007 cm (70 μm or 7 mg/cm^2) of nonviable skin cells that will be sloughed off, alpha particles will not penetrate this depth; alpha particles are thus not an external hazard.

An approximate general formula for the range of alpha particles in air is:

$$R_{air} = 0.56\,E \qquad\qquad E < 4\,MeV$$

$$R_{air} = 1.24\,E - 2.62 \qquad 4 < E < 8\,MeV$$

$$R_{air} \equiv cm$$

In other media, the range in the media (R_m, in units of mg/cm^2) is given by Cember[2] as

$$R_m = 0.56A^{1/3} R_{air}, \equiv mg/cm^2,$$

where A is the atomic mass number of the absorber.

The Bragg–Kleeman rule,[3] which addresses this issue, is stated as

$$\frac{R_2}{R_1} = \frac{\rho_1}{\rho_2} \frac{\sqrt{A_2}}{\sqrt{A_1}}$$

or

$$\frac{R_2 \rho_2}{R_1 \rho_1} = \frac{\sqrt{A_2}}{\sqrt{A_1}} = \sqrt{\frac{A_2}{A_1}}$$

Here ρ is the medium density. This is an empirical rule, originally purported to be accurate to within ±15%. The density of air (ρ_{air}) is

$1.293 - 10^{-3}$ g/cm^3 = 1.293 mg/cm^3. The density of human tissue (ρ_{tissue}), for comparison, is 1.04 g/cm^3.

Example

The atomic mass number of mylar (polyethylene terephthalate) is about 13, and its density is about 1.4 g/cm^3. What thickness of mylar would be needed to stop alpha particles from ^{241}Am (main alpha energy 5.49 MeV)?

The range of this alpha particle in air would be $(1.24 \times 5.49) - 2.62 = 4.2$ cm.

The range in mylar would thus be:

$$(\text{Cember}) \; R_{\text{mylar}} = 0.56 \times 13^{1/3} \times 4.2 = 5.53 \, \text{mg/cm}^2$$

The thickness of mylar needed would be

$$\frac{5.53 \frac{\text{mg}}{\text{cm}^2}}{1400 \frac{\text{mg}}{\text{cm}^3}} = 0.0040 \text{ cm} = 0.040 \text{ mm}$$

By comparison, the Bragg–Kleeman rule would give:

$$R_{\text{mylar}} = \frac{0.001293 \frac{\text{g}}{\text{cm}^3}}{1.4 \frac{\text{g}}{\text{cm}^3}} \frac{\sqrt{13}}{\sqrt{14.7}} 4.2 \text{ cm} = 0.0036 \text{ cm} = 0.036 \text{ mm}$$

4.3 Beta Particle Interactions

Beta particles are attenuated in an approximately exponential fashion, which thus appears as a linear function on a semilog graph. The range of any beta particle can best be determined by interposing successive layers of absorber materials between a beta source and a calibrated detector capable of detecting beta particles (see Figure 4.6, from Cember[2]). The maximum range can be established, which depends on the maximum beta energy in the decay spectrum. Recall that beta particles, unlike alpha particles or gamma rays, are emitted with a spectrum of energies ranging from zero to the maximum energy allowed for the transition. Therefore a beta source is continually giving off beta particles of different energies. The lowest energy betas will be easily stopped in small amounts of materials, whereas the highest energy particles will obviously need more, and it is these particles that determine the total thickness of material needed to stop all particles from a given source.

An approximate formula for beta particle range as a function of beta energy is given as

$$R = 412E^{1.265 - 0.0954 \ln E} \qquad 0.01 < E < 2.5 \, \text{MeV}$$
$$R = 530E - 106 \qquad E > 2.5 \, \text{MeV}$$

The range is expressed in units of mg/cm^2. This expression is a bit cumbersome to calculate by hand, but is easily implemented in any spreadsheet program or small program. A plot of the function is also very useful for making quick evaluations of the approximate range of electrons of a given energy (Figure 4.7). The range in this figure is given in units of g/cm^2; dividing by the density in g/cm^3 will give the range in cm. For human tissue, which has a density of 1.04 g/cm^3 and is thus well approximated using the density of water,

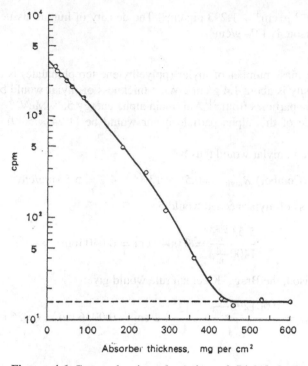

Figure 4.6 Curve showing absorption of Bi-210 beta particles by aluminum. (From Cember, with permission, McGraw-Hill, 1996.)

1 g/cm^3, the values on the abscissa can be just read off as the approximate range in cm.

Example

Given the experimental setup in the following figure, what thickness of an aluminum absorber (density 2.7 g/cm3) would be needed to stop beta particles from a ^{35}S source (maximum beta energy 0.167 MeV)?

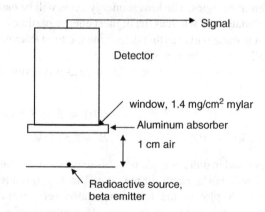

The beta particles will be stopped by the aluminum absorber, but also by the intervening air, and to a minor degree, by the detector's mylar window.

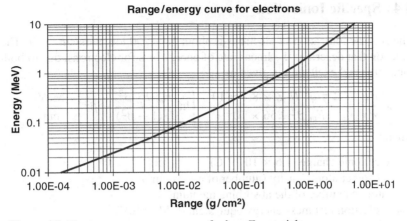

Figure 4.7 Electron range–energy curve for low-Z materials.

From the figure above, or the equations given, we find the approximate range of 0.167 MeV betas to be 0.0313 g/cm^2. The contributions from the other absorbers are:

Air: $1 \, \text{cm} \times (0.001293 \, \text{g/cm}^3) = 0.001293 \, \text{g/cm}^2$

Mylar: $0.0014 \, \text{g/cm}^2$

Aluminum needed: $0.0313 - 0.001293 - 0.0014 = 0.0286 \, \text{g/cm}^2$

$$\frac{0.0286 \, \frac{\text{g}}{\text{cm}^2}}{2.7 \frac{\text{g}}{\text{cm}^3}} = 0.0106 \, \text{cm} = 0.106 \, \text{mm}$$

Electrons lose energy in different media via interactions with orbital electrons in the atoms of the media. The interactions are inelastic collisions: the electric field of the beta particle interacts with that of an orbital electron and transfers energy to it, resulting in its expulsion from the atom (ionization) or it being raised to a higher energy level (excitation). In the case of ionization, the kinetic energy of scattered electrons is given by the difference in the energy of the incident beta particle (E_0) and the binding energy (E_b) of the atom with which it interacted:

$$E_k = E_0 - E_b.$$

This process continues until all of the beta particle energy is dissipated. The ionized electrons deposit their energy as they come to rest. As we learned earlier, because the mass of the beta particle and an orbital electron are the same, the individual interactions may cause the beta particle to lose a significant amount of energy in a given collision and have its path drastically altered, even to the point of being scattered back directly in the direction from which it came (180° scatter). If the energy of one of the scattered electrons is greater than about 1 keV, it has enough energy to itself cause other ionizations, and is called a *delta ray*. If one looks carefully at the tortuous beta ray tracks from Figure 4.3, one can observe some small delta ray tracks leading from the main track in a few locations.

4.4 Specific Ionization

The specific ionization of electrons is about $50 - 200$ ion pairs/cm in air. The function that describes the linear rate of energy loss for electrons, due to both ionization and excitation is given as

$$\frac{dE}{dx} = \frac{2\,\pi\,q^4\,N\,Z \times 8.1 \times 10^{37}}{E_m\,\beta^2\,2.56 \times 10^{-12}} \left\{ \ln \frac{E_m\,E_k\,\beta^2}{I^2(1-\beta^2)} - \beta^2 \right\} \frac{MeV}{cm}$$

where:

q = electronic charge (1.6×10^{-19} C)
N = absorber atom density (number of atoms/cm^3)
Z = atomic number of the absorbing medium
E_m = electron rest mass energy equivalent (0.511 MeV)
E_k = kinetic energy of the beta particle (MeV)
β = ratio of the beta velocity to the velocity of the speed of light (v/c)
I = mean ionization and excitation potential of atoms in the absorbing medium
  (MeV) (for air this is around 8.6×10^{-5}, for other substances it may be
  approximated as $1.35 \times 10^{-5} \times Z$)

The specific ionization is the ratio of dE/dx over the "work function" w, which is the energy needed to create an ion pair in the medium:

$$SI = \frac{dE/dx \;(eV/cm)}{w\;(eV/i.p.)} \equiv \frac{i.p.}{cm}.$$

A plot of the specific ionization for electrons in air is given in Figure 4.8. The range of beta particles in air can be up to several meters, for particles with energies of 1–2 MeV or so. The range in human tissue is of course shorter, up to ~1 cm for such energetic betas. For lower-energy betas, the range is of course lower, and some low-energy beta emitters such as ^{3}H do not have enough energy to penetrate the epidermal, or dead layer, of skin that covers our bodies (which has a variable thickness over the body, but averages around 70 μm, also cited frequently as 7 mg/cm^2).

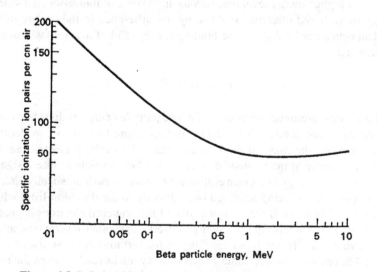

Figure 4.8 Relationship between beta particle energy and the specific ionization of air, from Cember, with permission, McGraw-Hill, 1996.

4.5 Mass Stopping Power

If we divide the energy loss rate by the density of the medium, we can define a new quantity, mass stopping power:

$$S = \frac{dE/dx}{\rho} \equiv \frac{MeV/cm}{g/cm^3} \equiv \frac{MeV - cm^2}{g}.$$

It is simple to also define the mass stopping power of one medium relative to that of another. A common comparison for a given medium is to calculate the relative mass stopping power of the medium, relative to the stopping power of air:

$$\rho_{medium} = \frac{S_{medium}}{S_{air}}.$$

Example

Faster (higher energy) beta particles have a lower rate of energy loss per unit pathlength. As they slow down, their rate of energy expenditure per unit pathlength increases. Media with more electrons per cubic centimeter for the beta particles to interact with have higher stopping power. Let's calculate the specific ionization and stopping power for a 0.1 MeV and a 1.0 MeV beta particle in air. Air has approximately 3.88×10^{20} electrons/cm^3.

To calculate the electron β (v/c) values, we need to use the relationship:

$$E_k = m_0 c^2 \left(\frac{1}{\sqrt{1 - \beta^2}} - 1 \right)$$

For the 0.1 MeV beta:

$$0.1 = 0.511 \left(\frac{1}{\sqrt{1 - \beta^2}} - 1 \right)$$

$$\beta^2 = 0.3005$$

$$\frac{dE}{dx} = \frac{2\pi(1.6 \times 10^{-19})^4 \times 3.88 \times 10^{20} \times (3 \times 10^9)^4}{0.511 \times 0.3005 \times (1.6 \times 10^{-6})^2}$$

$$\times \left\{ \ln \left[\frac{0.511 \times 0.1 \times 0.3005}{(8.6 \times 10^{-5})^2 (1 - 0.3005)} \right] - 0.3005 \right\}$$

$$\frac{dE}{dx} = 0.00481 \, \frac{MeV}{cm}$$

For the 1.0 MeV beta:

$$1.0 = 0.511 \left(\frac{1}{\sqrt{1 - \beta^2}} - 1 \right)$$

$$\beta^2 = 0.8856$$

$$\frac{dE}{dx} = \frac{2\pi(1.6 \times 10^{-19})^4 \times 3.88 \times 10^{20} \times (3 \times 10^9)^4}{0.511 \times 0.8856 \times (1.6 \times 10^{-6})^2}$$

$$\times \left\{ \ln \left[\frac{0.511 \times 1.0 \times 0.8856}{(8.6 \times 10^{-5})^2 (1 - 0.8856)} \right] - 0.8856 \right\}$$

$$\frac{dE}{dx} = 0.00215 \, \frac{MeV}{cm}.$$

The mass stopping powers are:

$$S_{\text{air}}(0.1 \text{ MeV}) = \frac{0.00481 \text{ MeV/cm}}{0.001293 \text{ g/cm}^3} = 3.718 \frac{\text{MeV} - \text{cm}^2}{\text{g}}$$

$$S_{\text{air}}(1.0 \text{ MeV}) = \frac{0.00215 \text{ MeV/cm}}{0.001293 \text{ g/cm}^3} = 1.66 \frac{\text{MeV} - \text{cm}^2}{\text{g}}$$

4.6   Linear Energy Transfer (LET)

The quantity specific ionization describes the rate of energy loss by the radiation in the medium. When we focus on the medium rather than the particle, we can describe the energy transferred to the medium. This is the Linear Energy Transfer (LET), and is given as

$$LET = \frac{dE_L}{dl}$$

where dE_L is the average energy locally imparted to a given medium by a charged particle passing through a length of medium dl. The units of LET are often given as keV/μm, although any units may be used. LET is an important parameter used in characterizing energy deposition in radiation detectors and in biological media in which we wish to study the effects of different types of radiation. As we show later, the LET of a type of radiation is often very important to understanding its effects on living cells.

4.7   Bremsstrahlung Radiation

A unique form of electromagnetic radiation is emitted when charged particles traveling at high speeds suffer rapid deceleration in the vicinity of an atomic nucleus:

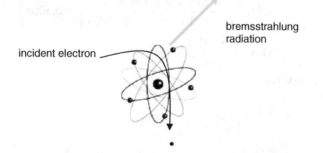

This electromagnetic radiation is identical in form and behavior to any other type of electromagnetic radiation. We have noted that the identities given to different forms of radiation in general, and to electromagnetic radiations in particular, depend on where the radiations originated, not necessarily their typical energies. X-rays originate from rearrangements of the orbitals of nuclei, whereas gamma rays originate from within the nucleus. X-rays are generally of lower energy than gamma rays, but there are low-energy gamma rays that have lower energies than some X-rays.

Bremsstrahlung radiation varies in energy from the maximum energy of the emitted beta particle down to zero (as beta particles as you recall are emitted with a spectrum of energies across this range). Strictly speaking, however, bremsstrahlung radiation is neither an X-ray nor a gamma ray because it was not produced in the ways defined for these radiations. Once formed, though, bremsstrahlung radiations will have the same interactions in matter as other photons (described next), will be shielded the same, and so on.

The total fraction of a beta emitter's energy that will appear as bremsstrahlung radiation has been described by a number of different empirical formulas. One useful formula is:

$$f = 3.5x10^{-4} \, ZE,$$

where E is the maximum (not average) beta energy (in MeV), Z is the average atomic number of the medium in which the interactions occur, and f is the fraction of incident beta energy converted to bremsstrahlung photons. A problem may be encountered with this formula at high values of Z and/or E: theoretically the fraction can exceed 1.0, which is not possible. A formula that avoids this possibility was given by Turner,[4]

$$f = \frac{6x10^{-4} \, ZE}{1 + 6x10^{-4} \, ZE}.$$

One should note also that these functions are only valid if the material is thick enough to completely stop the beta; they are not accurate in thin shield situations.

The fraction appearing as bremsstrahlung radiation is linearly related to the atomic number of the medium, so one should use low Z materials to shield beta sources. A common instinct is to use high Z materials, particularly lead ($Z = 82$) for shielding, as this is useful for shielding photons. Putting lead around a beta source will certainly stop the betas, but will also cause a very large production of bremsstrahlung radiation. One should thus use low Z materials to stop the betas, and then high Z material after that to attenuate any bremsstrahlung radiation formed. This idea is repeated and amplified in later chapters in which radiation shielding is treated in more detail.

4.8 Gamma Ray Interactions

The interactions and attenuation of photons are qualitatively different than for alpha or beta particles. Photons have no defined range in matter, as do alphas and betas. If we study a well-collimated, narrow beam source ("good geometry") and interpose different thicknesses of material (Figure 4.9), we find (for a monoenergetic source) that the intensity (here we mean the photon fluence rate, or exposure rate) of the measured radiation decreases exponentially with the placement of additional layers of shielding material:

$$I = I_0 e^{-\mu t}$$

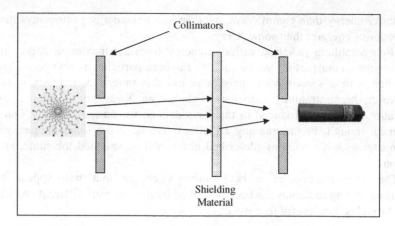

Figure 4.9 Gamma ray attenuation in shielding materials.

where:

I_0 = intensity (γ/s or MeV/s) with no shielding in place
I = intensity (γ/s or MeV/s) with shielding material present
t = thickness (cm) of shielding material present
μ = the linear attenuation coefficient (cm^{-1}) for photons in this material at the given photon energy

This *linear attenuation coefficient* μ gives the empirically observed rate at which the intensity of measured radiation decreases, according to the above formula. As noted in the definition, it is dependent on the incident photon energy and the material employed. Again, we revisit all of this in more detail in later chapters. One may also define a *mass attenuation coefficient*:

$$\mu_m = \mu/\rho \equiv cm^2/g.$$

Here ρ is the medium density, in g/cm^3. Figure 4.10 shows mass attenuation coefficients for gamma rays in water, and Figure 4.11 shows similar values for lead.

Example

Calculate the thickness of concrete or lead (Pb) to reduce the fluence rate of a beam of 0.5 MeV gamma rays to 10% of its initial intensity.

Reasonable values for the attenuation coefficients for these materials at this energy would be:

$\mu_{Concrete} = 0.252\,\text{cm}^{-1}$
$\mu_{Pb} = 1.8\,\text{cm}^{-1}$

$$I = I_0 e^{-\mu t}$$

$$\ln\left(\frac{I}{I_0}\right) = -\mu t$$

Concrete : $\ln(0.1) = 0.252t,$ $t = 9.1\,\text{cm}$
Lead : $\ln(0.1) = 1.8t,$ $t = 1.3\,\text{cm}$

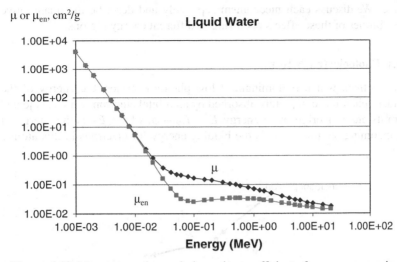

Figure 4.10 Mass attenuation and absorption coefficients for gamma rays in water. (Data from http://physics.nist.gov.)

4.9 Mechanisms

Photons interact with either orbital electrons or nuclei in passing through matter. Photons of low to moderate energy tend to interact with orbital electrons, whereas those of higher energy have interactions with atomic nuclei, with some interesting mechanisms and effects. For an incident beam of photons, even being purely monoenergetic, this is not an either/or proposition, however. Some of the reactions cause a shift from higher to lower photon energies, so some or even all of these effects may be observed to occur, and their cumulative effect is to reduce the radiation intensity as described by the equations

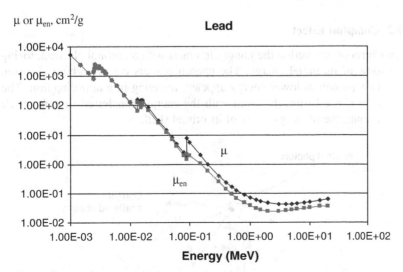

Figure 4.11 Mass absorption and attenuation coefficients for gamma rays in lead. (Data from http://physics.nist.gov.)

above. We discuss each mechanism separately and describe the quantitative probabilities of these effects occurring in different energy regions.

4.9.1 Photoelectric Effect

In this effect, which is dominant at low photon energies, the energy of the incident photon is completely absorbed by an orbital electron, which is ejected from its atom of origin with energy $E = E_\gamma - \phi$, where E_γ is the energy of the incoming photon and ϕ is the binding energy for electrons in this atomic shell.

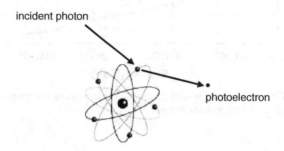

The incident photon thus completely disappears, and all of its energy is carried off by the photoelectron which acts in the medium just like any other electron or beta particle of that energy. It will ionize and excite other atoms, as described above, until all of its energy is dissipated. As an aside, this is why photons are one category of radiations termed *indirectly ionizing radiation*. Yes, the photon causes one ionization, but the vast majority of the ionization and excitation that occurs subsequently was caused by the photoelectron, not the photon itself. The cross-section (related to the probability of an interaction occurring) varies with Z^4/E^3, where Z is the mean atomic number of the medium and E is the photon energy.

4.9.2 Compton Effect

In this form of interaction the photon interacts with an orbital electron, losing only some of its initial energy. The photon scatters off the orbital electron, and a new photon of lower energy appears, traveling in a new direction. The electron is ejected from the atom with the energy transferred to it from the photon minus the binding energy of its orbital shell.

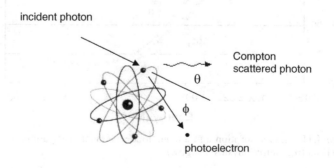

The energy (E) of the incident photon is given by

$$E = h\nu = \frac{hc}{\lambda}$$

Here h is Planck's constant (6.6262×10^{-34} J-s), ν is the frequency of the electromagnetic radiation (s^{-1}), c is the speed of light (3×10^8 m s^{-1}), and λ is the wavelength of the radiation (m). Treating the electron as being at rest just before the interaction, and conserving energy and momentum in the system:

$$\frac{hc}{\lambda} + m_oc^2 = \frac{hc}{\lambda'} + mc^2$$

The electron rest mass is m_0 and its mass at its new velocity v is m. If the photon is scattered from its original orientation through an angle θ and the electron is scattered in the opposite direction through an angle ϕ, we can also write:

$$\frac{h}{\lambda} = \frac{h}{\lambda'}\cos\theta + mv\,\cos\phi$$

$$0 = \frac{h}{\lambda'}\sin\theta + mv\,\sin\phi$$

Solving this system of equations, we can show that:

$$\Delta\lambda = \lambda' - \lambda = \frac{h}{m_0c}(1 - \cos\theta)$$

Now substituting $E = hc/\lambda$, we find that:

$$\boxed{E' = \frac{E}{1 + (E/m_0c^2)(1 - \cos\theta)}}$$

The photon will transfer the minimum possible energy by passing through an interaction with $\theta = 0°$. As you can imagine, this interaction is of little consequence in terms of biological consequences, shielding, and so on. The maximum energy transfer will occur at $\theta = 180°$, a backscattering collision. This will have an interesting consequence when we study radiation detection, in particular photon spectroscopy, later. As we study the spectrum of energies recorded from photon interactions in a medium, we show that this maximum energy of a Compton interaction causes there to be a region in the spectrum where no events will be recorded (between this maximum energy of transfer and the energy of the original gamma ray, which will have been recorded by a photoelectric effect). At this maximum energy, the energy of the scattered photon is given by

$$E' = \frac{E}{1 + (2E/m_0c^2)}$$

A complex formula was derived by investigators named Klein and Nishina, and gives the cross-section for photon scatter into a given solid angle Ω. The differential total scattering cross-section is given as

$$\frac{d\sigma_t}{d\Omega} = \frac{e^4}{2\,m_0^2c^4}\left[\frac{1}{1 + a(1 - \cos\theta)}\right]^2\left[\frac{1 + \cos^2\theta + a^2(1 - \cos\theta)^2}{1 + a(1 - \cos\theta)}\right]$$

In this formula, $a = h\nu/m_0c^2$, where ν is the photon frequency. In general, the probability of Compton interactions decreases with increasing photon energy and increasing atomic number of the medium. In low atomic-number media, Compton scattering is the predominant mode of interaction. Even in higher Z materials, at moderate energies, the Compton cross-sections are the highest, as we show shortly when we look at the combined cross-section curves that include all effects.

4.9.3 Pair Production

In this fascinating interaction, a photon of moderately high energy interacts with an atomic nucleus, and a positive and a negative electron are produced. Thus some of the quantum energy of the photon is converted to mass (according to $E = m_0c^2$), and the resulting particles are transported with any energy that remains. The photon energy must therefore exceed 1.02 MeV (the rest mass of an electron, 9.108×10^{-31} kg is 0.511 MeV in energy units). In practice, the process is really only apparent at energies above about 2 MeV.

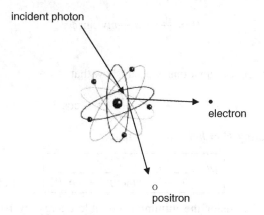

As with bremsstrahlung radiation, conservation of momentum considerations show that the positron and the electron will be projected in a forward direction, relative to the direction of the incoming photon. As with the photoelectric effect, the positron and the electron travel and lose energy in numerous ionization and excitation events, as would beta particles or positrons emitted from the nucleus of an atom. When the positron comes to "rest" it combines with an electron in the medium, they are annihilated, and two 0.511 MeV photons will be emitted (at 180° orientation from each other). So, consider for example a 7 MeV photon that interacts in a medium via the pair production mechanism. Each particle will carry (7 MeV – 2 × 0.511 MeV)/2, or about 3 MeV of kinetic energy. This energy will be dissipated in the medium eventually, and then two 0.511 MeV photons will be created from the positron/electron annihilation. These photons may then interact in the medium via Compton and photoelectric events.

4.9.4 Photodisintegration

In photodisintegration events, again most likely with higher-energy photons, a photon is captured by the nucleus of an atom, which subsequently de-excites with the emission of some particles, typically neutrons. This is a threshold phenomenon, as the energy added to the nucleus that absorbed the photon must be at least equal to the binding energy of a nucleon for that atom. These reactions generally have small cross-sections, ex. ^{9}Be (γ, n) ^{8}Be.

4.10 Photon Attenuation and Absorption Coefficients

When all of the photon interaction effects are combined, we can obtain plots of energy attenuation and absorption coefficients for photons of different energies in a given medium. The total attenuation coefficient (μ) is the sum of the coefficients for the interaction mechanisms discussed: Photoelectric (PE), Compton (C), and Pair Production (PP). So we can write that

$$\mu = \mu_{\text{PE}} + \mu_{\text{C}} + \mu_{\text{PP}}.$$

These coefficients have units of inverse length (e.g., cm^{-1}). They can be used to give the fraction of energy of a beam that is removed per unit distance in an attenuating medium. When looking at simply reducing the intensity of a source so that exposures on the other side of the medium from the source will be acceptable, we use these coefficients.

Another point of interest, however, is in the amount of energy that is actually deposited in the medium. This is important in evaluating biological effects, explaining the response of a radiation detector, and other applications. For this, we can also define a linear energy absorption coefficient. This coefficient includes only the energy absorbed in the medium from photoelectrons, Compton electrons, and the electron/positron pair. Energy carried away by scattered Compton photons, annihilation radiation (not subsequently absorbed in the medium), or bremsstrahlung radiation is not included. Whereas the attenuation coefficient is usually designated as just μ, the absorption coefficient is usually designated as μ_{en}. Both coefficients may be given as linear coefficients (with units of inverse length, e.g., cm^{-1}) or as mass coefficients (with units of, e.g., cm^2 g^{-1}), as shown above.

Tables of data are available from the National Institute of Standards and Technology (NIST; http://physics.nist.gov). Figures 4.10 and 4.11, above, show typical plots of attenuation and absorption coefficients for a pure element (lead, one used frequently in photon shielding) and for a composite substance (water, which is similar to human tissue).

4.11 Neutron Interactions

Treatment of neutron sources is often an unfamiliar subject to many practicing health physicists, as neutrons are not commonly encountered in many working environments. Important sources of neutrons are found in operating nuclear reactors and particle accelerator facilities. Reactors create significant neutron fluxes, and thus neutrons are an important contributor to ambient radiation fields around a reactor. Many particle accelerators create neutrons in capture reactions with various elements of the accelerator structural material (Table 4.1).

Neutron activation facilities house sources that emit significant quantities of neutrons for intentional activation of target materials to produce specific radionuclides. (Note: This is sometimes casually referred to as *isotope*

Table 4.1 Some representative nuclides produced by neutron reactions.

Target	Reaction	Product Use
^{98}Mo	^{98}Mo(n,γ) ^{99}Mo	^{99}Mo is the parent of ^{99m}Tc, and important nuclide in nuclear medicine.
^{235}U	^{235}U(n,f) ^{99}Mo	
^{124}Xe	^{124}Xe$(n,\gamma)^{125}$Xe $\rightarrow$ ^{125}I	^{125}I is used in nuclear medicine and as a biological tracer.
^{130}Te	^{130}Te$(n,\gamma)^{131}$Te $\rightarrow$ ^{131}I	^{131}I is used in nuclear medicine diagnosis and therapy.
^{235}U	^{235}U(n,f) ^{131}I	
^{59}Co	^{59}Co$(n,\gamma)^{60}$Co	^{60}Co is frequently used as a calibration source, among other uses.
^{191}Ir	^{191}Ir$(n,\gamma)^{192}$Ir	^{192}Ir is an important industrial radiography source.
^{58}Ni	^{58}Ni$(n,p)^{58}$Co	^{58}Co may be labeled to various substances and used as a biological tracer.
^{32}S	^{32}S$(n,p)^{32}$P	^{32}P is an important biological tracer and calibration source.
^{6}Li	^{6}Li$(n,\alpha)^{3}$H	^{3}H is used in nuclear weapons manufacture and is used as a biological tracer.

production.) In reactors and accelerators, materials become activated (i.e., nonradioactive material becomes radioactive) by capturing neutrons, and this is generally speaking an undesirable outcome, as new radiation sources are created that contribute radiation to the environment, in addition to the existing radiation field that exists. However, many reactors contain ports into which materials may be inserted for periods of time to intentionally activate the targets and make specific radionuclides. Instead of a full-power nuclear reactor, though, intense neutron sources (e.g., ^{252}Cf) can be housed in shielded facilities with access ports designed to permit the activation of specially prepared targets.

As we show in more detail later, these sources can also be used to identify elemental species in a sample by activating a nonradioactive atom and making it radioactive and then detecting the specific radiations it emits (neutron activation analysis). In addition, neutron calibration sources are available for calibrating neutron detection equipment. These sources often use a capture reaction as well, but this time another type of radiation, typically an alpha particle, is captured, and a neutron is released (Table 4.2).

Neutrons are generally categorized by their energy. Fast neutrons (although the strict definitions here may vary somewhat between sources) are those

Table 4.2 Some popular neutron sources.

Source Material	Reaction	Half-Life (Years)	Principal Alpha Energy (MeV)
^{226}Ra-Be	^{9}Be$(\alpha, n)^{12}$C	1,600	4.78−7.68
^{227}Ac-Be	^{9}Be$(\alpha, n)^{12}$C	22	5.71−47.44
^{239}Pu-Be	^{9}Be$(\alpha, n)^{12}$C	24,000	5.15
^{241}Am-Be	^{9}Be$(\alpha, n)^{12}$C	432	5.48

with energy >0.1 MeV, as initially created in a reactor. As neutrons interact with matter, they slow down to become intermediate neutrons (also called *resonance neutrons*). Eventually, the neutrons slow until they have reached thermal equilibrium with their environment (the energy level will depend on the temperature of the environment), and are called thermal neutrons. Most capture reactions occur with thermal neutrons, although, as we show later, specific high probabilities for certain reactions occur at intermediate energies, causing some neutrons to be captured before they reach thermal energies. These resonance energies are important to the overall function of a reactor, and to the design of neutron "poisons" (materials used to intentionally shut down nuclear reactions).

As neutrons penetrate matter, their number can be reduced by a number of interactions, including scattering and absorption. Instead of attenuation coefficients such as we use for gamma rays, the attenuation of neutrons is described by nuclear cross-sections in the following way.

$$I = I_0 e^{-\sigma N t}$$

where:
I_0 = original intensity of the neutron beam (n/s)
I = intensity after passing through matter of thickness t (n/s)
σ = nuclear cross-section (cm^2/atom)
N = the atom density of the material (atoms/cm^3)
t = the material thickness (cm)

This equation applies to a single pure material. If the material consists of several different materials, then this equation is expanded to include the effects of all absorbers.

The use of an "area-like" term to describe the probability of an interaction is unusual at first look. Neutron fields, however, are usually described in terms of the number of particles crossing a unit surface area per unit time (neutron *fluence rate*, although sometimes cited as *flux*). One can think therefore of a large number of baseballs or billiard balls being thrown by hundreds of people across a boundary. On the other side of the boundary imagine a wooden box turned so that its open side is towards the boundary. If the balls are thrown randomly, the chance that they will end up in the box is proportional to the cross-sectional area of the box opening; a larger box will simply intercept more balls.

As neutron interactions were first being studied, units for the interaction probabilities were considered, among them the Oppenheimer or the Bethe, to be named after important physicists who were working on problems involving neutron cross-sections in different materials. The story is told of a dinner on the campus of Purdue University in 1942, in which physicists M. G. Holloway and C. P. Parker were discussing a more easily repeated name for describing the size of atomic nuclei. The cross-sectional area of a uranium nucleus is about 10^{-24} square centimeters, which is tiny on the normal scale of size that we are used to dealing with but large compared with the size of many atomic particles. The barn, equal to exactly 1.0×10^{-24} square centimeters, is large compared with other farm buildings (as the dinner being in Indiana brought this mental image to mind), and thus came to be used to describe the size of atomic nuclei and the probability of nuclear interactions. In the above equation, if we multiply the cross-section per nucleus by the number of nuclei per cubic centimeter, we obtain a factor that is identical in function to a linear attenuation coefficient for a photon.

A very important fact to bear in mind, however, is that a large cross-section implies a large probability that an atom will have an interaction with a neutron: it does not relate to the physical size of the nucleus necessarily. Many nuclei of typical size have very large neutron capture cross-sections (e.g., cadmium, boron). Although we talk of large cross-sections, the reader should bear in mind that these are simply related to probabilities of interaction, not the physical sizes of the nuclei, even though they are expressed in units of square area. Cross-sections for scattering, as well as absorption, are expressed in barns. Important neutron interactions include the following.

4.11.1 Scattering

1. *Inelastic scattering*: In an inelastic scattering interaction, the neutron excites a nucleus in the medium, which subsequently emits a neutron and a photon. The threshold for such reactions is of the order of a few MeV. These interactions typically have low cross-sections.
2. *Elastic scattering*: "Billiard ball" type interactions, as were discussed above for other particulate radiations. A neutron may undergo a 100% energy loss with a H nucleus (from the mathematics shown above), or a maximum of a 22% loss with an O nucleus. Fast neutrons, as are formed during nuclear fission, scatter off various nuclei in slowing down, and may eventually reach thermal energies. Thermal neutrons are the most likely ones to be captured (see below). The scattered nuclei become ionizing particles, mostly H nuclei, which are now fast protons. So, like photons, neutrons are called indirectly ionizing particles, because they individually cause one other particle (such as an H nucleus), which in turn ionizes thousands of other atoms.

The distance that a neutron travels (i.e., the real distance: the length of track, which is always as long or longer than the depth of penetration) in reaching thermal energies is called the *fast diffusion length* or *slowing down length*. The distance traveled by a thermal neutron before being captured is related to the *thermal diffusion length*, which is defined as the thickness of a medium that attenuates a beam of thermal neutrons by a factor of *e*.

4.11.2 Absorption

Thermal neutrons are "captured" by nuclei with sufficiently large cross-sections. The nuclei that absorb the neutrons typically become radioactive, and thus various disintegration reactions follow. With the emission of a photon, the reaction is designated as (n, γ), with an alpha particle as (n, α), with a proton as (n, p), and so on.

A very special and important process is involved in neutron capture, and that is *neutron activation*. This involves the intentional production of a radioactive nuclide by absorption of a neutron. This has important applications in radionuclide production (to produce radiotracers, nuclear medicine compounds, etc.) and in the identification of unknown elements within a sample. The target or sample is irradiated with a flux of thermal neutrons. As more radioactive nuclei are produced, they are also decaying with their natural decay rate, so we can set up and solve the following equation.

$$\frac{dN}{dt} = \phi \sigma n - \lambda N$$

This equation just expresses the idea that the time rate of change of the number of atoms in a sample exposed to a constant neutron flux depends on the creation of new atoms over time (the first term) and their removal by radioactive decay. The terms in the equation are:

ϕ = neutron flux, neutrons/cm^2-s
σ = nuclear cross-section cm^2/atom
n = number of absorber atoms
λ = decay constant of radioactive species (s^{-1})
N = number of radioactive atoms

The solution of this differential equation is fairly easy and is:

$$\boxed{A = \lambda N = \phi \, \sigma n (1 - e^{-\lambda t})}$$

So the level of the created radioactive atoms builds up with a $(1 - e^{-\lambda t})$ form, to a constant value (eventually) of $(\phi \sigma n)$. At very early times (the term "early" being dependent on the value of λ), the curve is nearly linear. Then it begins to take the classic curvilinear inverse exponential shape and approaches the plateau value. Finally, at long times, it approaches this constant value of $(\phi \sigma n)$.

As with all $(1 - e^{-\lambda t})$ curves, it reaches one-half of the maximum value in one half-life $(0.693/\lambda)$ of the radioactive species produced, 75% of the maximum in two half-lives, 87.5% of the maximum in three half-lives, and so on. So when we wish to produce radioactive material by neutron activation, the question always arises, "How much time do we need to irradiate the target to get a specified activity level?" We often choose the target material because it has a good neutron cross-section in the first place, but after this, the only variables that we can manipulate to control the time of exposure are the neutron flux and the target mass. Then we have to just wait for the buildup of atoms to occur according to this formula.

We must always also keep in mind that after we remove the target from the neutron field, the radioactive atoms continue to decay. Thus if there is a significant delay from the time that irradiation stops and the material is

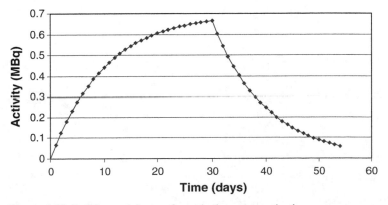

Figure 4.12 Buildup and decay of a typical neutron activation source.

delivered to the user, we need to produce more material than the user needs, to account for this loss of activity after the end of the irradiation period. Figure 4.12 shows a time-activity curve for a sample that will eventually reach a level of 0.7 MBq and has an approximately 7 day half-life. The sample is irradiated for 30 days (so it reaches only 95% of the saturation level) and is then removed from the flux and begins to decay.

Endnotes

1. G. F. Knoll, *Radiation Detection and Measurement*, 2nd ed. (Wiley, New York, 1989).
2. H. Cember, *Introduction to Health Physics*, 3rd ed. (McGraw-Hill, New York, 1996).
3. W. H. Bragg and R. Kleeman, On the alpha particles of radium and their loss of range in passing through various atoms and molecules, *Phil. Mag.* **10,** 318 (1905).
4. J. E. Turner, *Atoms, Radiation, and Radiation Protection* (Pergamon Press, New York, 1986).

5

Quantities and Units in Radiation Protection

Interaction of all kinds of radiation with matter ultimately results in the transfer of energy, through the processes of ionization and excitation. Quantifying the transfer of energy is important to the two most fundamental areas of radiation protection: radiation dosimetry and radiation instrumentation. Understanding these two areas is essential to the practice of health physics in almost every practical application. We have already studied the forms of these interactions. Now we discuss quantitative measures of these interactions. As discussed in Chapter 2, it is important to understand the distinction between a *quantity* (the parameter being measured) and a *unit* (the measure of the quantity). For example, velocity is the quantity that describes the time rate of motion of an object, and an example of the units that describe velocity is kilometers per hour (km/h).

It is also extremely important that one pay strict attention to the units in an expression. Making errors in the use and conversion of units is a common source of confusion in the quantitative assessment of a situation. One may easily give an expert assessment of a situation without quantities or units. For example, in a laboratory where a high intensity Cs-137 source is stored, it may be easy to guess that the dose rate at a meter from the source will be "high", even "dangerous" with no shielding applied to the source. That is a qualitative assessment. When asked, "What will the dose rate be to a person who stands at 1 m from this unshielded source?" we must then perform the appropriate calculations (outlined in Chapters 9 and 10), and provide a numerical answer that is correct and has the appropriate units.

As we soon show, many conversions that are used in the atomic world have very large or very small exponents (e.g., there are 1.6×10^{-13} joules in one megaelectron volt (MeV)). Thus, if one or more unit conversions are performed, involving multiplication and/or division, a small error in naming or citing the numerical value of a conversion constant may result in enormous errors in your final answer. Not only should the professional health physicist be diligent in the frequent checking of her units, she should also have a grasp on whether a calculated value is reasonable. The former should be practiced from day one of the study of health physics (and of course should be part of the routine discipline of any scientist); the latter comes only with experience. In early practice in real-life situations, the health physicist should be comfortable having another, more experienced health physicist check his work.

This author suggests another tip: when converting units in any reasonably complex situation, use horizontal lines to show and cancel units, rather than diagonal lines. This permits the clear view of which units are in the numerator and the denominator of the expression and thus correct cancellation of units. For example, compare the following two expressions:

$$\text{Activity} = \frac{10^{10}\ n}{cm^2 s}\ \frac{55\ b}{atom}\ \frac{10^{-24}\ cm^2}{b}\ \frac{\pi \times 0.5^2 \times 0.01\ cm^3}{}\ \frac{11.1\ g}{cm^3}\ \frac{mol}{60\ g}$$

$$\times \frac{6.025 \times 10^{23}\ atom}{mol}\ \frac{1\ trans}{n\ absorbed}\ \frac{1\ Bq-s}{trans} \times \left(1 - e^{\frac{-0.693 \times 24\ h}{2.9\ d \times 24\ \frac{h}{d}}}\right)$$

$$\text{Activity} = 1.02 \times 10^8\ Bq = 102\ \text{MBq}$$

$$\text{Activity} = 10^{10}\ n/cm^2 s \times 55\ b/atom \times 10^{-24}\ cm^2/b \times \pi \times 0.5^2$$

$$\times 0.01\ cm^3 \times 11.1\ g/cm^3 mol/60\ g \times 6.025 \times 10^{23}\ atom/mol$$

$$\times 1\ trans/n\ absorbed \times 1\ Bq - s/trans \times \left(1 - e^{\frac{-0.693 \times 24\ h}{2.9\ d \times 24\ h/d}}\right)$$

$$\text{Activity} = 1.02 \times 10^8\ Bq = 102\ \text{MBq}$$

The expressions are identical; the second was copied from the first and modified. It seems far easier to identify the correct units for cancellation in the first expression.

Finally, a word about significant figures. There are strict rules that apply to all formal calculations about significant figures. One should never have more significant figures in the result of a calculation than appeared in any individual value used in the calculation. You may know that the correct answer to "What is 9 times 9?" is not "81", but "80", in the problem as given. If you assumed that when I asked that question I was really using integer values in the question, so I was really asking, "What is 9.000...times 9.000...," then the answer "81" is correct under that assumption.

Some constants, like the number of hours in a day, have more significant figures than appear in their usage. But, given a problem in which a number of values are subjected to mathematical manipulation, it is important that the answer reflect a number of significant figures that is less than or equal to the number of significant figures in the value with the smallest number of significant figures. Then, common sense must be applied, and the final value should also be rounded to a sensible number of significant figures. In health physics, as in many forms of engineering, a number of modeling assumptions and simplifications are usually applied to solve a problem. Numerous uncertainties exist in the input values as well as the applicability of the values and the models used. Providing a dose rate near the above-mentioned Cs-137 source to 14 significant figures is just wrong, even in the unlikely event that all of the input data were this precisely known. Final answers in most health physics problems should be given to two or three significant figures, no more.

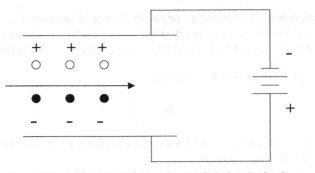

Figure 5.1 Schematic drawing of a parallel plate ionization chamber.

5.1 Exposure

As ionizing radiations interact with matter, they produce ions, which have charge. The simplest measurement of the quantitative effects of ionizing radiation with matter is the measurement of the number of ions that are created in air, using oppositely charged surfaces (typically charged metallic plates) to attract and count the ions formed. This approach was used in some of the earliest attempts to detect and quantify radiation field intensity and in the design of early radiation monitoring devices (Figure 5.1).

In all of the variables we study in this chapter, we cite the quantity that we are defining and give its exact definition, the units that are used to express numerical values of the quantity, and the special units that are used in health physics in common practice. For exposure, the definitions are:

Quantity: Exposure is defined as the sum of the electrical charges of all the ions of one sign produced in air by X-rays or gamma radiation when all electrons liberated by photons in a suitably small element of volume of air are completely stopped in air, divided by the mass of air in the volume element.

Units: Charge/mass of air, usually C/kg.

Special Unit: Roentgen (R) = 2.58×10^{-4} C/kg.

It is important to note that the quantity exposure is only defined for X and gamma radiation, and is only defined in air. Because radiation interactions produce ion pairs, and because these are easily collected by electronic devices, it is natural to evaluate the amount of ionization created in air by a radiation source as one's first evaluation of the intensity of a source and the amount of energy it is giving off to the environment. As is clear, the quantity exposure gives the amount of energy liberated in an air environment.

5.2 Absorbed Dose and Equivalent Dose

The quantity exposure may be useful for survey meter measurements (see Chapter 8 for an overview of radiation measuring devices), and gives us a measure of how much radiation is present in an environment. Ultimately we would like to know how much radiation was absorbed by objects in the environment, most importantly human beings. Chapter 6 gives an overview of the human body's response to radiation. The next quantity that we define moves us in this direction; it is *absorbed dose*. Absorbed dose is the energy absorbed per unit mass of any material. The definitions for this quantity are:

Quantity: Absorbed dose (D) is defined as:

$$D = \frac{d\varepsilon}{dm}$$

where $d\varepsilon$ is the mean energy imparted by ionizing radiation to matter in a volume element of mass dm.
Units: Energy/mass of any material. It can be erg/g, J/kg, or others.
Special Units:

$$rad = 100 \ erg/g$$
$$gray(Gy) = 1 \ J/kg$$
$$1 \ Gy = 100 \ rad.$$

The word *rad* was originally an acronym meaning "radiation absorbed dose". The rad is being replaced by the SI unit value, the gray (Gy), which is equal to 100 rad. Note that rad and gray are collective quantities; one does not need to place an "s" after them to indicate more than one. As emphasized above, absorbed dose may be defined for any kind of material at all. Given the exposure in air, we can calculate the absorbed dose in air (we do this shortly). We can calculate absorbed dose in brick, lead, water, human tissue, or any other material of interest. The practice of health physics is of course most directed towards the protection of human beings, so the dose to human tissue is of most interest.

As shown in Chapter 6, most biological effects of radiation can be related to an amount of absorbed dose. At very low doses, no effects may be observed. After dose reaches a particular threshold, effects may be observed, and will generally become more severe as more dose is received. As discussed in Chapter 6, however, when different experiments are performed in certain biological systems, using perhaps different kinds of radiation, or measuring different biological endpoints, different amounts of absorbed dose may be needed to observe a particular effect. This is particularly true for high LET radiations such as alpha particles and fast protons (from accelerators or as produced by neutron interactions in tissue). Another quantity has been traditionally defined to account for these differences; it is the equivalent dose. Equivalent dose is the absorbed dose modified by a factor accounting for the effectiveness of the radiation in producing biological damage.

Closely related to absorbed dose is the quantity *kerma*, which is actually an acronym meaning "kinetic energy released in matter". Kerma is given as

$$K = \frac{dE_{Tr}}{dm}.$$

dE_{Tr} is the sum of the initial kinetic energies of all the charged ionizing particles liberated by uncharged ionizing particles in a material of mass dm. Kerma has the same units and special units as absorbed dose, but is a measure of energy liberated, rather than energy absorbed. The two will be equal under conditions of charged particle equilibrium, and assuming negligible losses by bremsstrahlung radiation.

Quantity: Equivalent dose $(H_{T,R})$ is defined as

$$H_{T,R} = w_R D_{T,R},$$

where $D_{T,R}$ is the dose delivered by radiation type R averaged over a tissue or organ T and w_R is the radiation weighting factor for radiation type R.

Units: w_R is really dimensionless, so fundamentally, the units are the same as absorbed dose. Operationally, however, we distinguish using the special units:

Special Units:

$$\text{rem} = D(\text{rad}) \times w_R$$

$$\text{sievert (Sv)} = D(\text{Gy}) \times w_R$$

$$1 \text{ Sv} = 100 \text{ rem}.$$

Note that like rad and gray, the rem and sievert are collective terms; one need not speak of "rems" and "sieverts", although this may be heard in common speech and even observed in publications. Also note that units that incorporate a person's name (Roentgen, Gray, Sievert) are given in lower case when spelled out completely, but with the first letter capitalized when given as the unit abbreviation (e.g., "sievert" and "Sv").

Equivalent dose is defined for any kind of radiation, but only in human tissue. The recommended values of the radiation weighting factor have varied somewhat over the years, as evidence from biological experiments has been given and interpreted. The current values recommended by the International Commission on Radiological Protection (ICRP, see Chapter 7) are shown in Table 5.1.

So, for photons and electrons, numerical values of the absorbed dose in gray and equivalent dose in sievert are equal. For alpha particles and fast protons, the equivalent dose is a multiple of the absorbed dose. For neutrons, the radiation weighting factor is given as a function of energy, with a maximum value occurring in the range of 100 keV to 2 MeV. To adequately calculate the equivalent dose in a mixed neutron field, then, one must characterize the energy spectrum of the neutrons. Specialized survey meters have been designed for this purpose and are discussed in Chapter 8. Current U.S. regulations (Title 10, Code of Federal Regulations, Part 20, i.e., 10CFR20) apply the values shown in Table 5.2 for quality factors (another name for radiation weighting factors) in the routine practice of radiation protection.

Table 5.1 Radiation weighting factors recommended by the ICRP.

Type of Radiation	w_R
Photons, all energies	1
Electrons and muons, all energies (except Auger electrons in emitters bound to DNA)	1
Neutrons, energy:	
<10 keV	5
10 keV to 100 keV	10
>100 keV to 2 MeV	20
>2 MeV to 20 MeV	10
>20 MeV	5
Protons, other than recoil protons, $E > 2$ MeV	5
Alpha particles, fission fragments, heavy nuclei	20

Table 5.2 Quality factors given in 10CFR20.

Type of Radiation	Quality Factor (Q)
X, gamma, or beta radiation	1
Alpha particles, multiple-charged particles, fission fragments, and heavy particles of unknown charge	20
Neutrons of unknown energy	10
High-energy protons	10

Neutron Energy (MeV)	Quality Factor	Fluence per Unit Dose Equivalent (Neutrons cm^{-2} rem^{-1})
2.5×10^{-8}	2	980×10^6
1×10^{-7}	2	980×10^6
1×10^{-6}	2	810×10^6
1×10^{-5}	2	810×10^6
1×10^{-4}	2	840×10^6
1×10^{-3}	2	980×10^6
1×10^{-2}	2.5	1010×10^6
1×10^{-1}	7.5	170×10^6
5×10^{-1}	11	39×10^6
1	11	27×10^6
2.5	9	29×10^6
5	8	23×10^6
7	7	24×10^6
10	6.5	24×10^6
14	7.5	17×10^6
20	8	16×10^6
40	7	14×10^6
60	5.5	16×10^6
1×10^2	4	20×10^6
2×10^2	3.5	19×10^6
3×10^2	3.5	16×10^6
4×10^2	3.5	14×10^6

A simple analysis can show that the exposure in air can be related to the dose delivered to air. Assume that a source is giving off radiation such that 1 roentgen (1 R) is measured in a given time period. Knowing the definition of a roentgen, and that any ion pair carries 1.6×10^{-19} coulombs of charge of either sign, and furthermore that it takes about 34 eV of energy to create one ion pair in air, we can write:

$$1\ R = \frac{2.58 \times 10^{-4} \mathrm{C}}{\mathrm{kg}} \frac{1\ \mathrm{ion}}{1.6 \times 10^{-19} C} \frac{34\ \mathrm{eV}}{\mathrm{ion}} \frac{1.6 \times 10^{-19} \mathrm{J}}{\mathrm{eV}} \frac{\mathrm{Gy} - \mathrm{kg}}{\mathrm{J}}$$

$$1\ R = 0.00877\ \mathrm{Gy} = 0.877\ \mathrm{rad}$$

Table 5.3 Summary of important radiation protection quantities and units.

	Types of Radiation for Which It Is Defined	Types of Media in Which It Is Defined	Example of Generic Units	Example of Special Units
Exposure	X and γ	Air	C/kg	R (roentgen) = 2.58×10^{-4} C/kg
Absorbed Dose	All	Any	erg/g, J/kg	1 rad = 100 erg/g 1 Gy = 1 J/kg
Equivalent Dose	All	Human tissue	erg/g, J/kg	1 rem = 100 erg/g $\times w_R$ 1 Sv = 1 J/kg $\times w_R$

So, 1 R of exposure will deliver 0.877 rad to the air being irradiated in this time (assuming conditions of "electronic equilibrium", defined later). If a human being is standing in the air space, we would like to know how much dose is being received by the body tissues. The ratio of the mass absorption coefficients for human tissue/air is $\mu_{m-\text{tissue}}/\mu_{m-\text{air}} \approx 1.09$, so:

$$0.877 \text{ rad (air)} \times 1.09 = 0.95 \text{ rad (tissue)}$$

Thus 1 R measured in air will give a dose to tissue of about 1 rad, also 1 rem (as the w_R for photons is 1). This relationship has made conversions in the non-SI unit system particularly easy. One must be careful here, however: most radiation detectors give readings in counts/min or exposure. If you are really measuring photons (and the meter is suitably calibrated for the photon energies you are measuring; more on this in Chapter 8), your reading in mR/hr will give a good estimate of the absorbed dose rate or equivalent dose rate to persons in that area in mrad/hr or mrem/hr.

But you can also measure a beta source with a detector that has a thin window and get a reading in mR/hr. This usually means nothing in terms of dose. Survey meters can be calibrated to give quantitative data for particular beta emitters in particular geometries, using a translation factor to convert the meter reading in exposure readings to known correct values of beta dose rate. Simply waving a survey meter over an area contaminated with some beta emitter and assuming that the noted exposure rate is numerically equal to an absorbed dose rate may lead to significant errors.

As we study radiation dosimetry, we define some other specialized units of equivalent dose. For now, the quantities and units above and shown in Table 5.3 define most of what is needed for the practice of health physics.

5.3 Radioactivity

The reader is reminded of the other key quantity of activity, defined in Chapter 3.

Quantity: Activity is the number of nuclear transformations per unit time occurring in a given sample of radioactive material.
Units: Nuclear transformations/unit time.

Special Units:

The curie (Ci) $= 3.7 \times 10^{10}$ transformations/s

The becquerel, where 1 becquerel (Bq) $= 1$ transformation/s

Two other quantities are also defined:

Quantity: The radioactive decay constant (λ) is the rate constant for radioactive atoms undergoing spontaneous nuclear transformation.
Units: s^{-1}.

Quantity: The radioactive half-life is the time needed for one half of the atoms in a sample of radioactive material to undergo transformation. Mathematically the half-life is $\ln(2)/\lambda = 0.693/\lambda$.
Units: s.

5.4   Particle and Energy Field Units

In the measurement of radiation fields, particularly of neutrons, but also of photons, we define the number of particles or their energy in the field, with reference to a point or surface area in space.

Quantity: Particle flux is the quotient of the number of particles crossing a point in a given time interval.
Units: Particle flux is measured in s^{-1}. It is really particles per second, but "particles" is not a fundamental unit (mass, length, or time).

Quantity: Particle fluence is the quotient of the number of particles incident on a cross-sectional area, or incident on a sphere of a given cross-sectional area, and does not have time dependence.
Units: The unit for the fluence is m^{-2}, but is often shown as cm^{-2}.

Quantity: Particle fluence rate is the quotient of the number of particles crossing a cross-sectional are over a given time.
Units: Fluence rate is measured in $m^{-2} s^{-1}$.

Quantity: Particle radiance is the fluence rate of particles propagating in a specified direction within a stated solid angle.
Units: Radiance is measured in $m^{-2} s^{-1} sr^{-1}$.

We can also similarly define energy flux, fluence, fluence rate, and radiance with units of (for example) MeV s^{-1}, MeV m^{-2}, MeV $m^{-2} s^{-1}$, and MeV $m^{-2} s^{-1} sr^{-1}$.

Biological Effects of Radiation

6.1 Introduction: Background

On November 8, 1895, at the University of Wurzburg, Wilhelm Conrad Roentgen (1845–1923) was studying various physical phenomena and observed a glowing fluorescent screen in his laboratory. Studying cathode rays from an evacuated glass tube, Roentgen surmised that the fluorescence was from invisible rays originating from within the tube. The rays penetrated some opaque black paper wrapped around the tube. While considering this mystery, Roentgen realized the discovery of penetrating radiation, named "X-rays" because of its mysterious nature. This momentous event had an enormous and immediate impact on the fields of physics and medicine.

Antoine Henri Becquerel (1852–1908) was raised in a scientific environment. His grandfather studied electrochemistry and his father was a student of the fields of fluorescence and phosphorescence. Becquerel's proximity to this level of scientific inquiry and the minerals and compounds studied by his family allowed him to experiment in ways that might extend Roentgen's discovery. Becquerel chose to work with potassium uranyl sulfate. He exposed quantities of the material to sunlight and placed it on photographic plates wrapped in black paper. The developed photographic plates revealed images of the uranium crystals. Becquerel concluded that "the phosphorescent substance in question emits radiation which penetrates paper opaque to light."[1] He was operating under the assumption that the crystals were absorbing the sun's energy and then giving off X-rays. Serendipity played a role in his next observation. The 26th and 27th of February brought overcast skies to Paris. Becquerel's experiments were delayed, as he had no source of energy input for his crystals. Uranium-covered plates that were destined for experimentation were simply put away in a drawer. Several days later, on the first of March, he developed the photographic plates, not expecting much from the images. Astonishingly, the images were clear and strong. He surmised that the uranium itself was emitting some form of radiation with no external source of energy required. Becquerel had discovered radioactivity, the spontaneous emission of radiation by unstable substances.

Pierre Curie (1859–1906) and Marie Curie (1867–1934) worked together to discover other radioactive substances. Pierre was an established scientist, studying piezoelectricity (in which physical pressure applied to crystals results in the creation of an electric potential). He also had made contributions to the study of magnetism, including the identification of a temperature, named the "curie point", above which a material's magnetic properties are no longer observed. When he married Marie, they began investigating the recently divulged phenomenon of radioactivity discovered by Becquerel. Marie actually coined the term "radioactivity" to describe the interesting phenomenon. Following some chemical extractions that the Curies made of uranium from natural ore, Marie noted the residual material had a higher level of radioactivity than the natural ore. She concluded that the ore contained other elements that were also radioactive, indeed more radioactive than the original uranium. Through this, they discovered the elements polonium and radium; later they determined the separate chemical properties of each element. The Curies were awarded the Nobel Prize in Physics in 1903 for their work with radioactivity.

Three years later, while crossing a street in a rainstorm, Pierre was killed, trampled by horses drawing a carriage. Marie did an unprecedented thing and assumed Pierre's teaching position at the Sorbonne. No woman had ever taught there in its 650 year history. In Pierre's honor, the 1910 Radiology Congress chose the curie as the basic unit of radioactivity. It was defined as the quantity of radon in equilibrium with one gram of radium (we currently use the definition that $1 \text{ Ci} = 3.7 \times 10^{10}$ dps). One year later, Marie was awarded the Nobel Prize in Chemistry for her discoveries of radium and polonium; she was the first person to receive two Nobel Prizes. This remarkable scientist continued to study the use of radium as a possible treatment for cancer. She herself died on July 4, 1934, from pernicious anemia probably associated with her cumulative radiation exposure.

Very soon after the discovery of radiation and radioactivity it became evident that exposure to radiation could induce short- and long-term negative effects in human tissue. The first possible adverse effects of X-rays were observed by Thomas Edison, William J. Morton, and Nikola Tesla. These investigators independently reported eye irritations from experimentation with X-rays and fluorescent substances. The effects were first attributed to eye strain, or possibly, ultraviolet radiation from the long-term direct observance of fluorescence. Elihu Thomson (an American physicist) deliberately exposed the little finger of his left hand to an X-ray tube for several days, for about half an hour per day. This resulted in pain, swelling, stiffness, erythema, and blistering in the finger, which was clearly and immediately related to the radiation exposure. William Herbert Rollins (a Boston dentist) showed that X-rays could kill guinea pigs and result in the death of offspring when guinea pigs were irradiated while pregnant. In 1898, Henri Becquerel received a skin burn from a radium source given to him by the Curies that he kept in his vest pocket for some time. He carried the source with him on his travels, to use it in demonstrations during his lectures. He declared, "I love this radium but I have a grudge against it!"[2] The first death of an X-ray pioneer attributed to cumulative overexposure was C. M. Dally in 1904. It was later observed that radiologists and other physicians who used X-rays in their practices before health physics practices were common had a significantly higher rate of leukemia than their colleagues.

A particularly tragic episode in the history of the use of radiation and in the history of industrialism was the acute and chronic damage done to the

Figure 6.1 The Orange, N.J. radium dial painting factory in the mid-1920s.[3]

Figure 6.2 Close-up of the workshop, this one in Ottawa, Illinois.[3]

radium dial painters.[3] Radium was used in luminous paints in the early 1900s (Figures 6.1, 6.2). In factories where luminous dial watches were made, workers (mainly women) would sharpen the tips of their paint brushes with their lips, and ingested large amounts of radium. They had increased amounts of bone cancer (carcinomas in the paranasal sinuses or the mastoid structures, which are very rare, and were thus clearly associated with their exposures, as well as cancers in other sites) and even spontaneous fractures in their jaws and spines from cumulative radiation injury. Others died of anemia and other causes. The industry representatives (U.S. Radium) tried to deny the association. An organization called the Consumers League and journalist Walter Lippmann, an editor with the *New York World* took an active role in trying to publicize the plight of these women when it became clear that their maladies were associated with their ingestion of the radium.

One complication is that this occurred during a time when it was thought that moderate exposures to radiation in general and radium in particular could be beneficial. People visited mines and natural springs where radium and uranium concentrations were high. They intentionally ingested medical drinks sold over the counter until 1931, one of which, "Radithor," contained high levels of radium. They purchased special jars ("Revigators") into which they poured water and drank the water periodically.[4] The jars had sources of uranium, radium, or other materials that would allow radon gas to diffuse into the water so that when it was drunk it was "charged" with the invigorating radioactivity. Indeed, Surgeon General Dr. George H. Torney in about 1910 said that "Relief may be reasonably expected at the [radioactive] Hot Springs in... various forms of gout and rheumatism, neuralgia, metallic or malarial poisoning, chronic Brights disease, gastric dyspepsia, chronic diarrhea, chronic skin lesions, etc."[5]

Grace Fryer was an employee of U.S. Radium who worked with radium at one of the factories until 1920. About two years later, her teeth started falling out and her jaw developed a painful abscess, and she had significant systemic pain as well. X-ray photos of her mouth and back showed the development of serious bone decay, but her doctors did not understand the source of her problems. In July 1925, a doctor suggested that the problems may have been occupationally related. Along with four other women (Edna Hussman, Katherine Schaub, and sisters Quinta McDonald and Albina Larice), Grace Fryer decided to sue U.S. Radium in 1927. Each plaintiff asked for $250,000 in compensation for medical expenses and their suffering. The five became known through articles carried in newspapers in both the United States and Europe as "the Radium Girls." The case received considerable attention in the media due to the efforts of Lippmann and others, and was quite controversial. Marie Curie herself opined publicly about the matter.

By the time of the trial, Quinta McDonald and Albina Larice were bedridden. Grace Fryer had lost all of her teeth and could not walk or even sit up without the use of a back brace. When the first court hearing came up January 11, 1928, the women could not raise their arms to take the oath of testimony because of their advanced disease and pain.

Amelia Maggia, another former dial painter and also a sister of McDonald and Larice, died in 1922 from what was diagnosed to be syphilis. A New York dentist, Joseph P. Knef, who had treated Maggia, removed her decayed jawbone some time before she died. "Before Miss Maggia's death I became suspicious that she might be suffering from some occupational disease," Knef said.[4] At first he suspected that she might have the so-called "phossy jaw" necrosis, due to phosphorous intakes, common among workers in the matchmaking industry in the 18th century. Knef subsequently wrapped the jawbone in some unexposed dental film for a week and later developed the film. He also confirmed that the bone contained radioactivity using an electroscope and documented this in a published work with the Essex County medical examiner. A Newark, N.J. newspaper said "Dr. Knef then described how radium, hailed as a boon to mankind in treatment of cancer and other diseases, becomes a subtle death-dealing menace[4]". An autopsy requested by the Maggia sisters and Raymond Berry confirmed Knef's findings; her other bones were also highly radioactive. The original diagnosis of syphilis was thus refuted, and the dramatic exposed films were instrumental in the presented case. An out-of-court settlement was reached just days before the case was to go to trial. Berry and the "Radium Girls" agreed that each would receive $10,000 and a $600 per year while alive, and that all medical and legal expenses incurred would also be paid by the company.

6.2 Mechanisms of Radiation Damage to Biological Systems

Near the end of this chapter we look in more detail at some of the newer thinking that is ongoing about mechanisms of radiation effects on cells. The conventional wisdom until recently has been that damage to biological molecules, particularly the DNA, from ionization and excitation events, ultimately result in cell transformation or death. Cell transformation may be fatal, or may lead to expression of disease (typically cancer) at a later time. The damage to the DNA may be direct (i.e., the ionizing particle directly interacts with an atom in the DNA, causing damage to occur) or indirect (radiation interacts with water near the DNA, with subsequent attack of the formed free radicals on the DNA molecules).

Cember[6] shows, with a short mathematical example, why the majority of radiation damage, however, is not caused by direct radiation hits on the DNA, but by indirect effects. Consider that the $LD_{50/30}$ for the human for whole-body exposure (with no aggressive medical care) is about 4 Gy. Delivery of this level of energy to the body will result in the creation of 7.35×10^{17} ion pairs per kg of body weight:

$$\frac{4 \text{ Gy} \times 6.25 x 10^{18} \text{ eV kg}^{-1} \text{ Gy}^{-1}}{34 \text{ eV ion}^{-1}} = 7.35 x 10^{17} \text{ ions kg}^{-1}$$

If we assume that there are about 9 atoms excited for every one ionized (an assumption that Cember claims without support), and knowing that there are about 10^{26} atoms in each kg of tissue, we find that only about 1 atom in 10 million (10^{-7}) is directly affected by this deposition of energy in the body. Instead, the majority of damage comes from the creation of radicals due to ionization and excitation of water molecules. These created radicals diffuse for some distance, interact with other molecules, which may be of biological significance, and ultimately cause damage to cell structures, most notably DNA.

Radiation interactions with aqueous systems can be described as occurring in four principal stages (Figure 6.3, from Turner[7]):

1. Physical
2. Prechemical
3. Early chemical
4. Late chemical

In the *physical* stage of water radiolysis, a primary charged particle interacts through elastic and inelastic collisions. Inelastic collisions result in the ionization and excitation of water molecules, leaving behind ionized (H_2O^+)

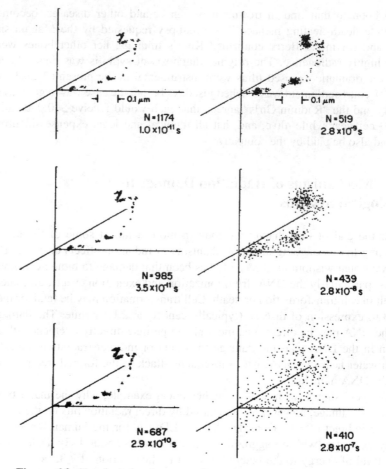

Figure 6.3 Chemical development of an electron track over the first $\sim 10^{-6}$ seconds after passage of the electron.[7]

and excited (H_2O^*) molecules, and unbound subexcitation electrons (e_{sub}^-). A *subexcitation* electron is one whose energy is not high enough to produce further electronic transitions. By contrast, some electrons produced in the interaction of the primary charged particle with the water molecules may have sufficient energy themselves to produce additional electronic transitions. These electrons may produce secondary track structures (delta rays) beyond those produced by the primary particle. All charged particles can interact with electrons in the water both individually and collectively in the condensed, liquid, phase. The initial passage of the particle, with the production of ionized and excited water molecules and subexcitation electrons in the local track region (within a few hundred angstroms), occurs within about 10^{-15} s.

From this time until about 10^{-12} s, in the *prechemical* phase, some initial reactions and rearrangements of these species occur. If a water molecule is ionized, this results in the creation of an ionized water molecule and a free electron. The free electron rapidly attracts other water molecules, as the slightly polar molecule has a positive and negative pole, and the positive pole is attracted to the electron. A group of water molecules thus clusters around the electron, and it is known as a "hydrated electron" and is designated as e_{aq}^-.

Table 6.1 Comparison of reaction rate coefficients and reaction radii for several reactions of importance to radiation biology.*

Reaction	k $(10^{10}\ \mathrm{M^{-1}\ s^{-1}})$	R (nm)
$H\cdot + OH\cdot \rightarrow H_2O$	2.0	0.43
$e_{aq}^- + OH\cdot \rightarrow OH^-$	3.0	0.72
$e_{aq}^- + H\cdot + H_2O \rightarrow H_2 + OH^-$	2.5	0.45
$e_{aq}^- + H_3O^+ \rightarrow H\cdot + H_2O$	2.2	0.39
$H\cdot + H\cdot \rightarrow H_2$	1.0	0.23
$OH\cdot + OH\cdot \rightarrow H_2O_2$	0.55	0.55
$2e_{aq}^- + 2H_2O \rightarrow H_2 + 2OH^-$	0.5	0.18
$H_3O^+ + OH^- \rightarrow 2H_2O$	14.3	1.58
$e_{aq}^- + H_2O_2 \rightarrow OH^- + OH\cdot$	1.2	0.57
$OH\cdot + OH^- \rightarrow H_2O + O^-$	1.2	0.36

*k is the reaction rate constant and R is the reaction radius for the specified reaction.

The water molecule dissociates immediately:

$$H_2O \rightarrow H_2O^+ + e_{aq}^- \rightarrow H^+ + OH\cdot + e_{aq}^-.$$

In an excitation event, an electron in the molecule is raised to a higher energy level. This electron may simply return to its original state, or the molecule may break up into an H and an OH radical (a *radical* is a species that has an unpaired electron in one of its orbitals; the species is not necessarily charged, but is highly reactive).

$$H_2O \rightarrow H\cdot + OH\cdot.$$

The free radical species and the hydrated electron undergo dozens of other reactions with each other and other molecules in the system. Reactions with other commonly encountered molecules in aqueous systems are shown in Table 6.1. Reactions with other molecules have been studied and modeled by various investigators as well.[8–11]

The *early chemical* phase, extending from $\sim 10^{-12}$ s to $\sim 10^{-6}$ s, is the time period within which the species can diffuse and react with each other and with other molecules in solution. By about 10^{-6} s most of the original track structure is lost, and any remaining reactive species are so widely separated that further reactions between individual species are unlikely.[7] From 10^{-6} s onward, referred to as the *late chemical* stage, calculation of further product yields can be made by using differential rate-equation systems that assume uniform distribution of the solutes and reactions governed by reaction-rate coefficients.

6.3 Biological Effects in Humans

There are two broad categories of radiation-related effects in humans, stochastic and nonstochastic. There are three important characteristics that distinguish them.

6.3.1 Nonstochastic Effects

(now officially called "deterministic effects", previously also called "acute effects") are effects that are generally observed soon after exposure to radiation. As they are "nonstochastic" in nature, they will always be observed (if the dose threshold is exceeded), and there is generally no doubt that they were caused by the radiation exposure. The major identifying characteristics of nonstochastic effects are:

1. There is a threshold of dose below which the effects will not be observed.
2. Above this threshold, the magnitude of the effect increases with dose.
3. The effect is clearly associated with the radiation exposure.

Examples include:
Erythema (reddening of the skin)
Epilation (loss of hair)
Depression of bone marrow cell division (observed in counts of formed elements in peripheral blood)
NVD (nausea, vomiting, diarrhea), often observed in victims after an acute exposure to radiation Central nervous system damage
Damage to the unborn child (physical deformities, microcephaly (small head size at birth), mental retardation

When discussing nonstochastic effects, it is important to note that some organs are more radiosensitive than others. The so-called *Law of Bergonie and Tribondeau*[12] states that cells tend to be radiosensitive if they have three properties:

• Cells have a high division rate.
• Cells have a long dividing future.
• Cells are of an unspecialized type.

A concise way of stating the law might be to say that "The radiosensitivity of a cell type is proportional to its rate of division and inversely proportional to its degree of specialization." So, rapidly dividing and unspecialized cells, as a rule, are the most radiosensitive. Two important examples are cells in the red marrow and in the developing embryo/fetus. In the case of marrow, there are a number of progenitor cells which, through many generations of cell division, produce a variety of different functional cells that are very specialized (e.g., red blood cells, lymphocytes, leukocytes, platelets) (Figure 6.4[13]). Some of these functional cells do not divide at all, and are thus themselves quite radioresistant. However, if the marrow receives a high dose of radiation, damage to these progenitor cells is very important to the health of the organism. As we show shortly, if these cells are affected, in a short period this will be manifested in a measurable decrease in the number of formed elements in the peripheral blood. If the damage is severe enough, the person may not survive. If the damage is not so severe, the progenitor cells will eventually repopulate and begin to replenish the numbers of the formed elements, and subsequent blood samples will show this recovery process.

   In the fetus, organs and systems develop at different rates. At the moment of conception, of course, we have one completely undifferentiated cell that becomes two cells after one division, then four then eight, and so on. As the rapid cell division proceeds, groups of cells "receive their assignments" and differentiate to form organs and organ systems, still with a very rapid

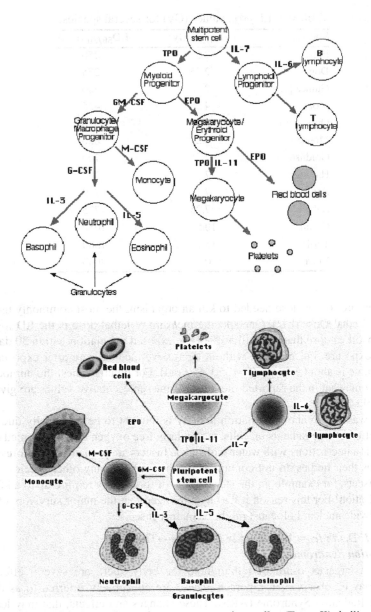

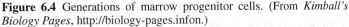

Figure 6.4 Generations of marrow progenitor cells. (From *Kimball's Biology Pages*, http://biology-pages.infon.)

rate of cell division. At some point, individual organs become well defined and formed, and cell division slows as the fetus simply adds mass. But while differentiation and early rapid cell division are occurring, these cells are quite radiosensitive, and a high dose to the fetus may cause fetal death, or damage to individual fetal structures. This is discussed further below. On the other hand, in an adult, cells of the central nervous system (CNS; brain tissue, spinal cord, etc.) are very highly specialized and have very low, or no, rate of division. The CNS is thus particularly radioresistant.

One important nonstochastic effect is death. This results from damage to the bone marrow (first), then to the gastrointestinal tract, and then to the nervous system. The dose levels involved are discussed shortly. There are a number

Table 6.2 $LD_{50/30}$ values (Gy) for several species.

Species	$LD_{50/30}$ (Gy)	$LD_{50/30}$ (rad)
Pig	2.5	250
Dog	2.75	275
Guinea pig	3	300
Monkey	4.25	425
Oppossum	5.1	510
Mouse	6.2	620
Goldfish	7	700
Hamster	7	700
Rat	7.1	710
Rabbit	7.25	725
Gerbil	10.5	1050
Turtle	15	1500
Newt	30	3000

of measures of the dose needed to kill an organism; the most commonly used is the Lethal Dose (LD). One specific measure of lethal dose is the $LD_{50/30}$, the absorbed dose that will kill 50% of the exposed population within 30 days after exposure. For humans, without aggressive medical care after exposure, this value is about 3.5–4.5 Gy (350–450 rad). For other species, the numbers vary somewhat in the reported literature. Some representative values are given in Table 6.2.

As was shown above, radiation toxicity is thought to be principally due to the action of free radicals on cells, particularly free oxygen radicals formed by radiation interactions with water molecules. Insects are particularly radioresistant, as their bodies do not contain as much water as many other species. To kill insects, for example, in the sterilization of foods, may require up to a kGy of radiation! For this reason it was hypothesized that the major survivors of a worldwide nuclear holocaust might be mainly insects.[14]

6.3.1.1 Death from Whole Body Exposure—The Acute Radiation Syndrome

After a large exposure of radiation to the body, there are several effects that may be observed in a person. These are collectively referred to as the Acute Radiation Syndrome (ARS). These changes and effects, that may lead to death, are broadly described in three general categories: the hemopoetic (or hematopoetic syndrome), the gastrointestinal syndrome, and the CNS syndrome. These are given in order of dose level at which they are observed; the hemopoetic syndrome will be observed at the lowest dose levels because of the radiosensitivity of the red marrow. In all cases, there are several phases of effect manifestation that are commonly observed:

- *Prodromal stage*: Immediately after exposure, subjects might experience a loss of appetite, nausea, vomiting, fatigue, and diarrhea. After extremely high doses, additional symptoms such as fever, prostration, respiratory distress, and hyperexcitability can occur. This phase occurs within minutes to days after the exposure.
- *Latent stage*: During this phase, the subject may look and feel healthy. In some cases, in fact, victims have reported feeling considerably better than

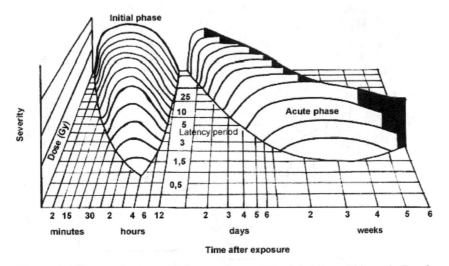

Figure 6.5 Phases of acute radiation syndrome. (Reprinted from Thierry de Revel, Menace terroriste approche médicale. *Nucléaire Radiologique Biologique Chimique*, 2005, with kind permission from Editions John Libbey Eurotext, Paris.)

usual, having a high energy level. This phase lasts for hours to a few weeks (depending upon the size of the radiation dose).

- *Manifest illness stage*: Here, the symptoms of the disease will manifest themselves. The severity and duration of this phase depends on the syndrome and again, the dose, but lasts for some hours to months after exposure.
- *Recovery or death*: If the subject will die from the acute exposure, death will occur within one to a few months. Complete recovery may continue for a week to years. Stochastic effects (cancer) may of course occur many years later, but if the subject survives this first period, he will generally recover to general good health eventually.

Bertho and colleagues[15] attempted to show the various phases of the ARS graphically (Figure 6.5).

6.3.1.1.1 Hemopoetic Syndrome: In the hemopoetic syndrome, as discussed above, damage to the progenitor cells of the red marrow causes changes in the levels of formed elements in the peripheral blood that can be observed through simple sampling of the blood and counting of the cells by a variety of methods. Changes in blood can be observed at whole-body doses of as low as 100–500 mGy (10–50 rad). The syndrome, with illness that may lead to death, occurs at $\sim$2 Gy (200 rad). The major characteristics of the syndrome are:

- A sudden sharp increase in granulocyte count, followed by decrease.
- A depression in platelet count, leucocyte count, and red blood cell count (after one to two weeks).
- Blood counts will reach a minimum (called the "nadir") and recover to normal levels within four to six weeks, or subject may die (>7 Gy).

Figure 6.6 shows typical responses for a subject exposed to radiation such that a significant change in blood counts was observed, but also such that the individual recovered.[6] Subjects will usually immediately express the NVD syndrome, then feel malaise and fatigue. Some epilation may also be commonly observed at some sites on the body.

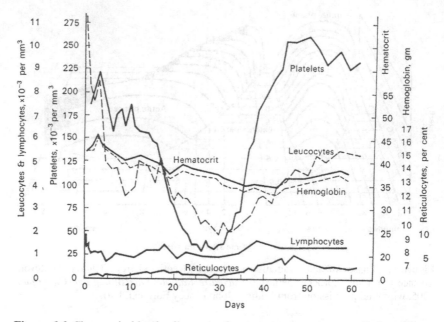

Figure 6.6 Changes in blood cell counts after an acute exposure to radiation. (From Cember, with permission, McGraw-Hill, 1996.)

6.3.1.1.2 Gastrointestinal (GI) Syndrome: In the GI syndrome, the damage is to the dividing "crypt cells" in the intestines. These are rapidly dividing cells at the base of the intestinal villi that constantly produce new villus cells, which migrate to the inner surfaces of the intestines and perform various functions. The villus cells themselves are not so radiosensitive, but the crypt cells are. They are specialized, but are also rapidly dividing. This syndrome manifests at doses of 10 Gy or more. As with the hemopoetic syndrome, some time is needed for the damage to be observed, as the existing generations of villus cells are not directly affected, but the loss of cell division capacity is seen over time. Due to damage to this important internal barrier, the symptoms include:

- Severe NVD effects
- Dehydration, electrolyte imbalance, risk of infection
- Death usually within several weeks

In addition, because of the dose level, the hemopoetic effect is also observed. It is rare, but not impossible for subjects to survive a dose of this level, with significant medical intervention (high doses of antibiotics to fight infections, infusion of fluids and electrolytes, and possibly blood and/or marrow transplants).

6.3.1.1.3 Central Nervous System (CNS) Syndrome: The CNS syndrome only occurs at doses of 20 Gy or more. Damage to the (quite radioresistant) CNS, as well as all other body systems, results eventually in system shutdown and death. Doses of this level cannot be survived with any currently known form of medical intervention. Immediately after exposure, subjects suffer agitation, apathy, disorientation, disturbed equilibrium, vomiting, convulsions, prostration, and may lapse quickly into a coma. Death occurs within hours to one to two days. We show a specific case where such an exposure occurred and the swift and severe effects that were observed in the subject, when we study neutron criticality.

6.3.1.2 Damage to Skin

Skin is a relatively radiosensitive organ, but not overly so. It is of frequent concern, as it will generally receive a higher dose than other tissues of the body in cases of external exposure. Photons are attenuated by tissue in an exponential fashion. If a beam of photons strikes the body, the skin will receive the highest dose, with deeper tissues receiving less dose, depending on their distance from the body surface. If electrons strike the skin, almost all of their energy will be deposited in the skin, with very little being deposited in other organs or tissues. Very low energy betas or electrons and alpha particles may not be able to penetrate the layer of dead cells that cover the entire body surface at different depths.

On average, the lowest depth at which viable cells exist below the skin surface is 70 μm. This depth is generally used in dose calculations. On the bottoms of the feet and major portions of the hands, the covering may be considerably greater. A number of accidents involving the skin of the hands have occurred, as people sometimes put their hands into beams of radiation while repairing or manipulating equipment. Damage to the skin is generally manifested some weeks after the exposure. As with marrow and the intestines, the damage is expressed only after several cycles of cell division have occurred. After high doses of radiation, however, there can be an early phase of *transient erythema* that is seen within about 24 hours. This erythema fades, however, and any serious injury that will be seen, including erythema due to damage to the skin cells appears later. The transient erythema is due to acute damage to the blood vessels supplying the skin; the radiation causes them to become temporarily inflamed, and the skin appears red for a short time while this effect is expressed. Other skin injuries have fairly well defined thresholds:

3 Gy: Epilation beginning around day 17
6 Gy: Erythema; distinguishable from thermal burn; minutes to weeks postexposure, depending on dose
10–15 Gy: Dry desquamation
20–50 Gy: Wet desquamation, two to three weeks postexposure, depending upon dose
>50 Gy: Radionecrosis, deep ulceration

The Centers for Disease Control categorize the stages of expression of cutaneous radiation injury:[16]

- *Prodromal stage* (within hours of exposure): This stage is characterized by early erythema (first wave of erythema), heat sensations, and itching that define the exposure area. The duration of this stage is from one to two days.
- *Latent stage* (one to two days postexposure): No injury is evident. Depending on the body part, the larger the dose, the shorter this period will last. The skin of the face, chest, and neck will have a shorter latent stage than will the skin of the palms of the hands or the soles of the feet.
- *Manifest illness stage* (days to weeks postexposure): The basal layer is repopulated through proliferation of surviving clonogenic cells. This stage begins with main erythema (second wave), a sense of heat, and

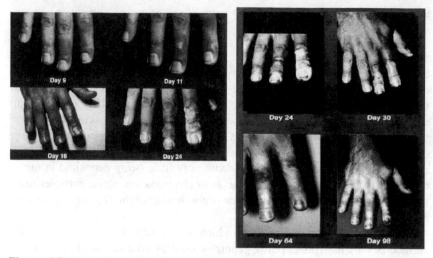

Figure 6.7 Progression of erythema in a patient involved in an X-ray diffraction accident, 9 days to 98 days postexposure (see color plate).

slight edema, which are often accompanied by increased pigmentation. The symptoms that follow vary from dry desquamation or ulceration to necrosis, depending on the severity of the CRI.

- *Third wave of erythema* (10–16 weeks postexposure, especially after beta exposure): The exposed person experiences late erythema, injury to blood vessels, edema, and increasing pain. A distinct bluish color of the skin can be observed. Epilation may subside, but new ulcers, dermal necrosis, and dermal atrophy (and thinning of the dermis layer) are possible.

- *Late effects* (months to years postexposure; threshold dose ~10 Gy or 1000 rads): Symptoms can vary from slight dermal atrophy (or thinning of dermis layer) to constant ulcer recurrence, dermal necrosis, and deformity. Possible effects include occlusion of small blood vessels with subsequent disturbances in the blood supply (telangiectasia); destruction of the lymphatic network; regional lymphostasis; and increasing invasive fibrosis, keratosis, vasculitis, and subcutaneous sclerosis of the connective tissue. Pigmentary changes and pain are often present. Skin cancer is possible in subsequent years.

- *Recovery* (months to years).

Skin injuries have been documented in dozens of cases in the last hundred years. Many accidents have occurred with radiation-producing machines, particularly when workers are unaware that high-dose beams have been activated, or are actively bypassing safety interlocks in order to save time, when the interlocks are there specifically to prevent such injuries. One accident occurred at an electron accelerator in Maryland in which safety precautions were bypassed, and a severe radiation injury was received by an operator. This person placed his hands, head, and feet in the beam path. The equipment was not energized in the normal sense, but the high voltage was still applied, and an electron *dark current* gave a dose of about 55–100 Gy to his hands. Several fingers from each hand were amputated about three months later. The man also suffered some minor epilation on his scalp two weeks after the exposure. Industrial

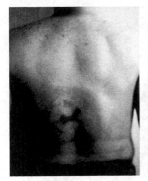

On March 29, 1990, a 40-year-old male underwent coronary angiography, coronary angioplasty and a second angiography procedure (due to complications) followed by a coronary artery by-pass graft.
Total fluoroscopy time estimated to be > 120 minutes. The image shows the area of injury six to eight weeks following the procedures. The injury was described as "turning red about one month after the procedure and peeling a week later." In mid-May 1990, it had the appearance of a second-degree burn.

Appearance of skin injury approximately 16 to 21 weeks following the procedures with small-ulcerated area present.

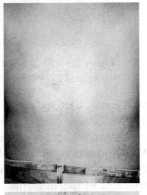

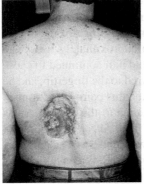

Appearance of skin injury approximately 18 to 21 months following procedures, evidencing tissue necrosis (and close-up of injury area)

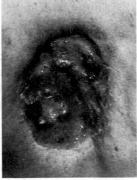

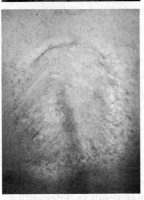

Appearance of patient's back following skin grafting procedure

Figure 6.8 Progression of skin damage in a patient who underwent a coronary angioplasty treatment that resulted in an overexposure to the skin on his back (see color plate).[18]

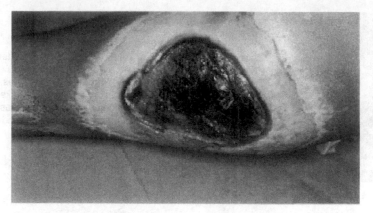

Figure 6.9 Exposed leg of worker in Yanango, Peru who encountered an unshielded Ir-192 industrial radiography source (see color plate).[19]

radiography sources, particularly ones that are lost during surveys and found by unsuspecting persons are other sources.

Figure 6.7 displays the progression of erythema in a patient involved in an X-ray diffraction accident, 9 days to 98 days postexposure. The day following the exposure (not shown), the patient displayed only mild diffuse swelling and erythema of the fingertips. On day 9, punctuate lesions resembling telangiectasis were noted in the subungal region of the right index finger, and on day 11, blisters began to appear. Desquamation continued for several weeks. The patient developed cellulitis in the right thumb approximately two years following exposure. The area of the right fingertip and nail continued to cause the patient great pain when even minor trauma occurred to the fingertip, and he required occasional oral narcotic analgesics to manage this pain. He continued to experience intense pain resulting from minor trauma to the affected areas for as long as four years postexposure. (from CDC Fact Sheet[17]). Figure 6.8 shows the progression of skin damage in a patient who underwent a coronary angioplasty treatment that resulted in an overexposure to the skin on his back[18].

In 1999 a worker in Yanango, Peru found an unshielded ^{192}Ir industrial radiography source and placed it in his rear pants pocket for several hours.[19] The severe lesion shown in Figure 6.9, with subsequent infection that spread through the leg and pelvic area, required amputation of the leg. The worker had a number of other lesions on the hands and legs from handling of the source, and members of his family were also exposed, as the source was inadvertently brought home. The pain and psychological impact from the disfiguration were significant, but the individual did survive. His family members did not receive life-threatening doses.

6.3.1.3 Gonads

Acute damage to the gonads may result in temporary or permanent sterility. We are referring here to the nonstochastic effects of radiation on the gonads; hereditary effects (in which damage to the gonads may result in deformations observed in the person's offspring) are discussed below. Recall that in the male gonads, there is constant cell division to produce new sperm whereas in the female, all eggs are formed at sexual maturity. Thus there is a difference in the thresholds at which effects are observed in the male and female gonads. As an aside, during sexual maturation, it is reasonable to expect heightened radiosensitivity, and we do see more breast cancer, for example, in women during puberty than at other ages, and the normal increase with age after about age 40.

Ovarian	
Dose (Gy)	Effect
0.6	No effect
1.5	Some risk for ovulatory suppression in women >40 years of age.
2.5–5.0	In women aged 15–40, 60% may suffer permanent ovulatory suppression; remainder may suffer temporary amenorhhea. In women >40 years of age, 100% may have permanent ovulation suppression. Menopause may be artificially induced.
5–8	In women aged 15–40, 60–70% may suffer permanent ovulatory suppression; the remainder may experience temporary amenorrhea. No data for women >40 years of age.
>8	100% ovulation suppression.

Testicular	
Dose (Gy)	Effect
0.1–0.3	Temporary oligospermia.
0.3–0.5	100% temporary aspermia from 4–12 months postexposure. Full recovery by 48 months.
0.50–1	100% temporary aspermia from 3–17 months post-exposure. Full recovery beginning 8–38 months.
1–2	100% temporary aspermia from 2–15 months postexposure. Full recovery beginning 11–20 months.
2–3	100% aspermia beginning at 1–2 months postexposure. No recovery was observed up to 40 months.

6.3.1.4 Cataract Formation

Radiation exposures have been definitively linked to the formation of cataracts in physicists who have worked at cyclotrons and who have had significant eye doses from exposure to beams of neutrons and gamma rays over extended periods. Some patients treated with X-rays for therapy have also had some incidence of cataract formation. Some of the Japanese bomb survivors also expressed some cataract formation. Radiation may damage any part of the eye, as with any other tissue in the body. Irradiation of the proliferating epithelial cells in the lens, however, may cause the formation of abnormal lens fibers, which may eventually form an opaque layer on the anterior surface of the lens. Interestingly, cataracts not caused by radiation will form on the posterior surface of the lens, so with careful diagnosis, radiation as a causative agent may be confirmed or ruled out. There appears to be a threshold of about 2 Gy (200 rad) for induction of cataracts by beta or gamma radiation. Neutrons, however, are much more effective at producing cataracts, based on data gathered in some animal studies. Here, the effective threshold seemed to be around 0.15–0.45 Gy (15–45 rad).

6.3.2 Stochastic Effects

Stochastic effects are effects that are, as the name implies, probabilistic. They may or may not occur in any given exposed individual. These effects generally manifest many years, even decades, after the radiation exposure (and were once called "late effects"). Their major characteristics, in direct contrast with those for nonstochastic effects are:

1. A threshold may not be observed.
2. The probability of the effect increases with dose.
3. The effect cannot be definitively associated with the radiation exposure.

Examples include:
- Cancer induction
- Genetic effects (offspring of irradiated individuals)

The most important and widely discussed effect here is cancer. It is not known at present whether there is a real threshold for cancer induction. When studying epidemiological data for many populations, a clear increase of the probability of contracting cancer is seen with increasing dose. Many different curve shapes may fit the data, but in many cases a linear relationship is not unreasonable. Whether this proportionality continues down to low doses and dose rates is unclear, and is the focus of considerable controversy. We will study this controversy more later. Many scientific advisory bodies have recommended, and many regulatory agencies have adopted, a model that assumes linearity to low doses for the purpose of setting regulatory limits for workers and others. This does not imply in any way that we know that a threshold does not exist for cancer induction; it is simply a prudent thing to assume for the present time. So this tenet of the nature of stochastic models is simply assumed. It is clear that the probability of contracting cancer increases with increasing dose above some level. If you contract a cancer, however, the severity is no different than at any other dose level (unlike nonstochastic effects, whose severity clearly depends on dose. The cancer is no more or less severe, and in other ways is as well indistinguishable from a cancer that another person contracted with no excess exposure to radiation (above natural background radiation, medical sources, etc.).

Radiation has been clearly shown to induce genetic effects in some animal species, but this effect has never been proven in humans. Nonetheless, as shown in the next chapter, worries about the possibility of inducing genetic effects in humans was the major driving force in the changes over the mid-1900s in our radiation regulations.

6.3.2.1 Cancer

The fact that ionizing radiation causes cancer is well established. Exposures of a number of populations, in addition to animal studies, have established clear causative links between radiation exposure and expression of a number of types of cancer. The quantitative relationship is sometimes fairly well established, and in other cases less well. At high enough doses, the rate of production clearly increases with increasing dose (i.e., probability increases with dose). A radiation-induced cancer is indistinguishable from a "spontaneous" cancer;

the causal link is established from the number of cancers induced in an exposed population in relation to that expected in that population otherwise. In the populations that have been studied to establish relationships between dose and risk, in most cases, the radiation doses themselves, and the rates of cancer, are subject to considerable uncertainty.

The most important population studied to date to determine these trends is the population of survivors of the Japanese nuclear bomb attacks. Several hundred thousand people died either instantly or within the first year after the attacks, from physical injuries and radiation sickness. The surviving population has been extensively studied over the years following the attacks. The most important single institution in this follow-up effort is the Radiation Effects Research Foundation (RERF),[20] with locations in both cities. The RERF (formerly the Atomic Bomb Casualty Commission) was founded in April 1975 and is a private nonprofit Japanese foundation. Funding is provided by the governments of Japan, through the Ministry of Health, Labor, and Welfare, and by the Department of Energy in the United States. Some 36,500 survivors who were exposed beyond 2.5 km have been followed medically continuously since the blasts. Of this population, about 4900 cancer deaths have been identified, including about 180 leukemia deaths and 4700 deaths from cancers other than leukemia. Of these, only about 89 and 340 deaths, respectively, appear to be attributable to radiation.

There are also a number of populations of individuals exposed to various medical studies using high levels of radiation (principally from the early 1900s, before radiation's dangers were fully appreciated). For different types of cancer, if we plot the cancer rate against the dose received, the data will show an upward trend (Figure 6.10). The question that none of the datasets clearly answers is that of the shape of the curve at low doses and dose rates. A controversy continues to rage over whether the relationship between dose and the absolute or relative number of induced cancers should be extrapolated to zero dose (i.e., all exposure to radiation, no matter how small, is associated with some risk of cancer) or if there is a threshold. There is evidence to support both views.

In fact, there is some evidence to support the controversial theory of *hormesis*, that exposure to low levels of radiation is associated with less cancer induction than in systems deprived of all radiation exposure.[21] The

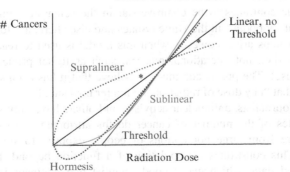

Figure 6.10 General principles of radiation risk models.

possible mechanism here is that exposure to radiation stimulates cellular repair mechanisms (the mechanism of "adaptive response"). In individual experiments, hormetic and adaptive response mechanisms have been demonstrated. Cells clearly have mechanisms for repairing DNA damage. If damage occurs to a single strand of DNA, it is particularly easy for the cells to repair this damage, as information from the complementary chain may be used to identify the base pairs needed to complete the damaged area. "Double strand breaks" are more difficult to repair, but cellular mechanisms do exist that can affect repair here also.

On the other hand, recent experiments suggest the expression of radiation damage in cells that have received no radiation exposure themselves, but are adjacent to, or somehow in communication with, exposed cells, the so-called "bystander effect".[22] Cultures of cells selectively irradiated with microbeams of alpha particles have been shown to demonstrate DNA damage in cells other than those that received any radiation. Similarly, if a population of irradiated cells is mixed with a batch of cells that were not irradiated, at some time later, both irradiated and unirradiated cells may have demonstrable DNA damage. So it appears that signaling or other sharing of information between cells allows damage from radiation to become more widespread than the direct impact might suggest.

Another mechanism, called "genomic instability"[23] also suggests that the effects from radiation may be felt in cells other than those directly irradiated. Cells irradiated with radiation have been shown to not have observable radiation damage, but subsequent generations of these cells may show DNA damage. Thus, although evidence from studies that purport to show hormesis and adaptive response suggest that low levels of radiation may not be harmful and may even be beneficial, experiments showing the bystander and genomic instability effects suggest that radiation's effects may spread considerably in cellular systems exposed to radiation.

The National Academy of Science's Committee on the Biological Effects of Ionizing Radiation (BEIR), perhaps the most influential scientific body writing on this matter, has, at the time of this writing, concluded that all of this evidence is at present not conclusive for either proof of a threshold or hormetic effect of radiation nor how the bystander and genomic instability evidence affects models predicting cancer effects at low doses.[24] Relying heavily on epidemiological data from the populations discussed above, this group has concluded that the most prudent model to use at present is still the Linear, No Threshold (LNT) model.

This issue continues to be controversial in the scientific community, but most regulatory bodies, in the United States and elsewhere, are following this advice. Problems arise, however, when this model is used to reach scientific conclusions (i.e., not operational reasons, such as to set prudent limits on radiation dose). The public continues to believe that it has been scientifically concluded that "any dose of radiation, no matter how small" can cause cancer. Scientists, journalists, antinuclear activists, and others have published numerical estimates of the number of cancer deaths attributed or to be expected in the future from large populations of people exposed to small doses of radiation. This constitutes extrapolation of a function beyond the limits of the observed data, which every good scientist and engineer knows to be improper.

One very important point to remember is that if radiation induces cancer in a population, the cancers are always expressed at some long time after the exposure (thus the early name "late" effects). This period of time between the exposure and the expression of the disease is called the "latent period". It is the shortest for expression of radiation-induced leukemia, being as short as 5 to 10 years after the exposure. With most solid cancers, the latent period is more like 20 years. After exposure of a population to radiation, a number of erroneous efforts (whether intentionally misleading or not) have been made to show increases in cancer rates in selected populations within months after the exposure.

6.3.2.2 Leukemia
Acute myelogenous leukemia (AML), and to a lesser extent, chronic myelogenous or acute lymphocytic leukemia, are among the most likely forms of malignancy resulting from whole-body exposure to radiation. Increases in leukemia incidence have been clearly shown in a number of populations exposed to high levels of radiation, including:

Early radiologists
Japanese bombing survivors
British children irradiated in utero

As noted above, leukemia has the shortest latent period of any form of cancer, as short as five to ten years. The last study cited remains controversial. In 1958 Alice Stewart, a British epidemiologist, and colleagues published an observation of a possibly statistically significant link between exposure to diagnostic levels of ionizing radiation and childhood leukemia.[25] This and other findings by expert committees of the National Academy of Sciences and the British Medical Research Council led to a significant reduction in unnecessary medical radiation exposures. Her results and methods were challenged by many, but the influence on the international scientific community was, and continues to be, significant. Dr. Stewart became a passionate crusader against nuclear power, as she was convinced that low levels of radiation of any kind may cause cancer.

6.3.2.3 Bone Cancer
In addition to the acute damage to the skeletons of the radium dial painters, as was discussed above,[4] a number of osseous tumors were confirmed in this population as well. Radium, strontium, and barium are chemically similar to calcium and so are taken up in bone in significant quantities. Plutonium also concentrates in the periosteum, endosteum, and trabecular bone. Rare earth elements such as cerium and praseodymium have also been demonstrated to have significant uptakes in skeletal tissues of animals. These species have been studied extensively in controlled experiments in animals.[26,27] Interestingly, no statistically significant incidence of bone cancer has been demonstrated in the Japanese bomb survivors.

6.3.2.4 Lung Cancer
Lung cancer has been clearly seen in the Japanese bomb survivors, but was first observed historically as early as the sixteenth century, in populations of miners exposed to high levels of the radioactive gas ^{222}Rn.[28] This species is a member of the ^{238}U decay series (see Chapter 3). Many epidemiological

studies have been performed on workers from uranium mines, exploited to provide uranium for reactor fuel and nuclear weapons. These earlier miners just happened to work in mines associated with particularly high levels of uranium deposits. A special unit of "exposure" has been derived to describe exposure to radon: the *Working Level* (WL), defined as "any combination of short-lived radon daughters in 1 L of air that will result in the ultimate emission of 1.3×10^5 MeV of alpha energy." As an extension, the *Working Level Month* (WLM) is defined as exposure resulting from inhalation of air with a 1 WL concentration for 170 hours (a typical working month).

Cumulative radon exposures have been definitively associated with statistically significant increases in lung cancer, and particularly in miners who also smoke. The effects of radiation exposure and smoking appear to be synergistic. It is difficult to see increases of cancer above those already associated with smoking, but both in these populations, using epidemiological methods, as well as in controlled animal studies, radiation exposure clearly causes an increase in lung cancer. As with other cancers, extrapolation to low levels of radon exposure, as are common in many homes, is uncertain, but has been an area of concern.

In certain areas of the United States, for example, several northeastern states, Tennessee, Florida, and some western states such as Colorado, there are significant deposits of uranium in natural rock formations. In Florida and the western states, in addition, uranium or phosphate mining may cause redistribution of naturally occurring radioactivity in a way that enhances the possibility for human exposure to radiation (termed TENR, Techologically Enhanced Natural Radiation)[29]. After mining activities, homes may be built on reclaimed lands where material with higher than average uranium concentrations has been discarded. In Tennessee and the northeast, it may be just certain geological formations that contain a high level of uranium. When homes are constructed over such formations, radon gas is continually produced and may diffuse through the ground and into homes. Basements tend to have the highest concentrations, but significantly enhanced concentrations have definitively been found in the more occupied portions of homes in these areas as well. As homes are often designed to be tightly sealed to facilitate air conditioning, radon gas (and other toxins as well) may become concentrated, with human exposures being increased. The levels of exposure are far less than those at which demonstrable increases in cancer have been observed, but concerns remain as we are still uncertain about whether such effects are proportional to dose even at low levels.

Another population of subjects in which increased levels of lung cancer have been observed are ankylosing spondylitis patients treated with X-rays.[30] This is a condition in which joints and bones of the spine become fused together. It is a painful, progressive, rheumatic disease that mainly affects the spine but which can also affect other joints, tendons, and ligaments and involve other body organs as well. In the early part of the twentieth century, radiation treatments were used to treat the inflammation of the spine in patients with ankylosing spondylitis. The treatments were effective, however, they were discontinued when physicians observed the acute damage to the bone marrow and later development of cancers caused by the radiation exposures.

Figure 6.11 Image of the nuclear blast near the Marshallese Islands. (From http://www.rmiembassyus.org/Nuclear%20Issues. htm.)

6.3.2.5 Thyroid Cancer

Thyroid cancer has been clearly associated with radiation exposure. Children irradiated with X-rays for enlarged thymus glands, ringworm of the scalp, and acne (again in the early days of radiation use) showed significant rates of thyroid cancer.[31] And again, such uses were quickly discontinued when the deleterious effects of radiation were noted. The thyroid concentrates free iodine in the body, so intakes of radioactive iodine, particularly ^{131}I, have been linked with increases in thyroid cancer. In 1954, an above-ground test of a nuclear weapon in the Pacific Ocean went awry, with weather conditions not matching expectations and the weapon yield being considerably greater than expected (Figure 6.11). A number of people living on nearby Rongerik, Rongelap, Ailinginae, and Utirik atolls (the Marshall Islanders) were blanketed in fallout from the weapon, resulting in considerable whole-body exposures, skin lesions from radiation exposure, and intakes of ^{131}I.[32] Although this was a relatively small population, a significant number of thyroid cancers (and other thyroid malignancies, as well as cancers in other body tissues) were observed, and the subjects were followed for some time. Eventually, their descendants were compensated by the U.S. government for this tragic event.

In April 1986, the most significant disaster in the nuclear power industry's history occurred at the Chernobyl nuclear power plant in the Ukraine.[33] The event was the product of a flawed Soviet reactor design coupled with serious mistakes made by the poorly trained plant operators. The accident destroyed one reactor and killed 31 people, 28 due to their radiation exposures. Another 100–150 cases of acute radiation poisoning were confirmed (the subjects recovered). There were no off-site acute radiation effects, but large areas of Belarus, Ukraine, Russia, as well as areas of Eastern Europe were contaminated to varying degrees. Many cases of thyroid cancer in children and adolescents in the contaminated areas were linked to their exposure to the ^{131}I releases from the site (other shorter-lived isotopes of iodine may also be involved in the radiation doses received by the children's thyroids).

6.3.2.6 Hereditary Effects

Radiation-induced hereditary effects have been clearly demonstrable in animal experiments involving mice and fruit flies, but never in any human population, including the Japanese bomb survivors, medical populations, and populations affected by the Chernobyl disaster. As with cancer, there is a spontaneous rate of mutations that is ongoing in the human population, with no excess exposure to chemicals, radiation, or other mutagenic agents. About 1 in 200 pregnancies involve a baby with a chromosomal abnormality, and about 3–4% of all pregnancies result in some abnormality being expressed in the child. Increases above this baseline, for radiation-induced genetic effects are expressed in a unique term called the Doubling Dose. The Doubling Dose is the radiation dose to the gonads that will eventually lead to a doubling of the expression of hereditary defects, over the "spontaneous" rate in a given population. In general cancer risk models, we usually express the absolute or relative risk (defined shortly) of a given dose of radiation. Here we refer to a specific dose that brings about this specific endpoint, namely a doubling of the baseline rate of defects.

In the next chapter, we show how concern about genetic effects from cumulative radiation exposure was the driving force behind the reduction in radiation dose limits over the years. The animal data suggested that significant increases of hereditary effects would be seen in human populations exposed to radiation, and this was the primary cause for the pressure to lower early dose limits for workers and the public. Risk weighting factors, which we study in Chapter 8, have also been the largest for the gonads, and are tied to risks of hereditary effects, as derived from animal studies. In the most recent recommendations of the ICRP, however, the risk weighting factor has decreased substantially, perhaps reflecting the fact that more time has gone by and no significant effects have been demonstrated in human populations.

6.3.2.7 Mathematical Models of Cancer Risk

This brief chapter cannot treat this subject in great detail, but an overview of the approaches and a few results of mathematical models to express cancer risk are given. The most influential group writing on this subject is the U.S. National Academy of Sciences Committee on the Biological Effects of Ionizing Radiation. They have issued several important reports periodically over the last 50 years; the latest is BEIR VII.[23] We can describe a generalized dose–response curve with the function:

$$F(D) = (\alpha_0 + \alpha_1 D + \alpha_2 D^2) \exp(\beta_1 D - \beta_2 D^2).$$

Here $F(D)$ is the incidence rate at dose D, α_0 is the natural or spontaneous rate, and the other parameters are fitted to the data for different cancer types (some may be zero). Generally these coefficients α and β are positive, as we have noted that the rates of observed cancers increase with increasing dose. There is one negative β coefficient in the equation; this accounts for a decrease

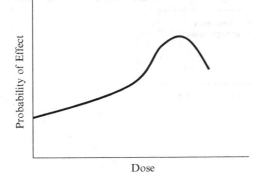

Figure 6.12 General curve for induction of stochastic effects.

in expressed cancers at high doses where cell killing begins to occur, and cells do not have the opportunity to express cancer (Figure 6.12).

A continuing effort goes into the assessment and reassessment of the Japanese bomb survivors' actual dosimetry and in the fitting of the observed cancer rates as a function of the radiation dose received. In its most recent assessment, the BEIR Committee has concluded that for all cancers except leukemia and genetic effects, data are "not inconsistent with" a linear no-threshold model (Figure 6.13). The leukemia data are best fit with a linear quadratic function (i.e., both α_1 and α_2 terms are nonzero).

The incidence may also be expressed as *absolute* risk (which gives the absolute difference in event rates between the exposed and control groups, or *relative* risk (which expresses the probability of an event in the exposed group divided by the probability of the event in the control group).

Additive (absolute) risk model:

$$\gamma(d) = \gamma_0 + f(d)g(\beta).$$

Multiplicative (relative) risk model:

$$\gamma(d) = \gamma_0 \left[1 + f(d)g(\beta)\right].$$

$f(d) =$ some function of d:

  i. Linear model (e.g., lung, breast, digestive tract) $f(d) = \alpha_1 d$.
  ii. Linear-quadratic model (leukemia) $f(d) = \alpha_1 d + \alpha_2 d^2$.

$g(\beta) =$ some function of other variables.

  i. For example, respiratory tract: $g(\beta) = \exp[\beta_1 \ln(T/20) + \beta_2 I(S)]$.
  ii. $T =$ age since exposure, in years.
  iii. $I(S) = 1$ if female, 0 if male.

$\gamma_0 =$ Age-specific background risk of death due to a specific cancer.
$d =$ Equivalent dose (Sv). The αs and βs are fitted parameters.

Current data support the use of relative, rather than absolute risk models. Extrapolation of observed cancer occurrence in the Japanese population to the general case is thought to require two important modifications:

- Application of a Dose and Dose Rate Effect Factor (DDREF). The population was exposed to high doses at a very high dose rate, whereas most exposures that we are concerned with routinely involve lower doses and dose

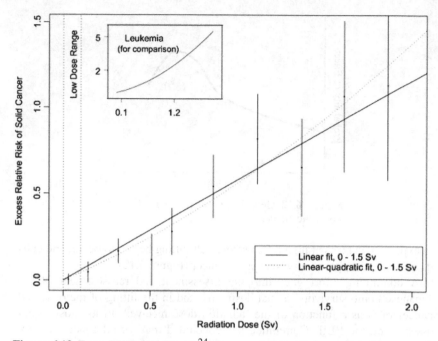

Figure 6.13 From BEIR VII Report.[24] Excess relative risk is equal to (the rate of a disease in an exposed population divided by the rate in an unexposed population) minus 1.0.

rates. The BEIR Committee thought that this factor should be between 1.1 and 2.3, and used a value of 1.5 (see Table 6.3).

• The Japanese population has different baseline risks of some cancers, so extrapolation of their absolute or relative risks to other populations was performed with some mathematical adjustments.

The BEIR VII lifetime risk model predicts that approximately one individual in 100 will develop cancer from a dose of 0.1 S*v*, whereas 42 of 100 persons will develop solid cancer or leukemia from other causes unrelated to radiation exposure.

6.4   Cell Survival Studies

So far we have looked at gross effects on the organism or individual tissues of the organism after exposure to radiation. Much information on the biological effects of radiation has been obtained for many years through the use of direct experiments on cell cultures. It is, of course, far easier to control the experiment and the variables involved when the radiation source can be carefully modulated, the system under study can be simple and uniform, and the results can be evaluated over most any period of time desired (days to weeks, or even over microseconds, such as in the study of free radical formation and reaction). After exposure of a group of cells to radiation, the most common concept to study is that of cell survival. Typically, the natural logarithm of the surviving fraction of irradiated cells is plotted against the dose received (Figure 6.14).

Table 6.3 The Committee's preferred estimates of the lifetime attributable risk (LAR) of incidence and mortality for all solid cancers and for leukemia with 95% subjective confidence intervals. Number of cases or deaths per 100,000 exposed persons.

	All solid cancer		Leukemia	
	Males	**Females**	**Males**	**Females**
Excess cases (including non-fatal cases) from exposure to 0.1Gy	800 (400,1600)	1300 (690, 2500)	100 (30, 300)	70 (20, 250)
Number of cases in the absence of exposure	45,500	36,900	830	590
Excess deaths from exposure to 0.1 Gy	410 (200, 830)	610 (300, 1200)	70 (20, 220)	50 (10, 190)
Number of details in the absence of exposure	22,100	17,500	710	530

The simplest survival curve is a single exponential:

$$S = S_0\, e^{-D/D_0}.$$

Here S is the surviving fraction, S_0 is the original number of cells irradiated, D is the dose received, and D_0 is the negative reciprocal of the slope of the curve, and is called the mean lethal dose. When cells receive dose D_0, the surviving fraction is 0.37, which is $1/e$. This dose may also be referred to as the D_{37} dose, just as we have defined above the LD_{50}, or lethal dose of radiation that will kill half of a population. Generally speaking, particles with a high LET will show this form of a survival curve, and those of low LET will have a more complicated curve, of the form:

$$S = S_0\,[1 - (1 - e^{-D/D_0})^n].$$

Here n is the assumed number of targets that need to be hit in order to inactivate a cell. If $n = 1$, the equation reduces to the simpler form shown above. The usual curve, however has a "shoulder," indicating that a certain amount of dose must be received before any significant effect on cell survival is seen. At higher doses, the curve attains the usual linear shape with slope $-1/D_0$. If the linear portion is extrapolated back to zero dose, it will intercept the y-axis at the

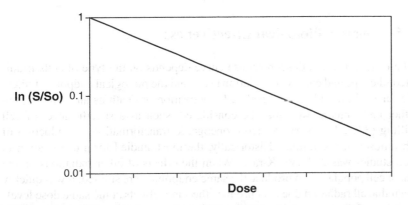

Figure 6.14 Cell survival curve: high LET radiation, single exponential.

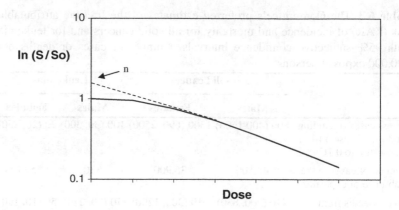

Figure 6.15 Cell survival curve: low LET radiation, demonstrating "shoulder" and extrapolation number.

extrapolation number n, which is numerically equal to the number of targets assumed to be relevant to the cells' survival (Figure 6.15).

Several factors affect the shape of the dose–response function other than the LET of the radiation, including:

- *Dose rate*: The LD_{50} of a population of cells will clearly increase as the dose rate at which a fixed dose D is delivered is decreased. Cells have a considerable capacity to repair radiation damage, and if time is allowed for repair, more radiation can be tolerated.
- *Dose fractionation*: If cells are given a cumulative dose D, but instead of being delivered all at once, it is delivered in N fractions of D/N each, the cell survival curve will show a series of shoulders linked together, because cellular repair is again ongoing between fractions. This is a strategy used in radiation therapy procedures to allow healthy tissues time for repair while still delivering an ultimately lethal dose to the tumor tissues.
- *Presence of oxygen*: Dissolved oxygen in tissue causes the tissue to be sensitive to radiation. Hypoxic cells have been shown to be considerably more radioresistant. The effect of oxygen is sometimes expressed as the *Oxygen Enhancement Ratio* (OER), which is the ratio of the slope of the straight portion of the cell survival curve with and without oxygen present.

6.5 Relative Biological Effectiveness

The exact form of a dose–response curve depends on the type of cells irradiated, the type and energy of radiation used, and the biological endpoint studied. So far we have only mentioned cell inactivation or death as an endpoint, but other endpoints as well may be considered, such as a specific level of cell killing (37%, 10% survival, etc.), oncogenic transformation, or induction of chromosomal aberrations. Historically, the most studied form of radiation in cell studies was 250 kVp X-rays. When the effects of other radiations on the same cell population to produce the same endpoint were studied, it was quickly seen that all radiation does not produce the same effects at the same dose levels as this "reference" radiation. If a dose D' of a given radiation type produces the same biological endpoint in a given experiment as a dose D of our reference radiation, we can define a quantity called the Relative Biological Effectiveness (RBE) as

$$RBE = \frac{D}{D'}.$$

So, for example, if a dose of 1 Gy of the reference radiation produces a particular cell survival level, but only 0.05 Gy of alpha radiation produces the same level of cell killing, we say that the RBE for alpha particles in this experiment is 20.

RBE is quite dependent on radiation LET. High LET radiations generally have high RBEs; the reader should note that 250 kVp X-rays are generally considered to be low LET radiation. The relationship of the two variables is not directly linear, but there is clearly a positively correlated relationship of RBE with LET, until very high LET values are reached, where "overkill" of cells causes the RBE not to increase as quickly.

The reader may have noted that in the numerical example chosen above, the RBE for alpha particles is exactly equal to the currently recommended value of w_R, the radiation weighting factor used in radiation protection. This was quite intentional. Values of w_R are very closely tied to RBE values, however, they are not exactly equal. Generally, conservative values of RBE were used to set the values assigned for w_R values (also formerly called "quality factors," you may recall). The important thing to remember about RBE values is that they are highly dependent on the experimental conditions (cell type, radiation type, radiation dose rate) and the biological endpoint defined for study. Radiation weighting factors, on the other hand, are single values to be applied to a type of radiation in all situations. Radiation weighting factors are operational quantities, used to solve a practical problem (how to best protect radiation workers from routine exposure to radiation), whereas RBEs are more scientific quantities relevant to the study of radiation biology.

Endnotes

1. http://www.accessexcellence.org/AE/AEC/CC/historical_background.html
2. http://www.physics.isu.edu/radinf/chrono.htm
3. Ross Mullner. Deadly Glow. The Radium Dial Worker Tragedy. American Public Health Association, Washington, DC, 1989.
4. http://www.radford.edu/~wkovarik/envhist/radium.html
5. http://www.orau.org/ptp/articlesstories/quackstory.htm
6. H Cember. (1996) Introduction to Health Physics, 3rd Ed. McGraw-Hill, New York, NY.
7. Turner J. Atoms, Radiation, and Radiation Protection. Pergamon Press, New York, NY 1986.
8. S. M. Pimblott and J. A. LaVerne. (1997) Stochastic simulation of the electron radiolysis of water and aqueous solutions, J. Phys. Chem. A 101, 5828–5838.
9. D. Becker, M. D. Sevilla, W. Wang and T. LaVere. (1997) The role of waters of hydration in direct-effect radiation damage to DNA, Radiat. Res. 148, 508–510.
10. H.A. Wright, J.L. Magee, R.N. Hamm, A. Chatterjee, J.E. Turner, and C.E. Klots. (1985) Calculations of physical and Chemical Reactions Produced in Irradiated Water Containing DNA. Radiat Prot Dosimetry 13: 133–136.
11. Turner JE, Hamm RN, Ritchie RH, Bolch WE. Monte Carlo track-structure calculations for aqueous solutions containing biomolecules. (1994) Basic Life Sci. 63:155–66.

12. Bergonie J, and Tribondeau L. (1906), De quelques resultats de la Radiotherapie, et esaie de fixation d'une technique rationelle. Comptes Rendu des Seances de l'Academie des Sciences, 143, 983–985.

13. Kimball JW. Kimball's Biology Pages, http://users.rcn.com/jkimball.ma.ultranet/BiologyPages/B/Blood.html#formation, 2006.

14. J. Greene, D Strom. Would the Insects Inherit the Earth? And Other Subjects of Concern to Those Who Worry About Nuclear War. (1988) Pergamon Press, McLean, VA.

15. Jean Marc Bertho, Nina M. Griffiths And Patrick Gourmelon. (2004) The Medical Diagnosis And Treatment Of Radiation Overexposed People. Eleventh Congress of the International Radiation Protection Association, Madrid, Spain.

16. Cutaneous Radiation Injury: Fact Sheet for Physicians. Centers for Disease Control and Prevention, 1600 Clifton Rd, Atlanta, GA 30333, USA. http://www.bt.cdc.gov/radiation/ criphysicianfactsheet.asp

17. http://www.bt.cdc.gov/radiation/criphysicianfactsheet.asp

18. Thomas B. Shope. Radiation-induced Skin Injuries from Fluoroscopy. US. Food and Drug Administration, Center for Devices and Radiological Health, Washington, DC, 1997. http://www.fda.gov/cdrh/rsnaii.html.

19. International Atomic Energy Agency. The Radiological Accident In Yanango. International Atomic Energy Agency, Vienna, Austria, 2000.

20. For example, Pierce DA, Shimizu Y, Preston DL, Vaeth M, Mabuchi K. Studies of the mortality of atomic bomb survivors. Report 12, Part I. Cancer: 1950–1990 Technical Report No. 11–95. Radiat Res 146:1–27, 1996.

21. Edward J. Calabrese and Linda A. Baldwin. Hormesis: The Dose-Response Revolution. Annu. Rev. Pharmacol. Toxicol. 2003. 43:175–97.

22. Eric J. Hall. The Bystander Effect. Health Phys. 85(1):31–35; 2003.

23. William F. Morgan. Non-targeted and Delayed Effects of Exposure to Ionizing Radiation: II. Radiation-Induced Genomic Instability and Bystander Effects In Vivo, Clastogenic Factors and Transgenerational Effects. Radiation Research 159, 581–596 (2003).

24. National Academy of Sciences. Health Risks from Exposure to Low Levels of Ionizing Radiation: BEIR VII Phase 2. The National Academies Press, Washington, DC, USA, 2006.

25. Stewart AM, Webb KW, Hewitt D (1958) A survey of childhood malignancies BMJ i 495-508, 1958.

26. Raabe, O.G., S.A. Book, N.J. Parks, C.E. Chrisp, and M. Goldman. (1981) Lifetime studies of 226Ra and 90Sr toxicity in beagles - A status report. Radiat. Res. 86:515-528.

27. Lloyd RD, Mays CW, Atherton DR. (1976) Distribution of injected ^{226}Ra and ^{90}Sr in the beagle skeleton. Health Phys. Feb;30(2):183-9.).

28. For example, B Grosche, M Kreuzer, M Kreisheimer, M Schnelzer, A Tschense. Lung cancer risk among German male uranium miners: a cohort study, 1946–1998. British Journal of Cancer 95, 1280–1287, 2006.

29. Henry Hurwitz Jr. The Indoor Radiological Problem in Perspective. Risk Analysis 3(1), 63–77, 1983.

30. N Feltelius, A Ekbom, and P Blomqvist. Cancer incidence among patients with ankylosing spondylitis in Sweden 1965–95: a population based cohort study. Ann Rheum Dis. 2003 December; 62(12): 1185–1188, 2003.

31. Raymond J. Carroll. Thyroid Cancer Following Scalp Irradiation: A Reanalysis Accounting for Uncertainty in Dosimetry. Biometrics 57 (3), 689–697, 2001.

32. Steven L. Simon, André Bouville, Charles E. Land. Fallout from Nuclear Weapons Tests and Cancer Risks. American Scientist, Volume 94 Number 1 Page 48, 2006.

33. Jim Smith and Nicholas A. Beresford. Chernobyl: Catastrophe and Consequences. Springer, New York, 2005.

7

The Basis for Regulation
of Radiation Exposure

The development of radiation protection standards has occurred rather gradually over the last century. Lauriston Taylor[1] has defined ten time periods in the development of radiation protection standards that characterize changes in thinking. We look at the development of these standards within the context of his time framework.

7.1 Period 1: 1895–1913

After the discovery of X-rays in 1895 by Roentgen and of radioactivity by Becquerel in 1896, Taylor notes that there was an acknowledgment of the need for protection against radiation, but no groups attempted to formulate a standard. In 1913, the German Radiological Society on X-Ray Protection Measures outlined nine steps for protection of radiologists from radiation injury. The requirements specified thickness of lead for protection of personnel (although they did not specify the tube voltage, etc. to which this corresponded).

7.2 Period 2: 1913–1922

In 1915, and again in 1921, the X-Ray and Radium Protection Committee of the (British) Royal Society of Medicine published a report outlining the risks and proposing safeguards for X-Ray operators and persons handling radium. Safety precautions included shielding of X-ray tubes, limitations on hours, use of protective gloves, use of proper equipment, caution with electrical equipment, providing proper ventilation, and use of procedures to increase distance and decrease time in use of radium sources. In 1922 George Pfahler addressed the American Radium Society and outlined a series of protective measures for radium users and X-ray operators that were essentially identical to the British recommendations.

7.3 Period 3: 1922–1928

The British X-ray and Radium Protection Committee published an expanded and revised set of recommendations. These (and their predecessor documents) can be pointed to as the first comprehensive set of recommendations that led to standards, as they were submitted to the Second International Congress on Radiology.

In Norway, certain safety committees were also looking into the formulation of protection standards. No licensing procedures were in place for purchase or use of X-ray equipment. Recommendations were drawn up for licensing of facilities and minimum requirements for medical students.

In 1925, Dr. A. Mutscheller proposed the concept of a tolerance dose. Based on some measurements made on several institutions, he arrived at a level corresponding to 1/100 of a threshold erythema dose per month. He later correlated this to a numerical value of "R-units," although it was not clear to whose "R-unit" this referred. This safety limit, however, was based solely on Mutscheller's judgment and did not correlate directly with a level correlated with the absence of observed biological effects.

At about the same time, and completely independently, R. Sievert of Sweden concluded that 1/10 of a skin erythema dose per year would be acceptable for radiation workers. Also, a British group suggested a value of 0.00028 of a skin erythema dose per day. It is interesting how all three of these independently derived guidelines were almost exactly the same.

There was considerable confusion during this time, however, regarding the quantities and units used by different experiments. In 1928, the International Commission on Radiological Units and Measurements (ICRU) first standardized the definition of the Roentgen (R) which has survived until now, except for its exclusion from the Système International (SI) system of units. Some researchers immediately recognized the need for a "quality coefficient" that accounted for the differing potential of various X-ray energies in producing erythema at a given exposure. At this point, studies proliferated looking at experiences of different departments with observed effects at measured exposure levels and attempts at defining an acceptable routine exposure level.

In 1928, an international committee was organized to derive standards for safe use of X-rays and radium. Originally called the International Committee on X-ray and Radium Protection (ICXRP), this group was really the first organization of the International Commission on Radiological Protection (ICRP). The original members were L. S. Taylor (United States), G. Grossman (Germany), I. Solomon (France), R. Sievert (Sweden), Dr. Ceresole (Italy), with Drs. Kaye and Melville from England as Honorary Secretaries.

7.4 Period 4: 1928–1934

In the United States, the perceived need for a local advisory body led to the organization of the Advisory Committee on X-ray and Radium Protection (ACXRP). This was chaired by L. Taylor and also included W. Coolidge, G. Failla, R. Newell, H. Pancoast, W. Werner, and F. Wood. Their agenda was to examine and evaluate existing regulations in the United States and develop a detailed set of new regulations. After several meetings and a few preliminary reports, the ACXRP proposed what might be the first exposure limit directly defined in measurable units. It was tied to Mutscheller's limit of 1/100 of a skin erythema dose per month, but was reduced by a factor of 2. This limit was 0.1 R/day, corresponding to about 25 R/year. This value was the first real radiation limit adopted by an advisory body, and was published in 1934.

Very soon thereafter, the ICXRP published a report recommending a limit of 0.2 R/day, because of the use of a different reduction factor with Mutscheller's proposed limit. Again, however, these limits were not tied to any biomedical observations about dose and effect, only the best judgment of a number of physicists of that day. Also, there was no link between tolerance dose for external sources of radiation and permissible body burden of internal emitters such as radium. Another interesting discussion, which arose during this time and continued for 25 years, was the use of extended vacations for radiation workers to counteract any effects from potential overexposures.

7.5   Period 5: 1934–1941

The recommendations regarding X-ray exposure were adapted to a number of other situations, including external exposure to radium gamma rays and use of fluoroscopic procedures. The first discussions about genetic risk also began. The belief of geneticists that low levels of radiation produced a cumulative effect in producing genetic damage led some in the ACXRP to propose a lowering of the limit from 0.1 R/day to 0.02 R/day (25 to 5 R/year). No action was taken on this because committee activities were suspended during the war. When the ACXRP was reorganized in 1946, it was then given the title National Committee on Radiological Protection (NCRP).

The other significant development during this time was the initial attempts to set a safe body burden for radioluminous compounds. In the late 1930s, the radiation-induced effects in the bodies of the radium dial workers became apparent. A report by the ACXRP in 1941 suggested (1) a limit of 0.1 μg of radium in the body (as indicated by radon exhalation), (2) a limit of 10 pCi/liter of radon in the workplace, and (3) a limit on whole-body exposure to gamma rays of 0.1 R/day.

The body burden of radium was not substantially changed in the next 30 to 40 years, and it became a reference point for evaluation of the body burden of other internal emitters.

7.6   Period 6: 1941–1946

The discovery of the sources and applications of atomic energy created a plethora of new problems in radiation protection. But these developments also generated a great wealth of experience as well as funding for research in this area. Out of the experience at the Manhattan Project came the basis for much of our radiation protection practice, much under the direction of H. Parker.

Activities at the Project opened up a variety of new safety problems including shielding for unusual nuclides and configurations (including consideration of "sky-shine"), exposure to β radiation, internal exposure to a variety of new radionuclides and compounds, waste disposal, and personal contamination. Development of units that would express the contribution of the different types of radiation also began here. The roentgen equivalent physical, or rep, was that amount of any type of ionizing radiation that would result in the deposition of 83 ergs/g in soft tissue. The roentgen equivalent man, or rem, was developed to equate biological effectiveness of the different types of radiation. Thus, the first quality factors were derived, although that term was not used.

The Project made use of the 0.1 R/day tolerance dose to limit external exposure and the 0.1 μg of radium limit to control internal exposures. They also had some separate limits for exposure of the skin and recognized that a body burden of 0.1 μg of radium resulted in about 0.16 rem per day to the bone. Thus we see the first steps in the evolution of differing standards for whole- and partial-body exposures.

Extensive work was also done in the study of standards for internal emitters, including study of routes of entry, fractional uptake, metabolism, and dosimetry. Body burdens of various species that would result in acceptable dose rates were derived. Also the first attempts to define a "standard man" were made.

7.7   Period 7: 1946–1953

Shortly after the reorganization of the NCRP, a subcommittee on Permissible External Dose prepared a new set of recommended limits for occupational exposure. The whole-body exposure limit for photons was recommended to be dropped from 0.1 R/day (0.5 R/week) to 0.3 R/week. The exposure of the hands was to be limited to 1.0 R/week of photons or 1.0 rep/week of β rays. The basis for reporting was switched to monthly so that the real proposed limits were 1.25 R/month photon whole-body and 4.2 R/month photon and 4.2 R/month for the hands. In addition, values for lifetime total exposure and accidental exposures were given (single exposures, whole-body, 25 R if under 45 and 50 R over 45; single exposures, hands, 150 R; accumulated exposure, whole-body, 300 R; accumulated exposure, hands, 1000 R). A limit was also specified for pregnant women (25 R in a single exposure) and workers under 18 (one-tenth of above limits). It was suggested that exposures to the gonads be kept as low as practicable under the 0.3 R/week level. Relative Biological Effectiveness factors were also specified (1 for x, γ, β; 5 for protons and thermal neutrons; 10 for alphas and fast neutrons).

A following meeting dealt with internal emitters and focused on K. Morgan's criteria from his work on the Manhattan Project. The various considerations for intake, metabolism, and dose to the critical organ led to specified concentrations of different radionuclides in air, water, and biological excreta that must be detectable in order to ensure the safety of the workers.

In July 1949, the state of California proposed a set of safety standards for radiation within a set of general safety considerations for all industries. This was the first such set of proposals seen by the NCRP. The NCRP responded to correct some misinterpretations in their document, but the real importance of the California action is that it led the NCRP to open a dialogue with the Atomic Energy Commission (AEC) to begin to promulgate a national safety standard. Up to now the NCRP had actively avoided having its recommendations formed into a regulatory framework, because of a fear of loss of flexibility and of the inclusion of technologically unsound considerations. It became evident by mid-1952 that some governmental bodies were going to formulate some kind of regulations, however, and the NCRP was asked to provide input.

Meanwhile, the ICRP had a meeting in 1950 to study international recommendations. In this set of recommendations, some new thinking emerged. First, they clearly stated a difference between carcinogenic and genetic effects. The concept of maximum permissible concentrations of radionuclides in air and water was acknowledged, as was the fact that the roentgen was not a suitable unit for measuring all types of radiations. Also, the concept of limiting exposures to produce the "lowest possible level" was stated. In a publication in 1953, many of the concepts and limits proposed by the NCRP appeared.

7.8 Period 8: 1953–1959

The struggle to formally develop a set of coherent international standards and national regulations continued. Geneticists continued to push for much lower exposure limits because of their perception of the risk of cumulative low-level effects. The good news out of all of this is that during this period, attempts were made to correlate the radiation protection standards to some real observable data regarding effects. Especially, observations published in a study by the National Academy of Sciences (NAS) on life-shortening due to cumulative radiation exposure began to heavily influence thinking in these groups.

As of 1957 the NCRP was still recommending a limit of 0.3 rem per week, but with a cap of 3 rem in 13 weeks and a cumulative limit of $5(N - 18)$ rem (where N is the person's age in years) and provided that no single annual exposure resulted in a person receiving over 15 rem. Permissible levels for internal emitters were to conform to these limits unless gonads were the critical organ, in which case the limits were to be reduced by a factor of 3. Nonoccupational exposure should not result in a person receiving more than 0.5 rem/year. Acceptable dose to the skin was to be twice that of the whole body. An emergency dose of 25 rem whole-body was also prescribed. These limits specifically exclude medical and background exposures.

Soon thereafter, the desire to limit a person's annual dose to 5 rem caused the NCRP to change its weekly limit to 0.1 R/week for new facilities. The real desire was to limit the cumulative exposure to age 30 to 50 rem and to limit exposures to 50 rem per decade thereafter.

The earliest proposed standard for the general public based on the NAS data was 14 million person rem per million population from conception to age 30, with one-third of that amount in each decade of life thereafter. Several other proposals were offered, with values between 9 and 12 million person rem per million. There was general agreement that background and medical exposures would contribute 4 and 3 rem, respectively, to this total, allowing for between 2 and 7 rem from manmade sources.

It was the ICRP, however, that first set the calendar quarter as the basis for occupational protection and moved away from the weekly limit. In the report of

ICRP Committee II published in the *Health Physics Journal* in 1960, the first comprehensive set of radiation dose limiting criteria appeared, which would directly influence national legislation in the United States. Their now familiar recommendations are as follows.

1. *Occupational exposure*: Dose to gonads or wholebody in a period of 13 consecutive weeks is not to exceed 3 rems and cumulative dose to age N shall not exceed $5(N - 18)$. The dose delivered to the skeleton in any 13 week period shall not exceed that from continuous content of 0.1 μg ^{226}Ra. The dose to any single organ shall not exceed 4 rem in 13 weeks or 15 rem per year, except for skin and thyroid which may receive 8 rem in 13 weeks, or 30 rem per year. The reduction in whole-body dose from 0.3 rem/week to 0.1 rem/week was to lessen the possibility of radiation-induced leukemia or life-shortening and the reduction in gonad dose was to limit potential deleterious genetic effects in future generations.

   If there is no external exposure, concentrations of radionuclides specified by maximum permissible concentration (MPC) values for air and water may be continuously tolerated during the 13 weeks. Concentrations may be permitted to vary over the 13 week period as long as total intake during the period based on continuous exposure is not exceeded. (Note this next statement.) If there is external exposure, MPC values must be *lowered* to bring the total RBE doses within the limits prescribed by the basic rules above.

2. *Special groups*: Adults who work in or near controlled areas but are not radiation workers were to be limited to 1.5 rem per year and concentrations equal to one-third of those for occupational workers. Members of the general public were to be limited to 0.5 rem per year and air and water concentrations equal to one tenth of the occupational values. Again, lower limits for MPCs were prescribed if significant external exposure resulted.

3. *Population exposure*: It is desired to limit the average dose equivalent to the gonads to 2 rem in the first 30 years of life (0.5 rem from external and 1.5 rem from internal). Note from the previous discussion on population exposures that this is the most conservative alternative offered.

7.9   Period 9: 1960–1965

These recommendations became the basis for our current regulatory structure in the famous (or infamous) 10CFR20. The thyroid was excluded from the permissible limit of 7.5 rem per quarter (30 rem per year), and an additional value of 18.75 rem per quarter (75 rem per year) for the extremities was added for occupational exposure. Individuals were allowed to exceed the basic requirements as long as dose equivalent in a calendar quarter did not exceed 3 rem and cumulative dose equivalent did not exceed $5(N - 18)$ rem. Under a separate heading, restrictions on concentrations of radioactive material in air and water, corresponding to the ICRP's MPC values were given. If you read carefully the language of the first section, it indeed implied what the ICRP intended, that is, that the individual's total dose equivalent from exposure to radiation or radioactive materials not exceed the basic limits. In practice, however, what was enforced was a two-part system, where a person's external exposure was subject to the basic requirements and in addition, his internal exposure to radioactive materials was governed by the MPC concept.

Persons under 18 were to be limited to dose equivalents equal to 10% of adult values and MPCs specified for the general public. Limits for the general public were as follows.

1. External radiation levels such that no person could receive more than 2 mrem in a hour or 100 mrem in seven consecutive days.
2. Concentrations in air and water separately specified. Standard man is assumed to take in 2200 cm^3 of water and 2×10^7 cm^3 of air in a 24 hour day. Because of higher activity at work, he is assumed to take in half of these values in the 8 hours at work and the other half during the 16 hours at home. To extrapolate occupational MPCs to the general public, the following factor is applied.

$$MPC_{GP} = MPC_0 \times 1/2 \times 5/7 \times 50/52 \times 1/10$$

$$MPC_{GP} = MPC_0 \times 1/30.$$

The factor of 1/2 is due to the consumption of twice as much air or water, the factors of 5/7 and 50/52 account for exposure on weekends and the two weeks of holiday assumed for the working population, and the factor of 1/10 accounts for the usual reduction used for the general public. So, with a few exceptions, MPCs for the general public are a factor of 30 lower than for workers.

These regulations thus became force of law and represented the thoughts of the international community of physicists and biologists based on early results on observed biological effects. Although they were perhaps more restrictive than some nuclear and medical industry people wanted and were less restrictive than many geneticists felt comfortable with, they were thought to provide a measure of safety comparable to, and probably better than, that afforded in most industries.

During this period, new biological studies were made and several other issues began to surface. The radiosensitivity of the lens of the eye was recognized by the ICRP and the radiosensitivity of the unborn child, which had been recognized previously, immediately became a bone of contention with some segments of the women's rights movement.

7.10 Period 10: 1966–Present

Although Dr. Taylor's period 10 ends in 1974, I continue it to the present to bring in what we know to date. Reports by the NAS and the United Nations Scientific Committee on the Effects of Atomic Radiation (UNSCEAR) enlarged and refined our understanding of radiation-induced effects in humans. A great knowledge gap still exists in the area where most regulations are concerned, however, at low doses and low dose rates. The proliferation of nuclear technology resulted in several serious radiation accidents involving a few individuals which tended mainly to confirm what we knew about acute whole-body or partial-body exposures to radiation. Two significant accidents at operating power plants, at TMI in the United States and at Chernobyl in the U.S.S.R., along with publicity about smaller accidents, heightened public concerns about the uses of nuclear technology.

During this period, several individuals with reasonable credentials began to influence the opinions of the public, selected members of Congress, and some segments of the judiciary that existing radiation protection standards were inadequate and should be lowered drastically. These individuals (e.g., J. Gofman, A. Tamplin, E. Sternglass, R. Nader, A. Stewart) mostly resorted to unreasonable and unscientific arguments to advance their arguments. Although they have from time to time managed to catch the ear of some legislators, their arguments have been widely recognized in the international community as fallacious and, fortunately, they have not been able (so far) to affect either the opinions of the ICRP or NCRP or of the existing regulatory bodies such as the NRC. In fact, in most cases, their voices are being heard less and less as their arguments have been discredited as "masterpieces of obdurate obfuscation and strange mixtures of facts used out of context and employing calculations based on unproven assumptions"[1].

So, although the concern of the public and several activists in Congress remains unnecessarily heightened, the regulatory process has so far escaped serious harm. The need for a revision to our current regulatory structure, however, has been widely recognized in the scientific community.

In 1977, the ICRP proposed a new system of radiation protection that was endorsed by the NCRP in 1987. It was formulated into a new set of regulations by the NRC and endorsed by President Reagan. They appeared in the *Federal Register* in early 1991 as the new 10CFR20. Licensees had a three year period to phase in compliance with the new regulations.

The new system is not radically different, either in the limits specified or the methods used to implement them. It represents mainly a refinement in the knowledge base and an attempt to rectify misapplications of the old recommendations.

The new system seeks to simultaneously limit stochastic effects (those for which probability is a function of dose without threshold, e.g., genetic effects and cancer) and nonstochastic effects (those for which severity is a function of dose, probably with a threshold, e.g., erythema). It more forcefully states that internal and external exposures must be summed so that total individual risk is limited. And it introduces one very new concept to both internal dosimetry and dose limitation: the effective dose equivalent concept, which attempts to allow whole- and partial-body exposures to be related on the basis of risk. The basic approach of the whole system is to eliminate nonstochastic effects and limit stochastic effects to "acceptable" levels.

To eliminate nonstochastic effects, no organ (including the skin and extremities) should receive more than 0.5 Sv (50 rem, note the switch in unit systems also), except for the lens of the eye, which should not receive more than 0.15 Sv in a year. To limit stochastic effects, the total whole-body dose equivalent should be limited to 0.05 Sv per year, when external and internal sources are considered and when partial-body dose equivalents are expressed in the equivalent whole-body dose (defined shortly). The limit for nonstochastic effects in based on the extensive data published in UNSCEAR reports regarding thresholds for observable effects in various organ systems.

The stochastic limit seeks to place the risk for occupational exposure to radiation in line with observed risk levels in other so-called "safe" industries.

The death rate in these industries averages about 110 per million workers per year. Based on extensive studies of cancer incidence and mortality data for a variety of cancers and over all ages and both sexes, a risk coefficient of 16,500 excess cancer deaths per million person sievert (lifetime) has emerged. Experience has shown that the average worker receives about 0.005 Sv per year under the dose limitation system which sets a maximum dose equivalent of 0.05 Sv per year. Multiplication of the above risk coefficient by the average dose equivalent of 0.005 Sv yields a total risk of about 80 expected deaths per million worker years, which is comparable to that for other industries.

The effective dose equivalent concept embodies knowledge about individual organ radiosensitivities and assigns dimensionless weighting factors to each organ in proportion to its sensitivity. The effective dose equivalent is numerically equal to the sum of the products of an organ's actual dose equivalent received and its dimensionless weighting factor:

Example

Organ	Actual Dose Equivalent Received (Sv)	Weighting Factor	Weighted Dose Equivalent (Sv)
Gonads	$0.0010 \times$	0.25	$= 0.00025$
Breast	$0.0020 \times$	0.15	$= 0.00030$
Lungs	$0.0020 \times$	0.12	$= 0.00024$
Red Marrow	$0.0015 \times$	0.12	$= 0.00018$
Thyroid	$0.0005 \times$	0.03	$= 0.000015$
Bone Surfaces	$0.0020 \times$	0.03	$= 0.00006$
Liver	$0.0030 \times$	0.06	$= 0.00018$
		Sum = Effective Dose Equivalent	$= 0.0012 \, Sv$

Because of the way in which the weighting factors are derived, this effective dose equivalent is the dose equivalent which, if uniformly received by the whole body, would result in the same risk as from the individual organs receiving these different dose equivalents.

Secondary and derived limits for inhalation or ingestion of radioactive materials are also specified. These limits only apply if the individual is exposed to that one radionuclide with no external exposure. The Annual Limit on Intake (ALI) limits the amount of a radionuclide that can be taken in during one year, considering both stochastic and nonstochastic effects. The Derived Air Concentration (DAC) is the concentration of the nuclide in air which would result in the intake of 1 ALI by inhalation.

Tests of compliance must now take the form:

$$\frac{H_{WB}}{0.05 \, Sv} + \sum_j \frac{I_j}{ALI_j} < 1$$

where H_{WB} is the external whole-body dose equivalent, I_j is the intake of nuclide j, and ALI_j is the ALI for nuclide j. One could also evaluate this expression using DAC – hours.

Superimposed on this dose limitation system is the concept of trying to keep all doses As Low As is Reasonably Achievable (ALARA) below the prescribed limits. Some have attempted to formulate ALARA into a regulation and penalize industries that do not practice it, but the question of what constitutes a reasonable reduction is not easy to resolve. The new 10CFR20 regulations actually include a statement that "The licensee shall use, to the extent practicable, procedures and engineering controls based upon sound radiation protection principles to achieve occupational doses and doses to members of the public that are as low as is reasonably achievable (ALARA)." Use of ALARA programs is widespread, but interpretations of how to implement this concept vary.

Limits for minors are similar to the previous ones. The limit for members of the general public, however, has been reduced to 1 mSv/year. Further, DACs for the general public are further reduced by a factor of 2, to account for the fact that the DACs derived for adults will be applied to persons of many ages, with perhaps different dose factors and radiosensitivities.

$$DAC_{GP} = DAC_{OCC} \times 1/2 \times 5/7 \times 50/52 \times 1/10 \times 1/5 \times 1/2$$

$$DAC_{GP} = DAC_{OCC} \times 1/300.$$

The NCRP and others have recommended the limitation of exposure of the unborn child to 0.005 Sv per gestation period, with no more than 0.0005 Sv to be received in any one month. The rationale for these limits is to protect the unborn child as a member of the general public and to limit exposure during the various radiosensitive periods of development. The new regulations include a provision for applying these limits, with some variations on the monthly dose equivalent, to workers who wish to declare their pregnancies. A 1991 Supreme Court ruling (United Auto Workers v. Johnson Controls) declared it unconstitutional to prohibit women from working with toxic substances, unless they choose to declare their pregnancies. The employer's rights to protect the "next generation" of persons and to protect themselves from possible litigation were outweighed by the employee's right to work in a nondiscriminatory workplace. Thus the dose limits above apply and are enforceable, as long as the woman officially declares her pregnancy to her employer. Furthermore, a woman may declare her pregnancy and at a later date choose to undeclare it, at which time the regulatory burden may not be enforced.

7.11 Period 11: The Future

The BEIR-V[2] report released in 1990 suggested to some that radiation limits for workers should be reduced by a factor of three to four. Such a reduction has been endorsed by the ICRP and has in fact appeared as revised recommended limits for exposures of workers (0.02 Sv/yr, i.e., 20 mSv/yr or 2 rem/yr) and intakes of radioactive materials in very recent ICRP publications. Many countries around the world have adopted the 20 mSv/yr limit and are acting in accordance with this limit. The NRC in the United States chose not to adopt either the 20 mSv/yr limit or the resultant new recommendations of the ICRP (see below) based on this limit. Thus, our current regulatory structure reflects the 1979 recommendations of the ICRP and is based on a 50 mSv/yr dose limit for workers.

The BEIR Committee recently released a new report, BEIR-VII[3] (2005), which basically reiterated their support for the linear, no threshold risk model and did not substantially change their cancer risk models. It will be interesting to see if any substantive changes in U.S. radiation protection regulations will occur in the foreseeable future. A new report to be issued by the ICRP in 2006 has a new provision: calculation of radiation dose to elements of the environment (flora and fauna). Our radiation protection policy for the last 100 years has been that if we protect humans, we will adequately protect the environment, from a dosimetric point of view. This new approach is not being widely accepted, as many still hold to this view, and believe that these new modeling and calculational approaches are not warranted, from the standpoint of wise use of societal resources, and because specific harmful endpoints in elements of the environment have not been demonstrated.

The future of the nuclear power industry, both fission and fusion, is highly uncertain. Public opinion has been very low for fission reactors and moderately good for fusion, the latter partially due to (yet another) misconception about the amount of radiation and radioactive materials that will be produced by such reactors. Concern over energy shortages, emissions of greenhouse gases, improvements in reactor designs (so-called "inherently safe" reactors), and reactor design standardization have given some hope that the nuclear industry will again expand in the coming years. In any case, concerns for this industry will certainly continue for the next 20 to 50 years during operation of existing reactors and decommissioning. Uses of radiation in the medical and other industries will continue into the foreseeable future. Changes in regulations are almost certain, given the tumultuous nature of our political processes and the discoveries about the effects of low-level radiation yet to be made. We can only hope that the integrity of the persons formulating such changes will be as high as those who have developed the science to its current status.

7.12 Radiation Regulations—An Acronym-onious History

7.12.1 Introduction

Significant sources of radiation and radioactivity in modern society include:[4]

- Nuclear reactors used: (1) to generate electricity; (2) to power ships and submarines; (3) to produce radioisotopes used for research, medical, industrial, space and national defense applications; and (4) as research tools for nuclear engineering and physics
- Particle accelerators used to produce radioisotopes and radiation and to study the structure of matter, atoms, and common materials
- Radionuclides used in nuclear medicine, biomedical research, and medical treatment
- X-rays and gamma rays used as diagnostic tools in medicine, as well as in diverse industrial applications, such as industrial radiography, luggage X-ray inspections, and nondestructive materials testing

- Common consumer products, such as smoke detectors, luminous-dial wrist watches, luminous markers and signs, cardiac pacemakers, lightning rods, static eliminators, welding rods, lantern mantles, and optical glass

Radiation protection is managed through an interaction between established regulatory bodies, scientific advisory bodies, and users of these technologies. The scientific advisory bodies were formed very early in the history of the use of radiation, and continue to function today. They have no "official" status, generally speaking. Some are appointed by a parent organization (e.g., the International Atomic Energy Agency (IAEA) was chartered in July 1957 by the United Nations), whereas others were formed as people perceived the necessity for them to exist, and their existence continues as long as some source of funding exists and there is a continued perceived need for their input. Some came into existence and were eliminated or replaced by other bodies over time. Scientific advisory bodies do not have authority to issue or enforce regulations. However, their recommendations often serve as the basis for the radiation protection regulations adopted by the regulatory authorities in the United States and most other nations.

7.12.2 Scientific Advisory Bodies

International Commission on Radiological Protection (ICRP) and National Council on Radiation Protection and Measurements (NCRP)
In 1928, at the Second International Congress of Radiology meeting in Stock-holm, Sweden, the first radiation protection commission was created. The body was named the International X-Ray and Radium Protection Commission (ICXRP). It was charged with developing recommendations concerning radiation protection. In 1950, to better reflect its role in a changing world, the Commission was reorganized and renamed the International Commission on Radiological Protection (ICRP). The ICRP is still very active today, and is considered to be the leading organization that develops recommendations for radiation protection, many of which are intended to (and do) influence the regulatory process in most countries.

In 1929, the U.S. Advisory Committee on X-Ray and Radium Protection (ACXRP) was formed. In 1964, the Committee was congressionally char-tered as the National Council on Radiation Protection and Measurements (NCRP). Both the NCRP and ICRP put out scientific documents, discussing the state of knowledge in a particular area of radiation protection science, or putting forth new knowledge or recommendations for practice. Several publications of the ICRP have formed the basis for specific regulations in the United States, in particular ICRP 2 (1959) and ICRP 30 (1979), which were directly used to write our most important radiation regulation, 10CFR20, discussed below. ICRP 60 (1991) proposed an update to this area, arguing for a reduction in overall dose limits from 50 mSv to 20 mSv. Many other countries have already adopted this, but the United States is still operat-ing under an ICRP 30-based system (which was only fully implemented in 1994).

International Atomic Energy Agency (IAEA)
The International Atomic Energy Agency (IAEA) was chartered in July 1957 as an autonomous intergovernmental organization by the United Nations

(U.N.). The IAEA gives advice and technical assistance to U.N. member states on nuclear power development, health and safety issues, radioactive waste management, and on a broad range of other areas related to the use of radioactive material and atomic energy in industry and government. As government bodies do not necessarily have to adopt the recommendations of the ICRP and NCRP, U.N. member states do not have to follow IAEA recommendations. If they choose to ignore the IAEA recommendations, however, funding for international programs dealing with the safe use of atomic energy and radioactive materials can be withheld and, in matters related to safeguarding special nuclear material, U.N. resolutions may be enforced legally, using government and even military intervention if needed. Many of the IAEA recommendations follow ICRP recommendations on radiation protection philosophy and numerical criteria. The IAEA has published a number of useful scientific documents, including tables of dose values for workers and the public, mostly drawing on results generated by the ICRP. The IAEA sponsors much international research, with funds mostly going to developing countries, in many areas of radiation research.

National Academy of Sciences (NAS)

The National Academy of Sciences (NAS) is a private, nonprofit, self-perpetuating society of distinguished scholars engaged in scientific and engineering research, dedicated to the furtherance of science and technology and to their use for the general welfare. Upon the authority of the charter granted to it by the Congress in 1863, the Academy has a mandate that requires it to advise the federal government on scientific and technical matters. Members and foreign associates of the Academy are elected in recognition of their distinguished and continuing achievements in original research; election to the Academy is considered one of the highest honors that can be accorded a scientist or engineer. The Academy membership is comprised of approximately 1900 members and 300 foreign associates, of whom more than 170 have won Nobel Prizes.

The NAS publishes on a variety of topics. Its most influential works in the area of radiation protection are its summaries of the biological effects of ionizing radiation. BEIR I, BEIR III, and BEIR V have all been highly influential in the setting of radiation protection standards, based on the total knowledge of radiation effects in humans and animals. BEIR VII was recently released, and the conclusions were relatively controversial, due to the current debates about low levels of radiation and health effects. The authors endorsed again the use of a Linear No Threshold (LNT) model for prediction of radiation carcinogenesis at low doses and dose rates. Some modification to specific model results were presented, based on new cancer data and new dosimetry analyses from the Hiroshima and Nagasaki bombings, but the ultimate conclusions were basically in agreement with those given in BEIR V. (Note: BEIR VI is a report strictly on dose/effect relationships for radon.)

United Nations Scientific Committee on the Effects of Atomic Radiation (UNSCEAR)

In 1955 the General Assembly of the United Nations established a Scientific Committee on the Effects of Atomic Radiation (UNSCEAR) in response to widespread concerns regarding the effects of radiation on human health and the environment. At that time, nuclear weapons were being tested in the atmosphere, and radioactive debris was dispersing throughout the environment, reaching the human body through intake of air, water, and foods. The Committee was requested to collect, assemble, and evaluate information on the levels of ionizing radiation and radionuclides from all sources (natural and produced by man) and to study their possible effects on man and the environment. The Committee is comprised of scientists from 21 member states. These member states are: Argentina, Australia, Belgium, Brazil, Canada, China, Egypt, France, Germany, India, Indonesia, Japan, Mexico, Peru, Poland, Russia, Slovakia, Sudan, Sweden, the United Kingdom, and the United States of America. The UNSCEAR Secretariat, which gives the Committee the necessary assistance in carrying out its work, is located in Vienna; it consults with scientists throughout the world in establishing databases of exposures and information on the effects of radiation. The Committee produces the UNSCEAR Reports, which are detailed reports to the General Assembly. The most influential of those are the reports on the sources and effects of ionizing radiation, which catalogue human exposure to natural background, occupational, and medical radiation sources worldwide.

International Commission on Radiation Units and Measurements (ICRU)

The International Commission on Radiation Units and Measurements (ICRU) was established in 1925 by the International Congress of Radiology. Since its inception, it has had as its principal objective the development of internationally acceptable recommendations regarding (1) quantities and units of radiation and radioactivity; (2) procedures suitable for the measurement and application of these quantities in diagnostic radiology, radiation therapy, radiation biology, and industrial operations; and (3) physical data needed in the application of these procedures, the use of which tends to assure uniformity in reporting. The ICRU endeavors to collect and evaluate the latest data and information pertinent to the problems of radiation measurement and dosimetry, and to recommend in its publications the most acceptable values and techniques for current use. The ICRU has published a series of useful documents, most importantly defining radiation quantities and units, but also discussing the state of the science in various applications of radiation in general protection and medicine.

Federal Radiation Council (FRC), Radiation Policy Council (RPC), and Committee on Interagency Radiation Research and Policy Coordination (CIRRPC)

The Federal Radiation Council (FRC) was established in 1959 by executive order of the president of the United States. The FRC was established as an official government entity and included representatives from all federal agencies concerned with radiation protection. The Council served as a coordinating body for all radiation activities conducted by the Federal Government and was responsible for:

... advising the President with respect to radiation matters, directly or indirectly affecting health, including providing guidance to all Federal agencies in the formulation of radiation standards and in the establishment and execution of programs of cooperation with States.... [5]

The FRC was a short-lived agency; it was dissolved in 1970, with its responsibilities being transferred to the newly created EPA. Another group, called the Radiation Policy Council (RPC) was founded in 1980 by President Carter to provide some similar interagency coordination. President Reagan disbanded this group a year later (!), creating in its place an entity called the Committee on Interagency Radiation Research and Policy Coordination (CIRRPC), which was to coordinate policy and resolve conflicts regarding radiation regulations and policies among the various governmental agencies in the United States. This group held some interesting meetings and put out a few not-so-influential publications before being disbanded in 1995.

7.12.3 Regulatory Bodies

Nuclear Regulatory Commission (NRC)

The Atomic Energy Act (AEA) of 1954[6] is the fundamental U.S. law regulating both the civilian and military uses of nuclear materials. Under the Atomic Energy Act of 1954, a single agency, the Atomic Energy Commission, had responsibility for the development and production of nuclear weapons and for both the development and the safety regulation of the civilian uses of nuclear materials. On the civilian side, it provides for both the development and the regulation of the uses of nuclear materials and facilities in the United States, declaring the policy that "the development, use, and control of atomic energy shall be directed so as to promote world peace, improve the general welfare, increase the standard of living, and strengthen free competition in private enterprise." The AEA requires that civilian uses of nuclear materials and facilities be licensed, and it empowered the AEC to establish by rule or order, and to enforce, such standards to govern these uses as "the Commission may deem necessary or desirable in order to protect health and safety and minimize danger to life or property." As we show later, the two functions (civilian and military) were later separated, and the civilian portion was given to the (then formed) Nuclear Regulatory Commission (NRC), which still functions today.

Agreement State Concept: Under Section 274 of the Act, the NRC may enter into an agreement with a state for discontinuance of the NRC's regulatory authority over some materials licensees within the state. States already regulate:

1. Naturally occurring radioactive materials (NORM)
2. Radiation-producing machines (medical and industrial X-rays, particle accelerators)
3. Radioactivity produced in accelerators

To become an agreement state, the state must first show that its regulatory program is compatible with the NRC's and adequate to protect public health and safety. The NRC retains authority over nuclear power plants but the agreement state then is given power to regulate within its borders the use of

1. Byproduct
2. Source
3. Special nuclear material (in small quantities).

Byproduct material is (1) any radioactive material (except special nuclear material) yielded in, or made radioactive by, exposure to the radiation incident to the process of producing or using special nuclear material (as in a reactor); and (2) the tailings or wastes produced by the extraction or concentration of uranium or thorium from ore.

Source material is uranium or thorium, or any combination thereof, in any physical or chemical form or ores that contain by weight one-twentieth of one percent (0.05%) or more of (1) uranium, (2) thorium, or (3) any combination thereof. Source material does not include special nuclear material.

Special nuclear material is plutonium, uranium-233, or uranium enriched in the isotopes uranium-233 or uranium-235.

The NRC periodically assesses the compatibility and adequacy of the state's program for consistency with the national program. Listed below are those states who currently are agreement states:

Alabama	Kansas	New York
Arizona	Kentucky	North Carolina
Arkansas	Louisiana	North Dakota
California	Maine	Oregon
Colorado	Maryland	Rhode Island
Florida	Mississippi	South Carolina
Georgia	Nebraska	Tennessee
Illinois	Nevada	Texas
Iowa	New Hampshire	Utah
New Mexico	Washington	

Low-Level Waste Disposal

The Low-Level Radioactive Waste Policy Amendments Act of 1985 (LLWPA) gave states the responsibility to dispose of low-level radioactive waste generated within their borders and allows them to form compacts to locate facilities to serve a group of states. The Act provides that the facilities will be regulated by the NRC or by states that have entered into agreements with the NRC under Section 274 of the Atomic Energy Act. The State Compact system for LLW is a complete mess at the moment.

There was once a nice scheme, dividing the country into several large regions, within which a site would be selected. This rapidly disintegrated into a political quagmire. At present, there are three operational LLW disposal facilities, but only two—at Richland, Washington near Hanford and Barnwell, South Carolina near Savannah River—are open to a wide variety of LLW. A third facility at Clive, Utah accepts a few limited categories of LLW. The future of other sites, in North Carolina, Texas, California, and elsewhere, is in question. When a site is chosen, an orchestrated protest is engaged by organized anti-nuclear groups, which generally is able to cause enough concern in the public to delay or stall the process of approval.

What is low-level waste? Here are the operational definitions of different types of nuclear waste. Waste is generally defined in categories based on its origins, not necessarily its present hazard level. A high activity ^{137}Cs source, definitely capable of delivering high doses if contacted, will be a type of "low-level" waste.

Category of Radioactive Waste	Definition
High-Level Waste (HLW)	1) *Spent Fuel*: irradiated commercial reactor fuel. 2) *Reprocessing Waste*: liquid waste from solvent extraction cycles in reprocessing. Also the solids into which liquid wastes may have been converted. Note: The Department of Energy defines HLW as reprocessing waste only, whereas the Nuclear Regulatory Commission defines HLW as spent fuel and reprocessing waste.
Transuranic Waste (TRU)	Waste containing elements with atomic numbers (number of protons) greater than 92, the atomic number of uranium (thus the term "transuranic," or "above uranium"). TRU includes only waste material that contains transuranic elements with half-lives greater than 20 years and concentrations greater than 100 nanocuries per gram. If the concentrations of the half-lives are below the limits, it is possible for waste to have transuranic elements but not be classified as TRU waste.
Low-Level Waste (LLW)	Defined by what it is not. It is radioactive waste not classified as high-level, spent fuel, transuranic, or byproduct material such as uranium mill tailings. LLW has four subcategories: Classes A, B, C, and Greater-Than Class-C (GTCC), described below. On average, Class A is the least hazardous and GTCC is the most hazardous.

10CFR20

10CFR20 is the main piece of legislation that governs radiation worker exposures. Other regulations may cover environmental releases, transportation of radioactive materials, and other issues, but the most important code is 10CFR20. Some important sections are given here (quoted directly from the published statute):

Dose Limits:

(a) The licensee shall control the occupational dose to individual adults, except for planned special exposures under §20.1206, to the following dose limits.

    (1) An annual limit, which is the more limiting of

        (i) The total effective dose equivalent being equal to 5 rems (0.05 Sv); or

        (ii) The sum of the deep-dose equivalent and the committed dose equivalent to any individual organ or tissue other than the lens of the eye being equal to 50 rems (0.5 Sv).

(2) The annual limits to the lens of the eye, to the skin of the whole body, and to the skin of the extremities, which are:

(i) A lens dose equivalent of 15 rems (0.15 Sv), and

(ii) A shallow-dose equivalent of 50 rem (0.5 Sv) to the skin of the whole body or to the skin of any extremity.

(b) Doses received in excess of the annual limits, including doses received during accidents, emergencies, and planned special exposures, must be subtracted from the limits for planned special exposures that the individual may receive during the current year.

(c) The assigned deep-dose equivalent must be for the part of the body receiving the highest exposure. The assigned shallow-dose equivalent must be the dose averaged over the contiguous 10 square centimeters of skin receiving the highest exposure.

(d) Derived air concentration (DAC) and annual limit on intake (ALI) values are presented in Table 1 of Appendix B to Part 20 and may be used to determine the individual's dose and to demonstrate compliance with the occupational dose limits.

(e) In addition to the annual dose limits, the licensee shall limit the soluble uranium intake by an individual to 10 milligrams in a week in consideration of chemical toxicity.

(f) The licensee shall reduce the dose that an individual may be allowed to receive in the current year by the amount of occupational dose received while employed by any other person. . . . The licensee shall demonstrate compliance with the dose limits by summing external and internal doses.

The annual occupational dose limits for minors are 10 percent of the annual dose limits specified for adult workers.

The licensee shall ensure that the dose equivalent to the embryo/fetus during the entire pregnancy, due to the occupational exposure of a declared pregnant woman, does not exceed 0.5 rem (5 mSv).

The total effective dose equivalent to individual members of the public from the licensed operation does not exceed 0.1 rem (1 mSv) in a year, exclusive of the dose contributions from background radiation, from any medical administration the individual has received . . .

The dose in any unrestricted area from external sources, . . . does not exceed 0.002 rem (0.02 millisievert) in any one hour.

Posting of Signs:

Posting of radiation areas. The licensee shall post each radiation area with a conspicuous sign or signs bearing the radiation symbol and the words, "CAUTION, RADIATION AREA."

Posting of high radiation areas. The licensee shall post each high radiation area with a conspicuous sign or signs bearing the radiation symbol and the words, "CAUTION, HIGH RADIATION AREA" or "DANGER, HIGH RADIATION AREA."

Posting of very high radiation areas. The licensee shall post each very high radiation area with a conspicuous sign or signs bearing the radiation symbol and words, "GRAVE DANGER, VERY HIGH RADIATION AREA."

Posting of airborne radioactivity areas. The licensee shall post each airborne radioactivity area with a conspicuous sign or signs bearing the radiation symbol and the words, "CAUTION, AIRBORNE RADIOACTIVITY AREA" or "DANGER, AIRBORNE RADIOACTIVITY AREA."

Posting of areas or rooms in which licensed material is used or stored. The licensee shall post each area or room in which there is used or stored an amount of licensed material exceeding ten times the quantity of such material specified in Appendix C to Part 20 with a conspicuous sign or signs bearing the radiation symbol and the words, "CAUTION, RADIOACTIVE MATERIAL(S)" or "DANGER, RADIOACTIVE MATERIAL(S)."

Opening and Receiving Packages

The licensee shall immediately notify the final delivery carrier and the NRC Operations Center by telephone, when removable radioactive surface contamination exceeds certain limits, or external radiation levels exceed certain limits.

Notification of the NRC

(a) Immediate notification.
   (1) An individual to receive
       (i) A total effective dose equivalent of 25 rems (0.25 Sv) or more; or
       (ii) A lens dose equivalent of 75 rems (0.75 Sv) or more; or
       (iii) A shallow-dose equivalent to the skin or extremities of 250 rads (2.5 Gy) or more; or
   (2) The release of radioactive material, inside or outside of a restricted area, so that, had an individual been present for 24 hours, the individual could have received an intake five times the annual limit on intake.
(b) Twenty-four hour notification.
   (1) An individual to receive, in a period of 24 hours
       (i) A total effective dose equivalent exceeding 5 rems (0.05 Sv); or
       (ii) A lens dose equivalent exceeding 15 rems (0.15 Sv); or
       (iii) A shallow-dose equivalent to the skin or extremities exceeding 50 rems (0.5 Sv); or
   (2) The release of radioactive material, inside or outside of a restricted area, so that, had an individual been present for 24 hours, the individual could have received an intake in excess of one occupational annual limit on intake.

Release of Wastes

1. By transfer to an authorized recipient as provided, or
2. By decay in storage, or
3. By release in effluents within specified limits (Excreta from individuals undergoing medical diagnosis or therapy with radioactive material are not subject to these limitations).

Some Important Definitions

Airborne radioactivity area means a room, enclosure, or area in which airborne radioactive materials, composed wholly or partly of licensed material, exist in concentrations

(1) In excess of the derived air concentrations (DACs) specified in Appendix B, or

(2) To such a degree that an individual present in the area without respiratory protective equipment could exceed, during the hours an individual is present in a week, an intake of 0.6 percent of the annual limit on intake (ALI) or 12 DAC-hours.

Committed effective dose equivalent ($H_{E,50}$) is the sum of the products of the weighting factors applicable to each of the body organs or tissues that are irradiated and the committed dose equivalent to these organs or tissues.

Controlled area means an area, outside of a restricted area but inside the site boundary, access to which can be limited by the licensee for any reason.

Declared pregnant woman means a woman who has voluntarily informed the licensee, in writing, of her pregnancy and the estimated date of conception. The declaration remains in effect until the declared pregnant woman withdraws the declaration in writing or is no longer pregnant. See the section below on the establishment of this unique regulatory consideration.

Deep-dose equivalent (H_d), which applies to external whole-body exposure, is the dose equivalent at a tissue depth of 1 cm (1000 mg/cm^2).

High radiation area means an area, accessible to individuals, in which radiation levels from radiation sources external to the body could result in an individual receiving a dose equivalent in excess of 0.1 rem (1 mSv) in 1 hour at 30 centimeters from the radiation source or 30 centimeters from any surface that the radiation penetrates.

Member of the public means any individual except when that individual is receiving an occupational dose.

Minor means an individual less than 18 years of age.

Radiation area means an area, accessible to individuals, in which radiation levels could result in an individual receiving a dose equivalent in excess of 0.005 rem (0.05 mSv) in 1 hour at 30 centimeters from the radiation source or from any surface that the radiation penetrates.

Restricted area means an area, access to which is limited by the licensee for the purpose of protecting individuals against undue risks from exposure to radiation and radioactive materials. Restricted area does not include areas used as residential quarters, but separate rooms in a residential building may be set apart as a restricted area.

Shallow-dose equivalent (H_s), which applies to the external exposure of the skin of the whole body or the skin of an extremity, is taken as the dose equivalent at a tissue depth of 0.007 cm (7 mg/cm^2).

Total Effective Dose Equivalent (TEDE) means the sum of the deep-dose equivalent (for external exposures) and the committed effective dose equivalent (for internal exposures).

Very high radiation area means an area, accessible to individuals, in which radiation levels from radiation sources external to the body could result in an individual receiving an absorbed dose in excess of 500 rads (5 grays) in 1 hour at 1 meter from a radiation source or 1 meter from any surface that the radiation penetrates.

For the complete text of the code, see http://www.nrc.gov/reading-rm/doc-collections/cfr/part020/index.html.

The Declared Pregnant Worker

This unusual regulatory status arose from a 1990 Supreme Court decision, "International Union, United Automobile, Aerospace & Agricultural Implement Workers of America, UAW v. Johnson Controls, Inc." This case involved facilities involved in battery manufacturing. Workers were occupationally exposed to lead, which involves health risks, including to the fetus, in the case of a pregnant worker. After eight employees of Johnson Controls became pregnant and had blood lead levels exceeding that suggested as "safe" by the Occupational Safety and Health Administration (OSHA) Johnson Controls formulated a policy barring all women, except those whose infertility was formally documented, from work situations involving actual or potential lead exposure exceeding the OSHA guidelines. A class action suit was filed in District Court, claiming sex discrimination. The court granted summary judgment for Johnson Controls, and the Court of Appeals affirmed. The Court of Appeals held that the petitioners had failed to satisfy their "burden of persuasion" in the case, but the Supreme Court disagreed, stating that

By excluding women with childbearing capacity from lead-exposed jobs, respondent's policy. creates a facial classification based on gender and explicitly discriminates against women. . . . The Policy is not neutral because it does not apply to male employees in the same way as it applies to females.

Thus, women were to be given full access to jobs involving materials or agents possibly harmful to the embryo/fetus, and were granted the authority to make risk/benefit decisions for themselves and their unborn children, should this situation arise. The debate over how to implement this was strong, as employers were concerned about their responsibilities to protect the unborn (seeing them as members of the public for radiation protection purposes) and to protect themselves against litigation, should a child be born with deformities after being exposed in an occupational setting.

The compromise was that women need to be informed of the risks of radiation exposure, to themselves and their potential unborn children, and be allowed to declare their pregnancies if they so choose, and at this point be subject to the more strict radiation limits noted above. If they do not choose to declare their pregnancy (formally, in writing), they must be treated equally to any other radiation worker. The highly unusual part of this is that a woman can declare her pregnancy to her employer one day, and then undeclare it the next, if she so chooses. If the pregnancy is declared, and then later "undeclared," the worker returns to "nonpregnant" status in the eyes of the law.

Department of Energy (DOE)

By the Energy Reorganization Act of 1974, the AEC ceased to exist. Its functions became the core of the Energy Research and Development Administration (ERDA), to which its laboratories were transferred, and the Nuclear Regulatory Commission (NRC). Other functions, and some single-program laboratories, were transferred to ERDA from the Department of the Interior (fossil fuels), the National Science Foundation (thermal and geothermal energy support), and the Environmental Protection Agency (alternative automobile power systems). This Act and the Federal Non-Nuclear Energy Research and Development Act of 1974 provided ERDA with a broad mandate to conduct

R&D in all energy areas and conduct demonstration projects and promote commercialization of energy technologies.

National energy-related problems continued into the administration of President Carter who requested the rapid establishment of a Department of Energy. Congress responded with the Department of Energy Organization Act in 1977 (P.L. 95-91) which transferred ERDA and all of its laboratories, and the Federal Energy Administration, to the new department. The Department of Energy was responsible for the development and production of nuclear weapons, promotion of nuclear power, and other energy-related work, and the NRC was assigned the regulatory work, not including regulation of defense nuclear facilities. The DOE thus gained broad responsibility for national energy policy, although major energy programs remained in the Department of the Interior. The DOE still shares much nuclear policy responsibility with the NRC.

The DOE provides regulations for DOE National Labs and test facilities. Their portion of the CFRs is *10CFR part 200-1000*. DOE Labs include:

Advanced Computing Laboratory (ACL)
Ames Laboratory
Argonne National Laboratory (ANL)
Bonneville Power Administration
Brookhaven National Laboratory
Center for Computational Sciences
Continuous Electron Beam Accelerator Facility (CEBAF)
Environmental Measurements Laboratory (EML)
Fermi National Accelerator Laboratory (Fermilab)
Hanford Reservation
Idaho National Engineering Laboratory (INEL)
Kansas City Plant (AlliedSignal Inc.)
Ernest Orlando Lawrence Berkeley Laboratory (LBNL)
Lawrence Livermore National Laboratory
LLNL High Energy Physics
Los Alamos Area Office
Los Alamos National Laboratory (LANL)
National Energy Research Supercomputer Center
National Renewable Energy Lab
Oak Ridge Associated Universities
Oak Ridge National Laboratory
Office of Scientific and Technical Information
Pacific Northwest National Lab
Pantex
Princeton Plasma Physics Lab
Sandia National Laboratories
Savannah River Site (Westinghouse)
Stanford Linear Accelerator Center
Superconducting Super Collider (SSC)
University of California (Lab Management)

The Department of Energy's (DOE's) rule on occupational radiation protection (10 CFR 835, Rev. 1) requires labs to have a formally approved Radiation Protection Program (RPP). The RPP is a legally binding document between lab and the DOE. The CFR contains all the elements of a comprehensive radiation protection program (i.e., management and administrative requirements; radiation dose limits; and requirements for monitoring individuals and areas, posting and labeling, recordkeeping, radioactive material control, training, and emergency response). DOE regulations were generally similar to those of the NRC, but there were some important differences in daily practice. DOE suffered through a number of episodes of criticism from external sources and complaints from unhappy employees in several safety-related matters, including radiation exposures (possibly linked to health effects). Until recently, DOE labs were entirely self-regulated. But now, authority for radiation protection practices at DOE labs is being transferred to the NRC, under HR 3907.

Environmental Protection Agency (EPA)

The Environmental Protection Agency was formed in 1970, and has authority to issue regulatory standards regarding radiation hazards from a number of different situations, including underground mining, uranium fuel cycle operations, management of uranium and thorium mill tailings, airborne emissions of radionuclide, and management and disposal of spent nuclear fuel and high-level and transuranic radioactive wastes.[4]

The crossover of authority between the EPA and NRC is sometimes difficult to understand, and for users to comply with. The EPA does have authority to "issue Federal guidance to limit radiation exposures to workers, as well as to the general public," whereas the NRC is principally responsible for limiting radiation dose to workers, and has oversight of releases of radioactive materials from many facilities. The EPA has published a number of guidance documents on calculating dose and risk; these are not regulations per se, but often they practically become law, as users generally must follow their recommendations, either in the absence of any other data, or because they represent the only numbers that the government will accept (similar to the situation with NRC Regulatory Guides and NUREG documents).

The Atomic Energy Act (AEA) of 1954 established the existence and authority of both the NRC and the EPA, and is the cornerstone of current radiation protection activities and regulations. It granted the EPA the authority to establish generally applicable environmental standards for exposure to radiation.[7] In 1977, the EPA issued standards limiting exposures from operations associated with the light-water reactor fuel cycle.[8] These standards, under 40 CFR Part 190, cover normal operations of the uranium fuel cycle. The standards limit the annual dose equivalent to any member of the public from all phases of the uranium fuel cycle (excluding radon and its daughters) to:

25 mrem to the whole body
75 mrem to the thyroid
25 mrem to any other organ

The standards also set normalized emission limits for some long-lived radionuclides into the environment. The dose limits imposed by the standards cover all exposures resulting from radiation and radionuclide releases to air and water from operations of fuel-cycle facilities. The development of these standards took into account both the maximum risk to an individual and the overall effect of releases from fuel-cycle operations on the population, and attempted to balance these risks against the costs of the technology that would be needed to control them.

In 1969, Congress passed the National Environmental Policy Act,[9] which defined national policy that encouraged harmony between the need for progress in technology and the impacts of those technologies on the public and the environment. The Act established a Council on Environmental Quality (CEQ) to assist the President in determining the status of environmental quality in the United States and in developing environmental policy initiatives. The Act also mandated that detailed Environmental Impact Statements (EISs) be submitted for any major action proposed by a federal agency or for legislation that would significantly affect the quality of the environment. The EIS must describe any adverse environmental effects that the proposal would cause, alternatives to the proposed action, effects of the project on the long-term productivity of the environment, and any irreversible and irretrievable commitment of resources involved in the proposed action.

The Safe Drinking Water Act (SDWA) was enacted to assure safe drinking water supplies and to protect against endangerment of underground sources of drinking waters.[10] Under the authority of the SDWA, the EPA issued regulations (40 CFR Part 141, Subpart B) covering the permissible levels of a number of contaminants (including radium, gross alpha, manmade beta, and photon-emitting contaminants) in community water supply systems.[11] Limits for radionuclides in drinking water are expressed as Maximum Contaminant Levels (MCLs), as for other hazardous chemical substances.

In December 1979, the EPA designated radionuclides as hazardous air pollutants under Section 112 of the Clean Air Act (CAA) Amendments of 1977.[12] In April 1983, the EPA proposed standards regulating radionuclide emissions from four source categories, one of which included DOE facilities. The rule established annual airborne emission limits for radioactive materials and specified that annual doses resulting from such emissions should not exceed 25 mrem to the whole body and 75 mrem to any critical organ (for members of the general public). A series of legal actions followed, causing the EPA to withdraw, revise, and resubmit its regulations, using a risk-based criterion, which the agency did in 1989, considering a lifetime risk to an individual of approximately 1 in 10,000 as acceptable.

Congress passed the Nuclear Waste Policy Act (NWPA) of 1982 to provide for the development of repositories for the disposal of high-level radioactive waste and spent nuclear fuel, and to establish research to demonstrate the feasibility of the project.[13] The Act established a schedule for the siting, construction, and operation of repositories that would provide a reasonable assurance that the public and environment would be adequately protected from the hazards posed by high-level radioactive waste. The Secretary of Energy was charged with nominating candidate sites for a repository and following a number of steps through a process of presidential and congressional approval, site characterizations, public participation, and hearings.

The Act also required the Secretary to adhere to the NEPA in considering alternatives and to prepare an EIS for each candidate site. Initially the Act called for the development of two mined geologic repositories. The first repository was to be selected from nine candidate sites in western states; the second repository was to be located in the eastern United States in crystalline rock. The EPA was charged with the responsibility of promulgating generally applicable standards for the protection of public health and the environment from off-site releases from radioactive material in repositories. The NRC, in turn, was responsible for promulgating technical requirements and criteria consistent with the EPA's standards to serve as the basis for approving or disapproving applications regarding the use, closure, and postclosure of the repository.

The Act also discussed interim waste storage requirements, as well as the payment of benefits to affected states and tribal groups to allow them sufficient resources to participate fully in the process.[1] This entire process has been bogged down for the past 20 years, more in the political struggles over the transportation and final disposal of these wastes than in engineering and scientific studies. The latter were completed in a reasonable time and showed the Yucca Mountain site (in Nevada) to be the best alternative. Construction has been ongoing, but political gamesmanship by anti-nuclear activists, the governor of the state of Nevada, the Clinton administration, and others has delayed the startup of the use of this site to the point that the nuclear utilities have now sued the federal government for 2–60 billion dollars in damages caused by the delays, which a federal judge has ruled will be paid by all American taxpayers. The future of the storage of high-level wastes is still in question.

Some problems have arisen with overlapping jurisdictions between the EPA and NRC, often leaving licensees laboring under conflicting or duplicate regulatory frameworks. Coordination of various agency functions was attempted by a committee called the Committee on Interagency Radiation Research and Policy Coordination (CIRRPC), but I cannot tell that they have had much of an impact. Negotiations over overlapping jurisdictions are fought somewhat case-by-case, at times with one or the other agency giving some ground, if public health and safety concerns can be shown to be adequately protected.

Some have compared the method of regulation of the two agencies, noting that the NRC rules from a "top-down" approach (setting upper limits on dose, with licensees expected to maintain doses below these levels in normal operating conditions), whereas the EPA operates from a "bottom-up" approach (setting restrictive limits to keep public exposures low, and permitting flexibility in specific circumstances when operators have trouble maintaining the limits).

Department of Transportation (DOT)

The Department of Transportation regulates the shipment of radioactive materials in the United States. Their portion of the CFRs for radiation protection can be found under in 49CFR Parts 170–175. Specifically, 49CFR173.403 defines levels of activity permitted in different categories of packaging, shipping papers, vehicle placarding, and general safety procedures to be followed in the shipping of radioactive materials. Tables establish Type A1 and A2 quantities of material. Type A1 is "special form," meaning that

1. It is either a single solid piece or is contained in a sealed capsule that can be opened only by destroying the capsule.
2. The piece or capsule has at least one dimension not less than 5 millimeters (0.2 inch).
3. It satisfies certain physical testing requirements.

If it is not special form, or if it is defined as Low Specific Activity (LSA) material (having activity concentration limits) or is a Surface Contaminated

Object (SCO; a solid object that is not itself radioactive but which has Class 7 (radioactive) material distributed on any of its surfaces), tabulated values of A2 will be used to determine if the package is a Type A package; otherwise similar values are given for A1 materials.

If the quantity does not exceed the activity value in the table for that nuclide, the package may be designated as Type A. Type B quantity means a quantity of material greater than a Type A quantity. Both the A1 and A2 quantities may use type A packages. Anything that does not qualify as Type A material is transported in a Type B package. A Type B package is an accident-resistant package. Finally, there are Type C packages that are designed to withstand severe air crashes and may have to be used if the shipment is to be made internationally by air.

Signage on the vehicle must specifically indicate that the material is radioactive, and show the Transport index (TI), a dimensionless number (rounded up to the next tenth) placed on the label of a package to designate the degree of control to be exercised by the carrier during transportation. The transport index is determined as 100 × the maximum radiation level in milliSievert(s) per hour at 1 m (3.3 feet) from the external surface of the package (unless the package contains fissile material, and a different definition may apply).

Some additional minor players (shhh, don't ever tell them I said that) are as follows.

Food and Drug Administration (FDA)

The Food and Drug Administration sets standards for the use of lasers (21CFR) and other nonionizing radiations, food irradiation, and pharmaceuticals. Their portion of the CFRs for radiation protection can be found under 21CFR 1000.

Occupational Safety and Health Administration (OSHA)

OSHA regulates the control of radioactive materials in the workplace that are not controlled by the DOE or the NRC. They also control overall safety of the workplace, including physical, chemical, sound, and light hazards including nonionizing radiation. Their portion of the CFRs for radiation protection can be found under 29CFR Parts 1910, 1926.

Department of Defense (DOD)

The DOD regulates some of the control of radioactive material in the military.

Endnotes

1. L. S. Taylor, *Organization for Radiation Protection: The Operations of the ICRP and NCRP, 1928–1974* (Office of Technical Information, U.S. Department of Energy, Washington, DC 1974).
2. National Academy of Sciences. Health Effects of Exposure to Low Levels of Ionizing Radiation: BEIR V. Committee to Assess Health Risks from Exposure to Low Levels of Ionizing Radiation, National Research Council, Washington, DC, 1990.
3. National Academy of Sciences. Health Risks from Exposure to Low Levels of Ionizing Radiation: BEIR VII Phase 2. Committee to Assess Health Risks from Exposure to Low Levels of Ionizing Radiation, National Research Council, Washington, DC, 2006.
4. USEPA. History of radiation protection in the United States and current regulations. USEPA, in *Public Health and Environmental Radiation Protection Standards for Yucca Mountain, Nevada* (Chapter 2, June 2001) EPA 402-R-01-005.

5. Federal Radiation Council, Radiation Protection Guidance for Federal Agencies, Federal Register, 25 FR 4402-4403 (May 18, 1960).
6. Atomic Energy Act, Public Law 83-703, as amended, 42 USC 2011 et seq. (1954).
7. The White House, President R. Nixon, Reorganization Plan No. 3 of 1970, Federal Register, 35 FR 15623-15626 (October 6, 1970).
8. U.S. Environmental Protection Agency, Environmental Radiation Protection Standards for Nuclear Power Operations, 40 CFR Part 190, Federal Register, 42 FR 2858-2861 (January 13, 1977).
9. National Environmental Policy Act of 1970, Public Law 91-190 (January 1, 1970).
10. Safe Drinking Water Act, 42 U.S.C. s/s 300f et seq., (1974).
11. U.S. Environmental Protection Agency, National Interim Primary Drinking Water Regulations, EPA 570/9-76-003 (1976).
12. U.S. Environmental Protection Agency, National Emission Standards for Hazardous Air Pollutants, ANPRM, Federal Register, 44 FR 46738 (December 27, 1979).
13. Nuclear Waste Policy Act of 1982, Public Law 97-425 (January 7, 1983).

8

Health Physics Instrumentation

Our natural senses do not detect radiation, even at its most intense levels. A possible exception might be that at very high exposure rates, degradation of oxygen molecules can result in the formation of ozone, which can be perceived. In such a situation, however, survival of the organism is unlikely, so the detection of the hazard may not be helpful. In the systematic measurement of radiation we mostly use electronic instruments that are designed to exploit the types of interactions that radiation has with matter to produce a signal that can be detected and quantified. The basic types of detectors available for routine use have changed little in the several decades since most of the technologies were first made. Some significant changes have occurred in the sophistication of the computer-related accessories, use of global positioning technologies, and the use of computer programs for analysis of data.

The most important reactions that ionizing radiation has with matter are ionization and excitation. These interactions produce measurable changes in the materials affected:

- *Heat*: Interactions of radiation with matter deposit energy, and some of this energy is dissipated as heat.
- *Chemical*: Changes in the chemical state of some species can cause reduction of film, or release of chemicals whose presence can be measured.
- *Electrical*: Ionization produces ion pairs, which can be collected and measured.
- *Light*: Some specialized substances emit light when excited by radiation (actually almost all materials in the world demonstrate this property, but only a few do so to a degree that is sufficient to produce a measurable signal).
- *Thermoluminescence*: This is a special category of light production that we discuss in detail.

8.1   Thermal Reactions

Only particularly high radiation levels will produce enough heat energy to be measured. *Bomb calorimeters* have been used to measure subtle temperature changes in closed systems being irradiated. This is a rarely used technology.

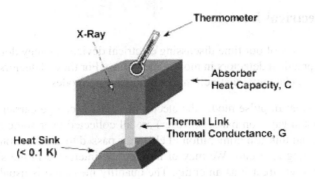

Figure 8.1 X-ray microcalorimeter.

A dose of about 5000 Gy (500,000 rad) is needed to raise the temperature of water by about 1°C, so use of these devices is limited to characterization of very high intensity beams of radiation, such as are used in external beam radiotherapy. This was also demonstrated to be useful in X-ray spectroscopy (Figure 8.1[1]).

"In 1982, a new approach to non-dispersive X-ray spectroscopy, based on the measurement of heat rather than charge, was proposed by Moseley et al. (1984). The detector, an X-ray microcalorimeter, works by sensing the heat pulses generated by X-ray photons when they are absorbed and thermalized. The temperature increases indicate the photon energy. This invention combines high detector efficiency with high energy resolution. Any microcalorimeter must have a low-heat-capacity mass to absorb incident X-ray photons, a weak link to a low-temperature heat sink which provides the thermal isolation needed for a temperature rise to occur, and a thermometer to measure change in temperature"[1].

8.2 Chemical Reactions

The *Fricke dosimeter* is a dose-measuring device based on the work of Fricke and Hart in 1962[2] and based on changes in the oxidation state of iron atoms that produce optical changes in a chemical solution that can be quantified to estimate the energy deposited, and thus the dose, in the solution. The method has a strong dependence on LET, so it has limited practical applications. The usual solution is a ferrous sulfate mixture: 1 mM $FeSO_4$ in 0.8 N H_2SO_4. Ionizing radiation oxidizes Fe^{2+} to Fe^{3+}, which generates a blue color in the solution that can be quantified with a spectrophotometer. The detector continues to have a number of specific applications in biology and medicine, even being suggested for use in calculating three-dimensional magnetic resonance imaging of dose distributions from radiation therapy.[3]

Another device is the *film dosimeter*. The discovery of radiation (Chapter 6) was based on radiation's well-known ability to expose photographic film. Instead of an oxidation reaction, radiation causes the reduction of silver halide in the film to grains of metallic silver. We discuss film dosimeters in more detail later.

8.3   Electrical Devices

We spend much of our time discussing electrical devices, as they dominate the available practical detectors in most common use. For these detectors, we need to make a fundamental distinction in their operating modes:

- *Pulse mode*: In pulse mode, the electronic species (charge carriers of some kind, not always ion pairs, however) are all collected over some fixed time period, and this conglomeration of charge is passed to the electronic system for counting as a unit. We may or may not be interested in the size of the pulse, but we treat it as an entity. The quantity measured is usually charge (e.g., coulombs, C).
- *Current mode*: In current mode, the electronic species are streamed continuously through the charge collection system, and the current generated is related to the rate of exposure or dose being delivered to the detection apparatus. The quantity thus measured is typically a current (e.g., amperes, C/s).

8.3.1 Gas-Filled Detectors

Many detectors work on the principle of the collection of ion pairs (charge carriers) in a gas. One of the earliest methods for quantifying radiation exposure was to apply a potential across parallel metal plates that had just air between them, expose the air to ionizing radiation, and collect the formed charges on the plates and then pass them through a measuring circuit (Figure 8.2). Charges are collected by applying a potential across a region containing a gas (usually a cylinder with a wire in the center; the wire is the anode (positive potential) and the cylinder is at a negative potential). Basically, the electrons will be attracted towards the anode and the positive ions towards the cathode.

The collected charge is sent to an analysis circuit (not shown in the diagram) for processing. The circuitry will contain both resistor and capacitor elements, and is generally referred to as *RC circuitry* (Figure 8.3). The RC circuitry is designed so that the circuit time constant is less than that needed to collect all of the ions created by the original ionizing particles. Pulses are then passed by the circuit to shaping, amplification, and counting devices (Figure 8.4[4]).

Different gas-filled detectors operate at different voltages (potential differences) across the collecting chamber. If the voltage across the chamber is very low, many of the ions formed may recombine, and thus not be counted. As

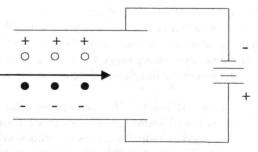

Figure 8.2 Simplified drawing of a gas-filled detector.

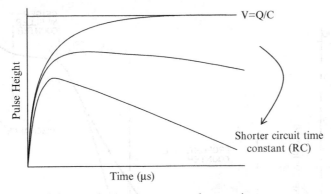

Figure 8.3 Plot of pulse shape versus detector time constant.

the voltage increases, more and more of the ions formed are collected, until basically all of them are collected and processed. If voltage is increased still further, the ions may be accelerated towards the collectors (let's concentrate on the electrons: they are lighter, and thus more easily accelerated, and are collected more quickly). These fast-moving electrons may themselves produce secondary ionizations.

If voltage is increased even further, the secondary electrons can produce more electrons, that produce more, and so on, and a massive "avalanche" of particles may be produced every time that an interaction occurs. Further

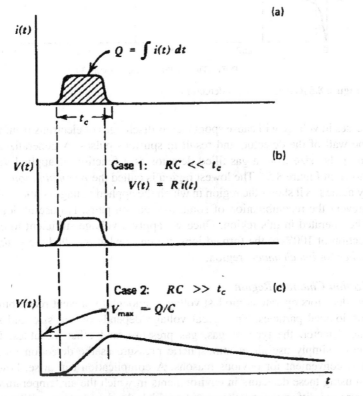

Figure 8.4 Long versus short detector time constant and effect on pulse shape.

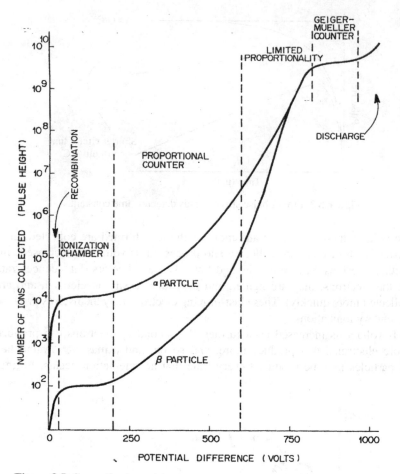

Figure 8.5 Generalized gas detector curve.

increases in voltage will cause spontaneous discharges of electrons from atoms in the wall of the detector, and result in spurious pulses. A generalized plot of the pulse size from a gas-filled detector as a function of applied voltage is shown in Figure 8.5.[4] The lowest region is called the *recombination region*, aptly named as it shows the region in which the applied voltage is not sufficient to prevent the recombination of some formed ion pairs. No useful detectors can be operated in this region. Once we apply a voltage sufficient to ensure collection of 100% of the formed ions, we enter what is called the *ionization chamber* (or *ion chamber*) region.

Ionization Chamber Region

These detectors operate at modest voltages, enough to prevent recombination of the ionized particles. The actual voltage depends on the size and shape of the detector, the type of gas, gas pressure, and other variables. Many detectors simply use air at atmospheric pressure as the detection medium. This is convenient for obvious reasons. A complication can arise, however, when using these detectors in environments in which the air temperature and pressure are different from those under which the detector was calibrated, as the number of ion pairs produced will vary with the number of air molecules

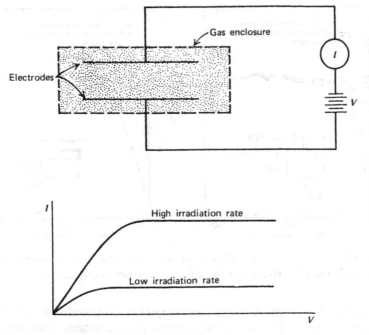

Figure 8.6 Ionization chamber with high or low irradiation rate.

encountered by the radiation, which is in turn dependent on the density of the air in the chamber. Higher pressures will result in higher densities, and higher temperatures will result in lower densities, if all other variables are held constant. Thus, if a chamber is calibrated at sea level at ambient temperature, but then used in a high-temperature environment or taken to a city at high altitude, the chamber reading will need to be corrected.

Fortunately, the correction is a simple linear ratio; one must only remember to use absolute temperatures (K), not relative temperatures (C or F) for the correction. If the temperature and pressure at which the measurements are made (T_2, P_2) are significantly different from those at which the calibration was performed (T_1, P_1), then the following correction factor should be applied to the measured values.

$$\text{Correction Factor} = \left(\frac{P_1}{P_2}\right) \times \left(\frac{273.2 + T_2}{273.2 + T_1}\right).$$

In this expression, the temperatures T_1 and T_2 are assumed to be in degrees Celsius.

Ionization chambers are generally operated in *current* mode, meaning that pulses are integrated over fixed times to give a continuous readout of the rate of ionization, which is easily converted to an exposure rate with proper calibration (Figure 8.6[4]). Ion chambers do not "click." The clicks that are heard from some survey meters are individual pulses being recorded and then being represented by an audible signal. For some meters that operate in pulse mode, such as Geiger counters and scintillation counters that we study shortly, audible clicks may be heard in the routine use of these meters.

Ionization chambers are generally quite energy-independent, over a broad energy range (Figure 8.7[4]). The energy dependence of a survey meter is an

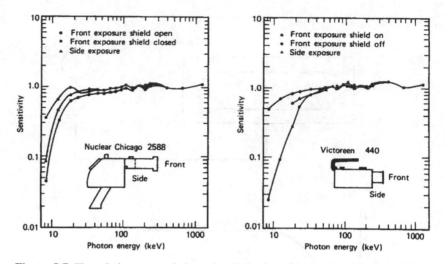

Figure 8.7 The relative energy-independent behavior of ionization chambers. (From Knoll.[4]) Contrast this with the later figure for Geiger counters (Figure 8.14).

important consideration. All survey meters must be calibrated with some chosen radiation source. The most common source chosen, because of availability, half-life, and other considerations, is ^{137}Cs. The principal photon energy from this nuclide has an energy of 0.662 MeV, which is a nice medium energy. If the sources being detected have an energy quite different from this, however (e.g., ^{99m}Tc, with a photon energy of 0.14 MeV), if the meter has an energy-dependent response, a correction factor may be needed to adjust the observed reading to a correct value. With ionization chambers (examples shown in Figures 8.8–8.10), the response is quite linear over a broad range, as is shown in Figure 8.7. Until we reach low energies, the reading taken directly from the meter will be useful for any individual emitter or a mixture of sources.

Many older ionization chambers were shaped somewhat like large "guns," which is practical for handling when performing surveys. The sensitive

Figure 8.8 Eberline portable ionization chamber.

Figure 8.9 Radiation detector (www.drct.com/specials/
eberline.htm) showing display and various scales (www.
drct.com).

chamber is typically cylindrical, and the readout is oriented towards the user.
Older meters will have analog displays, with a needle indicator showing the
measured exposure rates. Newer meters may also use an analog display, but
others may have only digital readouts (Figures 8.8 to 8.10). The exposure rate
typically will vary somewhat as measurements are made, due to variability in
the exposure rate at a point, which is in turn due to the stochastic nature of
radiation emission and interactions in the detector. We discuss some of the
statistical aspects to the behavior of radiation later in this chapter.

This author finds it much easier to visually estimate the average of a variable reading from a swinging needle than from varying numbers on a digital
display. One thing that can help is that some meters have a switch allowing

Figure 8.10 Image of an older "gun type" ionization chamber,
the Hanford "Cutie Pie." (From http://www.orau.org/ptp/collection/
surveymeters/hanfordcutiepie.htm.)

selection of either a slow or fast integration time. Changes in this setting affect which specific resistor and capacitance settings are used in the RC circuit and thus over what period of time the signals are averaged. For general survey measurements, usually a slow integration time is better, as more information is used to express an average value, and the readings do not vary as much. If one is searching for a lost source, however, using the fast integration setting may be better, as more sensitivity to variations in exposure rate are desirable, to avoid passing by the source and not noticing abrupt changes in the exposure rates. Several scales are typically provided (e.g., ×0.1, ×1, ×10). Scales may run from 0 to 10, 0 to 3, or other ranges. Thus the sensitivity can be varied in this way as well. At lower exposure rates, obviously a lower scale will be desirable, not only to show the variability in the readings, but to give a readable absolute value. At higher exposure rates, the meter will "peg" on the lower scales, and one must use a higher scale to obtain any reading at all.

Standard ionization chambers are used for moderate to high exposure rate readings (mR/hr to R/hr; i.e., not environmental monitoring, which is generally in the μR/hr to low mR/hr region). Geiger chambers, which we study shortly, overlap the ranges where we would use an ion chamber (mR/hr to R/hr). Specialized Geiger detectors are used to monitor the very highest exposure rates observed (e.g., close to nuclear reactor components, perhaps 10s to 100s of R/hr!), and to measure beta and other particulate emissions. Ion chambers are useful for general monitoring of photons in work environments. Sodium iodide scintillation detectors, which are also discussed shortly, measure low to moderate exposure rates, and are thus used for environmental monitoring through moderate workplace exposures.

One may, however, monitor much lower exposure rates through the use of pressurized ionization chambers (Figure 8.11). Large pressurized ionization chambers have been used for many years for passive monitoring of ambient exposures or for absolute calibration of other survey meters. More recently,

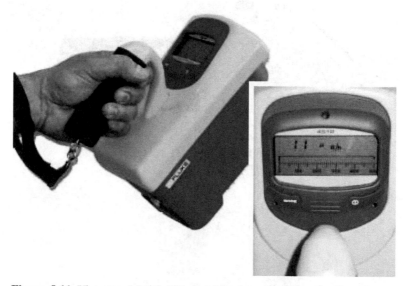

Figure 8.11 Victoreen Model 451 portable pressurized ion chamber. (http://www.flukebiomedical.com/rms/productDataSheets/451B-ds.pdf, with kind permission from Fluke Biomedical.)

small, portable pressurized ionization chambers have been designed that have sensitivities in the μR/hr region. Pressurized ionization chambers typically use high-pressure argon as the detection gas. Because there are many more molecules of gas in the detector, the chance for photon interactions is enhanced, and lower exposure rates may be reliably monitored. An advantage of the typical ionization chamber, which uses air at ambient pressures, is that the ionization gas cannot be lost, as it is in equilibrium and exchanging with the ambient atmosphere.

Proportional Counters

In the "proportional" region of the voltage region, the voltage is such that the electrons created by the initial ionization of the gas molecules are accelerated to energies sufficient that the electrons themselves become ionizing particles, thus proportional counters are operated at voltages that permit the triggering of secondary electron ionizations. The signal obtained is still proportional to the intensity of the original signal, but the number of ion pairs ultimately collected is greater than that originally formed. We therefore say that the "gas amplification factor" is thus >1.0. In an ionization chamber, the gas amplification factor is exactly 1.0; that is, 100% of the original ions are collected. Proportional counters do not use air as the ionizing gas. Air (mostly the oxygen component) has a significant "electron attachment coefficient," meaning that free electrons can become associated with the gas molecules and lost to ultimate collection.

Proportional counters operate in pulse mode and are typically operated with so-called proportional gases, such as P10, which is an argon/methane mixture (90% argon, 10% methane). The argon is the main ionizing gas of interest; the methane serves a special purpose, which is discussed in Knoll's book on radiation detection and measurement.[4] Because of the higher specific ionization of alpha particles, their pulses can be distinguished from those of beta particles. The pulse collected is considerably bigger for an alpha track than for a beta track, as more initial ion pairs are formed in the chamber. A count rate versus voltage curve for a proportional counter can be made with a mixed α/β source, to determine the "α plateau" and "$\alpha + \beta$ plateau," that is, the regions over which the count rate is fairly stable:

Proportional counters are not often used as survey meters, but are very useful for laboratory counting of α, β, or mixed α/β sources (Figure 8.12). They are often used to count filter papers that have been used to monitor surface contamination ("swipe" or "smear" samples), or that have been used in air

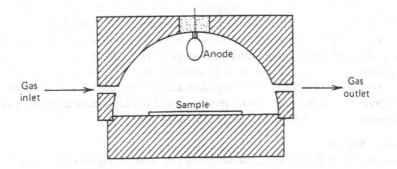

Figure 8.12 Diagram of 2-pi proportional counter.

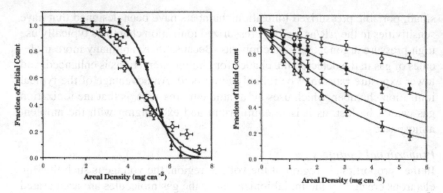

Figure 8.13 Alpha and beta self-absorption curves, from Luetzelschwab et al.[5] (Reprinted with permission from the Health Physics Society.)

samplers to collect air samples over fixed periods of time. These detectors must be operated with thin entrance windows (or no window at all) so that these weakly penetrating emissions can be registered in the counting chamber. Mylar is a good choice for a proportional counter window. These counters are often hemispherical in shape, with various geometries used for the cathode and anode materials.

Efficient detection of alpha and beta particles requires the use of thin entrance windows between the sample and the counting chamber. Typically, very thin mylar windows are used, which must be treated with care, as they can tear or break easily if touched, or if the chamber gas pressure is excessive. Another important consideration in proportional counter detection of particulate radiations is the sample self-absorption. Filter papers may build up small or large amounts of filtered materials. If water samples are evaporated onto metal planchets, similarly, there may be little or no residue on the planchet or large quantities if the sample contained surface waters or waters from a private consumer well, for example. In these cases, the particles may have a lot of material to pass through before they can enter the counting chamber, and may suffer considerable attenuation within the sample which will affect the numerical results observed. A correction based on previously counted samples with varying amounts of sample material must be applied to obtain the correct results (Figure 8.13[5]).

Another application of proportional counters is found in *position sensitive* detectors. In these detectors, multiple, criss-crossing anode wires are used to detect the signal, and the position at which an event has occurred is registered by the intensity of the signal in specific wires. These detectors are used in laboratory analysis of radiotracers in phosphoresis gels, in which the position of the tracer tells the experimentalist which species it was, as different species have different migration potentials in the gels. They also have application in the characterization of large areas potentially contaminated with radioactive material, study of neutron scattering, detection of "hot particles" (discussed in Chapter 9) in laundry and portal monitors, and the characterization of X-ray emissions of distant galaxies using X-ray telescopes.

Geiger Region
Note in Figure 8.5 showing the different gas detector regions that there is a region of the gas detector curve called the region of "limited proportionality."

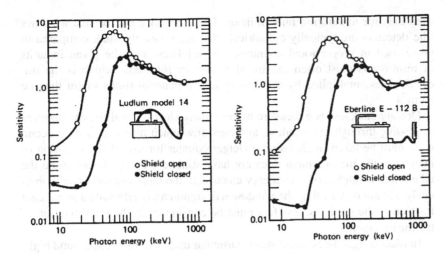

Figure 8.14 Relative energy-dependent behavior of Geiger counters.

In this region, no useful radiation detectors can be operated. Gas multiplication still occurs, but not in a way that can be predicted and controlled, thus the detected signal does not well represent the original ionizing events. If we continue to increase the detector voltage, however, we reach a region called the Geiger region, in which every pulse results in an avalanche that extends along the length of the anode. Proportional detectors create small manageable avalanches of charge multiplication in fixed regions of the detector gas. In a Geiger detector, each small avalanche starts another avalanche, which starts another, and so on, until one large pulse that involves the entire collecting anode is created. Geiger counters thus cannot distinguish among the various types of radiation that they detect. Every initiating event results in a huge avalanche of the same magnitude, be it from a particle of low or high LET.

Geiger counters are used in the laboratory for some applications, but more for survey work at moderate to very high radiation dose levels, as discussed above. Benchtop Geiger tubes can be used to count filter papers or other samples collected from the laboratory. Mostly they are designed to work with portable monitoring detectors, for measuring ambient exposure rates or surface contamination. Geiger tubes operate in pulse mode, as individual events are registered as single counts. The scales on a Geiger counter thus might be given as counts per minute, with various scales available to register low and high count rates. The individual events, however, may also be summed by the measuring circuit and converted to a time-varying count rate which is shown on the readout (analog or digital) as exposure rate in mR/hr, for example. Unlike ion chambers, Geiger counters are typically very energy-dependent in their response to photon fields.

The plots in Figure 8.14 show some typical Geiger counter calibration curves, and one notes immediately the overresponse at moderate energies and very poor response at low energies. Geiger tubes are usually operated using argon gas at negative pressures (see Knoll[4] for reasons why), and thus they need to have relatively thick entrance windows for passage of the radiation, if they have entrance windows at all. Some are windowless, and are used only for monitoring photon fields. Detectors have end or side windows (the detectors are typically cylindrical) that are relatively thick compared to those used in proportional counters. The thickness may be given in units of mm, but is most often expressed as *density thickness*, which is the linear thickness multiplied by the density of the material (units might then be g/cm^2).

Density thickness is a measure of how many electrons will be encountered in passing through the window, and thus how much attenuation may occur. Care must be taken in choosing a Geiger counter for use: if the radiation of interest to the measurement does not have sufficient energy to penetrate the tube window (alphas or low-energy electrons), the measurements made obviously will not be useful. If the window will significantly attenuate a significant fraction of the particles, this fact must be considered in the interpretation of the measurements.

In older Geiger tubes, there was a particular danger in their use around high-intensity photon sources. The detectors can saturate at very high count rates, and the counting rate may actually decrease as you approach a very intense source. This is due to the buildup of detector "dead time" (discussed soon), and the eventual paralysis of the detector as count rates increase. So, as you move towards a hot source, the recorded count rate or exposure rate may increase, increase, and then begin to decrease, causing one to think that one is moving away from the source. Newer Geiger tubes contain correction circuits that prevent this from occurring.

An important issue in the operation of Geiger tubes is that of *quenching*: as the positive ions strike the wall of the tube, they may cause excitation of atoms in the wall, with subsequent emission of ultraviolet (UV) radiation, which can initiate another (separate) avalanche. This can be reduced or eliminated by quenching; two strategies are given here:

- *Electronic quenching*: One may, using uniquely designed electrical circuits, lower the anode voltage for a short period of time after a pulse is received, to permit all of the positive ions to be collected.
- *Chemical quenching*: One may introduce a gas, typically an organic vapor, which is particularly prone to interactions with UV radiation. Should a UV event occur, this gas "catches" the event and dissipates the energy in a way that does not cause a new event to be registered in the detector. Typically, the energy goes into the actual dissociation of the halogen molecule. Some quench gases are slowly expended with tube use, requiring ultimate replacement of the tube; others are cleverly designed using molecules that, after dissociation, may recombine, thus extending the life of the tube.

Dead Time Evaluation

All radiation detectors need a certain amount of time to process the information from an ionizing event through their electronic circuitry. While this processing is ongoing, new information from a new separate pulse cannot be processed. The time during which the detector is not able to process a new pulse is sometimes called the detector *dead time* (to be precise, the total time is called the *resolving time*, which has a distinct definition from "dead time"; see Knoll's book for details;[4] the term "dead time" is more commonly used).

Geiger counters have particularly long dead times. You may not think of a microsecond as a long time, but many pulses may be arriving in a detector from sources of even moderate intensity, and if pulses occur close together in time by chance, it is clearly possible for many pulses to be lost. The dead time for a specific detector can be directly measured, generally by one of these methods:

1. *The split source method*: To perform this method correctly, a special source must be designed that has two halves, each of which has the same geometry, plus one needs another "half source" that has no activity. If both halves are placed near the detector, we will obtain a count rate $R_{1,2}$. If we use just one half, we will obtain rate R_1 (the "dummy" half of the source needs to be counted with R_1 in order to have the same counting and scattering geometry). Then, similarly, we count the other half of the source, with the "dummy" source in the other position, and obtain R_2. If the detector does not suffer dead time losses when counting $R_{1,2}$, then $R_{1,2} = R_1 + R_2$. But of course we design the sources so that there will be losses. The time during which the detector is not able to process pulses is given by the expression:

$$\tau = \frac{R_1 + R_2 - R_{1,2} - R_b}{R_{1,2}^2 - R_1^2 - R_2^2}$$

In this expression R_b is the ambient background counting rate. Then, the true counting rate for an unknown source T may be related to its observed counting rate C by the expression:

$$T = \frac{C}{1 - C\,\tau}.$$

If C is not very large, then the observed counting rate will be representative of the true counting rate.

2. *The decaying source method*: In this method, a source is counted whose initial count rate is such that the detector will suffer dead time losses, but with time, the source decays to a rate that does not overwhelm the detector. If we plot the ln of the observed count rate over time, we expect a straight decreasing line. The curve will depart from linearity where the count rate exceeds the detector's capability to process the pulses (Figure 8.15). Thus we can obtain an empirical correction that can be applied to high count rate values to obtain the true count rate.

8.3.2 Light Production: Scintillation Detectors

Whereas gas-filled detectors produce an electronic signal by collecting charged species from ionized gases, scintillation detectors convert the kinetic energy of the ionizing particle into visible light photons that are collected and converted to electric pulses via interaction with a *photocathode*, which is a device that releases electrons on interaction with the visible light photon. *Scintillation* is the emission of light by fluorescent substances after excitation by radiation. Fluorescence of rocks and other substances is a well-known phenomenon. The British scientist Sir George G. Stokes, named the phenomenon after fluorite, a known strongly fluorescent mineral. Stokes noted that fluorescence

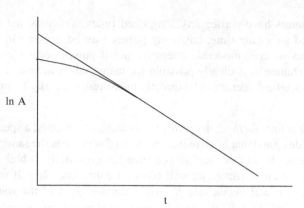

Figure 8.15 Decaying source curve to demonstrate dead time.

is induced in some substances after stimulation with ultraviolet light. He believed that the wavelength of the fluorescent light is necessarily greater than that of the exciting radiation (Stokes law), but exceptions to his law have been found. Various organic and inorganic substances will fluoresce after stimulation with ultraviolet light as well as visible light, infrared radiation, X-rays, radio waves, cathode rays, friction, heat, pressure, and other external stimuli.

Phosphorescence is different than fluorescence, in that it occurs later, and will continue after the external source of stimulation is removed. Phosphorescence is not a useful phenomenon in radiation detection, as we need for our detectors to gather the information about an interaction, process it, and produce a pulse in shorter times than those over which phosphorescence occurs. In radiation detection, we exploit the fluorescent qualities of a number of both organic and inorganic substances to produce a light signal in a short time (relative to that needed by the collecting circuit to process the information). Most of the substances are solids that are fabricated into forms suitable for measuring the activity in fixed sources (typically disk shaped); an important liquid scintillator system, however, is used to measure activity (mostly beta and alpha) in liquid samples that are mixed with the scintillator. This achieves very high (near 100%) sample efficiencies for counting.

As noted above, when radiation from a source strikes the scintillator material, the material gives off photons in the visible light region, which interact with a substance known as a *photocathode*, which absorbs the photons and gives off a number of electrons called *photoelectrons*. These electrons are then accelerated towards a *dynode*, which is a collecting device maintained at a positive potential relative to the photocathode. The dynode is designed so that it not only collects the electrons, but also releases a higher number of electrons, proportional to the number that it collected. Thus, each electron that strikes the dynode causes several more electrons to be ejected, which are accelerated towards a second dynode, where they are further multiplied, and so on, for ten or more stages, where the final pulse is collected, shaped, and amplified. The device that causes this successive multiplication of the initial photocathode signal is called a *photomultiplier tube* (PMT; see Figure 8.16). Each dynode

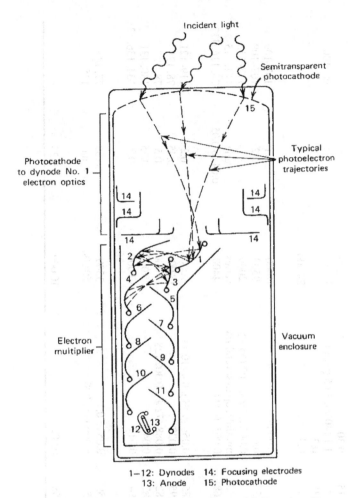

Figure 8.16 Schematic diagram of a photomultiplier tube. (From Knoll.[4])

causes a multiplication of the signal by a fixed amount δ, which is defined as

$$\delta = \frac{\text{number of secondary electrons emitted}}{\text{primary incident electron}}.$$

A PMT contains a series of dynodes at successively positive potential. The final multiplication after N dynodes is:

$$\text{Gain} = \alpha \cdot \delta^N.$$

Here α is the fraction of all photoelectrons collected and N is the number of dynode stages in the PMT.

Scintillators: Solid Scintillators

The most popular solid scintillator (see Table 8.1[4]), useful only for counting photons, is a crystal of sodium iodide (NaI), which intentionally includes an impurity, thallium (Tl), which is very efficient at converting photon energy to light. These detectors are called "thallium-activated sodium iodide" crystals or

Table 8.1 Properties of Common Inorganic Scintillators, adapted from Knoll[4].

	Specific Gravity	Wavelength of Max. Emission	Refractive Index	Decay Time (μs)	Abs. Light Yield in Photons/MeV	Relative Pulse Height Using Bialk. PM tube	References
Alkali Halides							
NaI(Tl)	3.67	415	1.85	0.23	38 000	1.00	—
CsI(Tl)	4.51	540	1.80	0.68 (64%), 3.34 (36%)	65 000	0.49	78, 90, 91
CsI(Na)	4.51	420	1.84	0.46, 4.18	39 000	1.10	92
LiI(Eu)	4.08	470	1.96	1.4	11 000	0.23	
Other Slow Inorganics							
BGO	7.13	480	2.15	0.30	8200	0.13	
CdWO$_4$	7.90	470	2.3	1.1 (40%), 14.5 (60%)	15 000	0.4	98–100
ZnS(Ag) (polycrystalline)	4.09	450	2.36	0.2		1.3[a]	
CaF$_2$ (Eu)	3.19	435	1.47	0.9	24 000	0.5	
Unactivated Fast Inorganics							
BaF$_2$ (fast component)	4.89	220		0.0006	1400	na	107–109
BaF$_2$ (slow component)	4.89	310	1.56	0.63	9500	0.2	107–109
CsI (fast component)	4.51	305		0.002 (35%), 0.02 (65%)	2000	0.05	113–115
CsI (slow component)	4.51	450	1.80	multiple, up to several μs	varies	varies	114, 115
CeF$_3$	6.16	310, 340	1.68	0.005, 0.027	4400	0.04 to 0.05	76, 116, 117
Cerium-Activated Fast Inorganics							
GSO	6.71	440	1.85	0.056 (90%), 0.4 (10%)	9000	0.2	119–121
YAP	5.37	370	1.95	0.027	18 000	0.45	78, 125
YAG	4.56	550	1.82	0.088 (72%), 0.302 (28%)	17 000	0.5	78, 127
LSO	7.4	420	1.82	0.047	25 000	0.75	130, 131
LuAP	8.4	365	1.94	0.017	17 000	0.3	134, 136, 138
Glass Scintillators							
Ce activated Li glass[b]	2.64	400	1.59	0.05 to 0.1	3500	0.09	77, 145
Tb activated glass[b]	3.03	550	1.5	~3000 to 5000	~50 000	na	145
For comparison, a typical organic (plastic) scintillator:							
NE102A	1.03	423	1.58	0.002	10 000	0.25	

[a]For alpha particles.
[b]Properties vary with exact formulation. Also see Table 15.1.
Source: Data primarily from Refs. 74 and 75, except where noted.

"NaI (Tl)". The crystal is optically coupled to the photocathode and PMT. The PMT output generally passes through a pulse-shaping preamplifier, an amplifier, and then is passed to a device that sorts the pulses by size. The size of the initial pulse before amplification is proportional to the energy of the recorded interaction. Amplification just changes the size of the pulse by some fixed amount, so proportionality is preserved. More energy deposited in the crystal means more light produced, which causes the production of more electrons, and thus creation of a bigger pulse.

This is a key issue in gamma ray *spectroscopy*, in which we sort the pulses by size in order to investigate the identity of the radionuclides in our sample. In many cases in radiation detection, we simply measure the exposure rate (with a survey meter) or the count rate (of a radioactive source). If we do not know the identity of the substance(s) emitting the photons, we have some information, but may need more to finish our analysis. Using spectroscopy, we can identify the radionuclide(s) in our sample as well as their absolute concentrations, after performing a suitable calibration of the counting system. A few other scintillator materials and their properties are shown in Table 8.1. NaI is by far the most common scintillator for photons. Solid organic scintillators for counting beta emitters exist, but are not in widespread use.

Scintillators: Liquid Scintillation

Use of the liquid scintillator toluene is a very popular way to detect low-energy beta emitters (H-3, C-14). The sample is dissolved in the liquid using some kind of solvent material. The solution also contains a fluor that scintillates when struck by radiation from the liquid sample. The two are in intimate contact, so the conversion efficiency is nearly 100%. The wavelength of the scintillation light is not always ideal for many PMTs, so sometimes the system contains an additional material used as a wavelength shifter.

The sample is placed between two PMTs (Figure 8.17). Because light is emitted in all directions, a coincidence circuit can be used to ensure counting of only those pulses that were detected by both detectors simultaneously. Liquid Scintillation Counters (LSCs) are encountered in almost every university, national laboratory, nuclear power plant, and other research and industrial settings.

8.3.3 Semiconductor Detectors

A semiconductor is a specially designed crystalline material that has properties intermediate between those of an insulator (a material that does not readily conduct electricity, such as glass, ceramic, or rubber) and a conductor (a material that can carry an electrical current). The most popular semiconductor materials used in radiation detection are germanium (Ge) and silicon (Si). These materials have a unique configuration of their valence and conduction bands such that, when properly configured, only a small amount of energy is required to create useful charge carriers ($\sim$1.1 eV).

Interaction of ionizing particles with semiconductors does not result in the production of ion pairs as in a classical gas-filled detector, but of *electron–hole pairs*. The holes are vacancies in the crystalline lattice structure where an electron should be. Similar to ion pairs, under an applied voltage, the electrons will migrate towards the positive terminal, and the holes "migrate" towards the negative terminal. The holes do not themselves move, but electrons jump

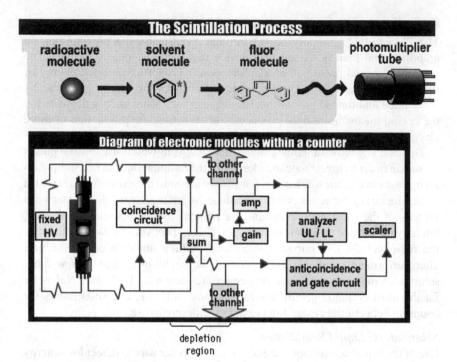

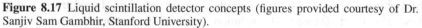

Figure 8.17 Liquid scintillation detector concepts (figures provided courtesy of Dr. Sanjiv Sam Gambhir, Stanford University).

from other positions in the lattice to fill the holes, and the position of the hole "moves" in the opposite direction.

The pure forms of the elements exist in a lattice structure joined by covalent bonds. Addition of an impurity, particularly one from the group V elements, results in the presence of an extra electron that can move easily about the crystal (this is called an *n-type semiconductor*). Addition of an impurity from the group III elements results in regions in the crystal with a deficiency of electrons (*p*-type semiconductor). Now, if a *p*-type semiconductor region is coupled to an *n*–type region, a *p–n junction* is formed. When a *p–n* junction is formed, some of the electrons from the *n*-region that have reached the conduction band are free to diffuse across the junction and combine with holes, forming a *depletion region* (Figure 8.18).

Such a detector with no bias applied to it can function as a radiation detector, but not a particularly good one. If we apply a bias across the two materials, we can affect the ability of electrons to move in the system. If we "forward bias" the system, we essentially open the floodgates to free movement of electrons across the junction and to the positive pole. This also does not represent a useful radiation detector, as current is flowing all the time, whether a radiation interaction has occurred or not. If we, however, "reverse bias" the detector, we accomplish a number of goals and produce a very sensitive radiation detector:

- First, we sweep out all of the charge in the depletion region.
- Then the direction of the electric potential actually causes an increase in the size of the depletion region. With more applied potential, we can continue to increase the size of the depletion region as far as the detector will allow before we cause electronic breakdown of the system. Some detectors, in fact,

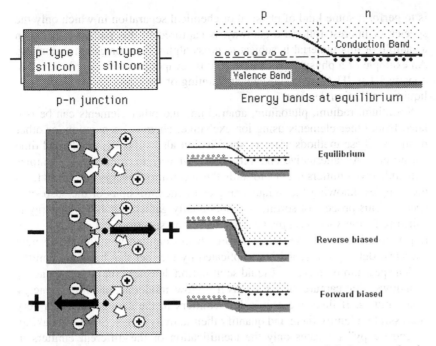

Figure 8.18 Schematic diagram of semiconductor detectors and their energy levels (http://hyperphysics.phy-astr.gsu.edu/hbase/solids/pnjun2.html#c2).

are designed to be *fully depleted* detectors, meaning that the entire detector is comprised of a depletion region.

- We set up a strong resistance to the flow of electrons so that no charge flows in the system unless an interaction occurs (depicted by the "hill" that the electrons have to "climb" in Figure 8.18).
- When radiation interacts with the depletion region, it creates a large number of electron–hole pairs that are rapidly swept away into the collection circuit. The reader will certainly understand intuitively why it is better to have a large number of charge carriers in any radiation detector. Shortly, we show quantitatively why this is advantageous, and why, when we perform gamma spectroscopy, semiconductor detectors have much higher energy resolution than NaI(Tl) scintillation detectors.

8.4 Alpha and Gamma Spectroscopy/Spectrometry

As briefly noted above, sorting the pulses that we gather in a radiation detector by energy can thus tell you much about the energy of the particles or photons interacting with the detector. Sometimes all we want to know is that an interaction occurred, take a count, and quantify the absolute activity in a sample. This is fine if we know the identity of all radionuclides in our sample, or if we wish just an estimate of total sample activity (e.g., in public water samples, we may wish simply to know the amount of gross alpha activity in the sample, without regard to the identity of the individual alpha emitters). It is very common, however, to have a sample that may have one or more unknown species for which we wish to know their specific identities as well as their absolute activities. One approach to identifying individual radionuclides is to perform some kind of physical or chemical separation in which only the species of interest follows a pathway, is captured in this pathway, and then a sample of this material is subject to gross alpha, beta, or gamma counting. An example of a physical pathway would be evaporation of water believed to contain ^{3}H as ^{3}H$_2$O, with subsequent counting of the reprecipitated vapor in a liquid scintillation detector.

Strontium, radium, plutonium, americium, and other elements can be isolated from other elements using ion exchange, chemical separation, or other methods. These methods are very specific, but also generally costly and time consuming. An alternative and very powerful way to "fingerprint" samples with unknown emitters may be to study the spectrum of energy emissions from the samples, knowing the unique energies of the emissions of the suspected species. This process of sorting the individually gathered pulses by energy is called radiation *spectroscopy*. Gamma and alpha spectroscopy are fairly easy to perform, because these radiations are always emitted with unique discrete energies. Beta spectroscopy is complicated by the fact that betas are emitted with a spectrum of energies. Liquid scintillation detectors set up broad energy "windows" that capture the emissions from low, moderate, and higher energy beta emitters. If there are only a few emitters in the sample, this can very successfully identify them and quantify their activity levels. Strictly speaking, "spectroscopy" indicates only the identification of the different emitters in any sample, whereas *spectrometry* means identification and quantification of the information from a spectroscopic analysis. In routine daily mention, these terms may be used interchangeably by many professionals.

In an ideal world, a gamma or alpha spectroscopy detector measuring a sample would show discrete delta functions at the energies of the radionuclides' important emissions. This does not occur, first, because there is always some statistical variability in the number of charge carriers produced by an ionizing event. Thus, at the photopeak energy, there will be a peak that is generally Gaussian in shape, with a finite width, around the photopeak energy. Now, in the radiation detector, we also do not always get all of the photon energy deposited with every ionizing event. Sometimes we obtain a photoelectric event, so that the entire photon energy has been absorbed by a photoelectron, and this photoelectron then gives up its energy in the crystal, producing the signal (light in a scintillation crystal or electron–hole pairs in a semiconductor).

In some cases, we will get only part of the initial photon energy deposited, in a Compton event, with the rest of the photon energy leaving the crystal as a scattered photon. This is not useful information, because there is nothing unique about the original photon (any photon energy may be scattered at any angle to give almost any lower energy). We may have a Compton event followed by a photoelectric event in the crystal, and all of the energy information is preserved. (These events happen quickly relative to the resolving time of the detector electronics, so this looks just like a single photoelectric event.)

If the initial photon energy is very high, a pair production event can occur. The energy of the electron and positron will be given up in the crystal, but the two 0.511 MeV photons created when the positron annihilates may or may not be captured. Thus we may get counts in the photopeak, and also at energies equal to the photopeak minus 0.511 MeV (one annihilation photon escapes the crystal) and at the photopeak minus 1.02 MeV (both escape). In addition, all radiation detectors will measure some radiation from the ambient radiation background in which we all live, so there will be photopeak and scattered photon events in the detector from natural sources (e.g., ^{40}K is present in nature and all human beings, and emits a photon at 1.46 MeV). Samples typically have more than one emitter, and most emitters have more than one photon emission. So the observed spectrum is complicated by these various features, and interpretation of gamma and alpha spectra takes a significant level of skill and training.

The various electronic pulses that we will collect are converted to digital values by a device called an Analog-to-Digital Converter (ADC). The pulses are then sorted into energy bins in another device known as a MultiChannel Analyzer (MCA). If we sort the events coming into the crystal by size, we can begin to display them on a plot of energy versus the number of events and thus characterize the photons being emitted by the sample we are measuring. Let's consider gamma spectroscopy and assume that we have only one radionuclide in our sample, and it has only one important gamma ray. After some time, a number of events will be logged at the energy of the gamma ray. There will also be many counts at lower energies, from Compton scattering events. Remember from Chapter 4 the concept of a maximum energy for transfer by scattering that occurs at $\theta = 180°$, a "backscattering" collision. Between the energy of this event and the photopeak energy itself, there is a region in which no events from this photon can possibly occur. This area is called the *Compton gap*, and the photon energy at which the backscatter event occurs defines the so-called *Compton edge*. Between this event and zero energy, there will be a distribution of events known as the *Compton continuum*. Thus a gamma spectroscopy plot from this sample could look like the plot in Figure 8.19.

As noted above, there are statistical fluctuations in the number of photo-electrons produced at the photocathode, and at each dynode. Therefore, energy pulses from a single monoenergetic photon will appear in a number of energy bins around the central value, so that the peak actually looks like a Gaussian curve. The width of the photopeak determines the resolution of the detector system. If the peak is wide (poor resolution), it will be harder to identify photopeaks that may be close together. The common quantity used to specify the resolution of a system is called the Full Width at Half Maximum (FWHM),

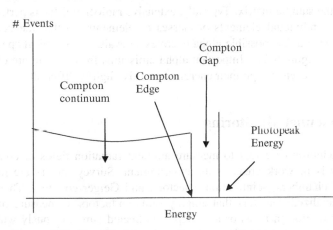

Figure 8.19 Simplified drawing of a gamma spectrum.

which is, as the name implies, the full width of the photopeak at a point where the counts are one-half of the peak maximum. One may also define the Full Width at Tenth Maximum, but this is more rarely encountered. The FWHM varies as a function of energy, so it must be specified at a particular energy value. The *percentage resolution* is defined as the FWHM as a percentage of the peak energy. Although the FWHM itself increases with energy in practical detector systems, the percentage resolution decreases with increasing energy.

Many more charge carriers are formed in a semiconductor crystal than in a scintillator for a given energy of interaction. We soon learn the exact mathematical reasons why, but in general, more charge carriers means better resolution. The resolution of a Ge detector is far superior to that of a NaI detector (FWHM of perhaps 1–2 keV vs. 30–40 keV). The classic plot in Figure 8.20 from Dr. Knoll's book[4] shows the difference in resolution between a germanium and NaI detector for a complicated photon spectrum, showing the clear superiority of germanium semiconductor detectors for such applications.

Germanium is considerably more expensive than NaI (so the detectors tend to be smaller and thus have lower absolute efficiencies). These detectors must be operated at very low (liquid nitrogen) temperatures, as electronic noise in the system can degrade the otherwise excellent signals that are produced. Thus, these detectors require significantly more "care and feeding," in the way of frequent refilling of liquid nitrogen containers, or electrical systems that are capable of cooling the detectors to this level. Sodium iodide detectors are not trouble-free to maintain. They tend to have more pronounced daily drifts in their energy calibrations and require more attention in this regard. One never worries about their operating temperatures, generally, and they are much cheaper to purchase initially and have higher efficiencies than most germanium detectors (note in Figure 8.20 how the NaI spectrum has more counts in any given channel than the Ge system). So, if one is really only going to see a few photons of interest in a sampling program, use of NaI(Tl) detectors may be preferred.

Spectroscopy can also be performed for sources of alpha radiation with specialized semiconductor detectors. The detectors have a thin layer of semiconductor material implanted into or diffused onto the detector unit. Samples to be studied must themselves be very thin, to reduce attenuation of the emitted alphas in the sample matrix. Typically, extensive radiochemistry is performed to separate individual elements or classes of elements, so that the detector will only have a few possible alpha energies to evaluate. A typical spectrum is shown in Figure 8.21[4]. Important alpha emissions from ^{241}Am are clearly distinguished, even though their energies are only slightly different.

8.5   Personnel Monitoring

We use radiation detectors to measure ambient radiation fields or contamination levels in work areas or the environment. Survey meters are mostly ionization chambers, scintillation detectors, and Geiger counters. There are a few specialized detectors that employ semiconductors to measure photon or alpha radiation in areas or air samples collected simultaneously with the detection process. We use other radiation detectors to measure the radioactivity

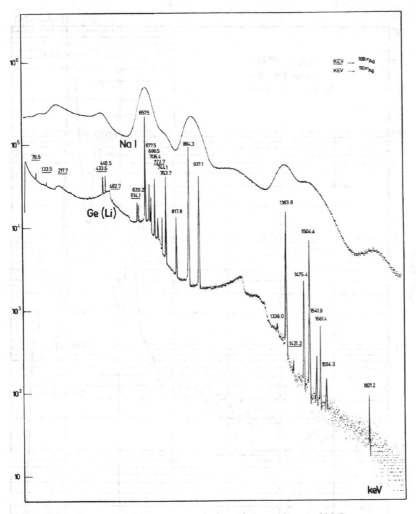

Figure 8.20 Contrast of the energy resolution of germanium and NaI spectroscopy detectors. (From Knoll.[3])

in samples of many kinds of materials. Such detectors are generally solid and liquid scintillation detectors, proportional counters, Geiger detectors, and semiconductor detector systems. The goal of all such monitoring is ultimately the protection of people and the environment. We may at times wish to directly measure the radiation received by workers, or even members of the public, by placing specialized radiation detectors on the person and measuring some collected signal after a fixed period of time. This section briefly describes some of the technologies used in this application.

The first technology we discuss is that of pocket dosimeters or pocket ionization chambers (see Figure 8.22). These are clever little devices that detect radiation using a passive approach. The detectors are comprised of a closed metallic chamber, generally filled with air. The chamber contains a central wire anode and a metal-coated fiber. The anode may be charged to a positive potential, such that the charge is distributed between the wire anode and the fiber. Electrostatic repulsion causes the fiber to be deflected. When radiation

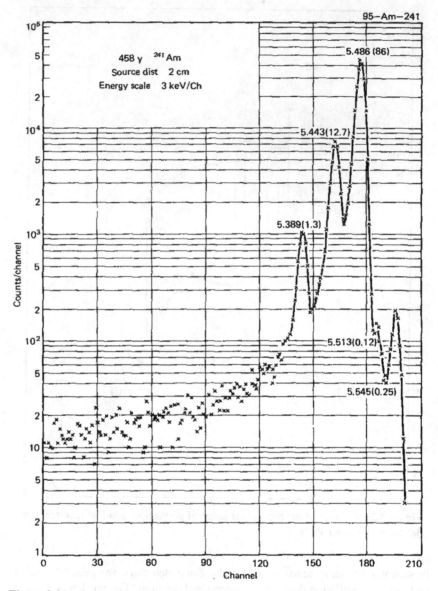

Figure 8.21 Alpha spectrum.

creates ion pairs in the chamber, the charge created neutralizes some of the charge placed on the fiber, causing the separation between it and the anode to decrease. The device is designed so that the distance the fiber is displaced is directly proportional to the quantity of radiation measured by the chamber. Self-reading devices can be read by looking down the barrel; others may be read in a charge reader. These devices can also be coated with boron to detect thermal neutrons. We discuss neutron detection in the next section, and the reason for the boron lining is explained.

For many years, most personnel monitoring was done with so-called *film badges*: these were just small pieces of photographic film. As we know, radiation reduces silver halide in film, causing darkening, and the amount of

Figure 8.22 Pocket ionization chamber. (From http://www. hsa.gov.sg/html/business/crp_consultancy_instrument.html.)

darkening can be related to the amount of radiation that the badge received, which is worn on the person in a place such that the reading is representative of the average dose received. Usually these badges are worn on the chest or waist, but specialized badges (finger badges) are also designed to monitor the dose received by the hands, if one is working with smaller intense sources of radiation with one's hands. Body badge holders were also designed to have several pieces of film, or a single large piece of film, covered by filters designed to block beta particles, low-energy photons, and other radiation types (Figure 8.23). An unfiltered region of the film would record the total radiation dose received, and the other filtered regions would measure other signals, that, when considered together, would explain different components of a radiation field to which a worker was exposed. Film badges cannot be reused. New badges with new film need to be issued periodically to the workers. The reading process is not destructive, but to maintain reasonable levels of detection, new unexposed film must be used each time. Theoretically, film dosimeters can give information about the direction of an incident radiation field from an accident.

An alternative to film badges is the use of specialized light-producing detectors which can be produced in many configurations, including small "chips," long fibers, loose powder, and other forms. These ThermoLuminescent Dosimeters (TLDs) operate using a unique phenomenon that occurs in the material's valence band (see Figure 8.24). Absorption of energy from ionizing radiation causes excitation of electrons, as with normal scintillation detectors. Instead of falling back quickly into their normal site in the valence band with the emission of visible light, these electrons may become held in "traps" within activator molecules introduced into the TLD material, or in the crystal lattice itself (Figure 8.24). These electrons stay in their "traps" until heating of the crystal causes release of the electrons from the traps, with the emission

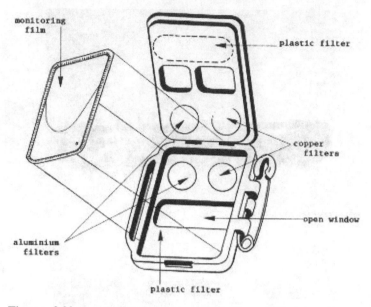

Figure 8.23 Sample film badge and holder. (From http://www. e-radiography.net/radtech/f/film_badge_holder.htm.)

of visible light. Thus, any radiation exposure will be integrated and held in the badge until it is desired to read it out. When heated, the emitted visible light is detected by a photomultiplier tube, which can be tuned to respond well to the particular spectral output of the TLD, often called a *glow curve* (Figure 8.25).

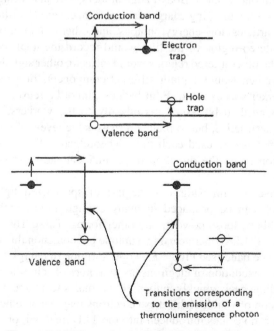

Figure 8.24 Thermoluminescent detector principles (from Knoll[4]).

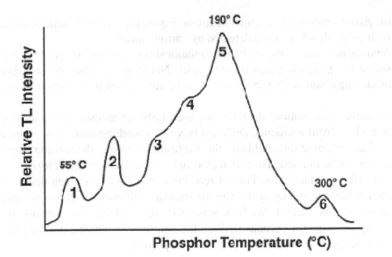

Figure 8.25 TLD glow curve. (From http://www.nukeworker.com/study/hp/tlds/tlds.shtml.)

The dosimeters may also be read by irradiating them with a laser beam, referred to as *Optically Stimulated Luminescence* (OSL). Most TLDs used for personnel dosimetry today use the material lithium fluoride (LiF), although other materials may be useful. TL material generally offers a very linear light output with energy and dose. Some fading of the stored signal is inevitable with any TL material. This effect is usually not so important in personnel monitoring, which is carried out on a monthly or quarterly basis in many facilities. TLDs may be put in place as well to monitor radiation doses over longer periods of time in environmental or workplace monitoring, and here fading may be more important. Just as with film badges, the use of filters in multi-TLD badges permit monitoring of different types and energies of radiation.

8.6 Neutron Detection

Neutrons, like photons, are *indirectly ionizing*; they do not produce many ion pairs themselves, but they react with the medium to produce the primary ionizing particles. In the case of photons, it is the photoelectron or Compton scattered electron that produces most of the ionization after a photon interaction. With neutrons, a number of other reactions are possible, and may be exploited to develop useful radiation detectors. In each case, we seek materials that have a high cross-section, or probability for entering into the kind of reaction desired.

1. (*n*, *α*) *reactions*: Thermal neutrons will react according to $^{10}B(n, \alpha)^{7}Li$; boron may be enriched in ^{10}B, and used as a gas-filled detector (BF$_3$) or as a lining that produces αs that are detected in different ways.
2. (*n*, *p*) *reactions*: High-energy neutrons will scatter well with low-Z materials, particularly hydrogen nuclei (*p*). The *p* is then detected (it has properties similar to an α particle: it travels in straight lines and is densely ionizing.

3. (*n, fission*) *reactions*: Thermal neutrons may be captured by fissile material, with the fission fragments detected by various means.
4. *Neutron activation*: We studied the relationships that describe neutron activation for producing radioactive nuclei. Neutrons can thus be absorbed, producing a radioactive species that can be detected by one of its emissions.

These latter (*n*-activation) detectors are used only for measuring higher dose rates, such as from accidents. Different types of absorbing materials are used, which have different thresholds for the reactions; therefore, the neutron energy spectrum can be defined. These detectors are called *threshold detectors*.

Now, after producing the kind of reaction that we want, we can detect the secondary particle in any of the various standard radiation detectors we have discussed in this chapter. So, in a sense, we are "tricking" the detector into detecting neutrons (or perhaps tricking the neutrons into being detected by a detector designed to measure another kind of particle)!

BF₃ Detectors

An important category of neutron detectors is gas-filled proportional counters. If we find a substance with high cross-section for a reaction that will introduce an ionizing particle into the chamber, we have a highly efficient detector. An important reaction that is widely exploited is the $^{10}B(n, \alpha)^7Li$ reaction. Several advantages to the use of this reaction include:

- ^{10}B has a 4010 b cross-section for this capture reaction.
- Two ionizing particles are created by the reaction: the α and the ^7Li recoil particle. Both are densely ionizing and easily distinguished from the environmental photon background. In most health physics applications we don't worry much about the recoil atom, but in this detector, the contribution to the signal is appreciable. In standard radiation detection of alpha particles by gas-filled proportional counters, the recoil atom is not present in the counting chamber.
- These devices are easy to calibrate to the thermal neutron dose equivalent rate. A flux of 680 n/cm^2-s for 40 hr is equivalent to 1 mSv (100 mrem) per week.

The BF₃ detector, which operates on the (n, α) reaction, sees only thermal neutrons. If we enclose the detector in a paraffin exterior, typically in a spherical configuration, neutrons of above thermal energies will be scattered by the paraffin and possibly slowed to thermal energies. If we use a series of spheres of different diameters, we can thermalize neutrons of low, medium, and higher energies; detect the thermal neutrons in the BF₃ detector; and infer the magnitude of the neutron flux at different energies. The principle of neutron detection using this series of thermalizing spheres was developed by Bramblett, Ewing, and Bonner in 1960;[6] the detectors are typically called *Bonner sphere* sets. The alpha particles may also be detected in a LiI(Eu) *scintillation counter* instead of a BF₃ gas-filled proportional counter. ^6Li also has a high cross-section for the (n, α) reaction, so a lithium iodide detector enriched in ^6Li also makes a good thermal neutron detector, and may also be surrounded by a Bonner sphere set for study of neutron energy spectra.

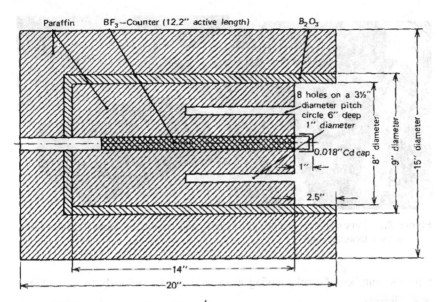

Figure 8.26 Long counter (from Knoll[4]).

Long Counter

A uniquely designed BF_3 detector (Figure 8.26) has been designed to have a fairly energy-independent response over a broad range of neutron energies. The BF_3 detector, which as we know operates on the (n, α) reaction, sees only thermal neutrons. This detector is enclosed in paraffin as are the Bonner spheres, and thus thermalizes fast neutrons, so that they are counted. Increasing the thickness of the paraffin will increase the detector count rate until the paraffin begins to absorb a significant fraction of the thermal neutrons. The optimum thickness depends on the neutron energy, but at about a 6 cm thickness, the detector has an energy-independent response over a wide energy range, thus the name "long counter".

Proton Recoil Counter

Most radiation doses that humans receive by exposure to neutrons are due to scattered protons (originally hydrogen nuclei) that are formed by the (n, p) reaction. The neutron causes one ionization event (this time removing the nucleus rather than an orbital electron, as is common with photons), and the proton is a densely ionizing particle that deposits its energy through ionization and excitation events with orbital electrons it encounters along its path as it slows down. Protons are most similar to alpha particles in their interactions with matter: they are densely ionizing, have a high LET, and travel in straight paths. Proton recoil counters exploit this process to detect neutrons. The counters may be filled with hydrogen, methane, or a hydrogen/argon mixture.

Track-Etch Detectors

Special foils can be marked by α particles emitted by our well-known $^{10}B(n, \alpha)^7Li$ reaction. The alpha tracks can be enhanced by a chemical treatment (etching in an alkaline bath). The resulting image can actually be visually scored by a trained analyst; more commonly, the image is analyzed digitally,

Figure 8.27 Personal bubble dosimeters (left) and dosimeter reader (right). (From http://www.bubbletech.ca.)

with the number of tracks having been calibrated previously to an ambient neutron flux.

Albedo Detectors

A unique personnel monitor uses an unusual strategy to detect neutrons. Neutrons interact with the body and are thermalized, and some will backscatter into a detector worn on the body. Thus, the detector uses the human body as a moderator. These detectors are difficult to calibrate; a correction for photon contribution is also needed.

Bubble Dosimeters

Most of the radiation detectors that we have studied in this chapter were basically discovered and developed in the early days of the development of the science of radiation protection (the Manhattan Project; see Chapter 7). Since then, detectors have become simpler, more portable, and more computer-based. Reading TLDs with a laser instead of a heated surface is a modification, an advance, not a revolution. A truly revolutionary and new idea in radiation detection, however, was given in the development of superheated drop detectors for neutron detection. In this detector, droplets of superheated liquid dispersed in a gel or polymer are vaporized by neutron interactions and form tiny explosions in the gel, leaving a visibly detectable bubble in the dosimeter (see Figure 8.27). These have been used in stationary detectors and are applied as well in some personal dosimeters.

8.7   Calibration Considerations

Radiation detectors give digital or analog signals from the steady-state currents that they generate based on continuous input to the sensitive region of the detector, or from counts based on isolated pulses that they form based on individual interactions. These signals must be related to absolute values of known activity or exposure rate from traceable reference sources, in order to allow the observed signal to be related to these absolute quantities.

8.7.1 Photons

Some photon detectors (NaI scintillation survey meters, ion chambers, Geiger (GM) counters), as shown above, have a high degree of energy-dependence and others have a lower degree. It is important to know the energy-dependence of any meter used for quantitative measurements. Calibration of meters with several different scales should be made at least one, and preferably two or more, points on each scale. The absolute value of the true exposure rate from a point source can be calculated from the well-known relationship (see also the derivation in section 9.2.1):

$$\dot{X} = \frac{A\Gamma}{r^2}$$

Here Γ is the specific gamma ray constant for the radionuclide of interest (e.g. R-cm^2/mCi-h), A is its activity (mCi), and r is the distance from the source (cm). The inverse square law may be used to obtain appropriate readings for each scale.

$$\frac{\dot{X}_2}{\dot{X}_1} = \frac{d_1^2}{d_2^2}$$

True point sources do not really exist. Small sources can be used and suspended from a point in a low scatter geometry. A small source will appear to the detector as a point source, and follow this rule, if the distance from the source to the detector is more than ten times the largest dimension of the source.

8.7.2 Electrons/Beta

Sources of electron and beta radiation may be detected with proportional counters operated on the $\alpha + \beta$ plateau, an end window or side window GM detector, or possibly (not ideally) with an ion chamber. These sources often represent removable activity from laboratory or other surfaces made with filter papers, water samples, air or water filter samples, or other such samples. The response of radiation detectors or survey meters is highly energy- and geometry-dependent. The detector window must be thin enough to permit entry of the electrons, of course. Thus most measurement of electrons/beta rays is for detection, not quantification. It is tempting to relate a reading from a meter (e.g., in R/hr) to an electron dose rate in rad or rem/hr, but this is completely inappropriate. NIST-traceable sources can be obtained, and an absolute calibration obtained that relates the count rate or exposure rate of a survey meter or laboratory detector to sample activity or dose rate, if quantification is desired.

8.7.3 Alpha

Most of the same arguments made for beta sources apply to alpha sources. Alpha survey instruments are generally used to measure surface contamination, so only count rates are registered. Detection is made with ZnS scintillation detectors. Electroplated alpha sources may be purchased for daily checking of the survey instruments. When surface contamination is detected, quantification is performed by taking wipe samples and quantifying the activity in a gas flow proportional counter, liquid scintillation detector, or perhaps an alpha spectrometer.

8.7.4 Neutrons

A "source" of neutrons is not easy to come by. Sources are created by using an (α, n) (not an (n, α)) reaction. As discussed in Chapter 4, PuBe, RaBe, and PoBe sources are the most popular. Once the neutron flux is known, it can be related to measurements and used to develop a calibration.

8.8   Counting Statistics

There are some important areas of radiation science that require consideration of their statistical nature. These are:

1. Radioactive decay itself, which thus influences the certainty with which we know the measurement of the activity of a sample.
2. The creation of charge carriers in radiation detectors, which influences the variability in the size of the pulse measured by the detector. This is generally important, but especially noteworthy when we perform photon or alpha spectroscopy. Similarly, the number of electrons liberated from photocathodes and multiplied at the dynodes in photomultiplier tubes (discussed above) varies statistically just as does the charge carrier production process.

To begin our study of this application of counting statistics, we consider a process that can be treated using the binomial distribution. In this distribution, the following rules apply.

- There are two possible outcomes for each trial: success and failure (success is not necessarily good; we may be counting fatal accidents at an intersection, cancers in a worker population, etc.).
- The probability of success is the same for each trial.
- There are n trials.
- The n trials are independent.

Now, if we define the probability of success as p and the probability of failure as q (where q is equal to $1 - p$, as the total probability for all outcomes is 1), it can be shown[7] that the probability of x successes in n trials, $P(x)$ is given as

$$\frac{n!}{x!(n-x)!}p^x(1-p)^{n-x}$$

For example, consider a fair die. Each face has a probability of 1/6 of appearing in any one trial. The probability of getting 3 sixes in 4 tries is

$$\frac{4!}{3!(4-3)!}0.1667^3\,(0.8333)^{4-3} = 0.0154$$

Thus the probability is about 15 out of a 1000, or about 1.5%. So if you had no life at all and wanted to just sit and roll a die four times, see if you got three sixes, then repeat that nine hundred and ninety-nine more times, you would expect that in about 15 of those trials you would actually get 3 sixes. Actually, this is just an expected value. If 20 losers sat around a room on a Friday night and did this experiment in parallel, everyone would not get exactly 15 sets of 3 sixes in every thousand trials, but this would be the long-term average of this process. Figure 8.28 shows the solution to the equation for the binomial distribution describing the fair die, for up to eight trials.

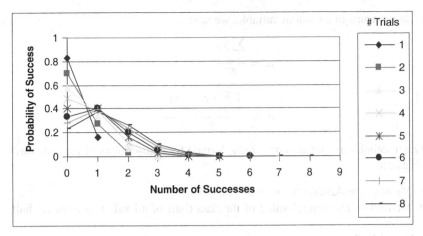

Figure 8.28 Binomial distribution results for different numbers of trials and successes.

8.8.1 Gaussian Distribution

For two trials, you see that the probability in one trial of getting a six is 1/6 and the probability of getting something else is 5/6, of course. The other probabilities are easily calculated with the formula, and you can see the case of 3 sixes in 4 trials, as we calculated. As the number of trials increases, though, the distribution begins to have a more interesting shape, where the probability of a small number of successes has a lower probability than that of a higher number, so that the distribution of probabilities first increases, then reaches a maximum, and then decreases. If we continue to increase the number of trials, or as n increases, the binomial distribution approaches a bell-shaped form, known as the Gaussian curve.

In Figure 8.29, μ is the average value and σ is the standard deviation. The mean μ gives the average of the distribution; σ expresses the spread of the distribution: how much uncertainty, or how much deviation from the average, could be expected in any one measurement. Mathematically, from a series of

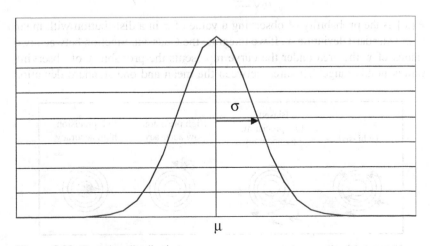

Figure 8.29 Gaussian distribution.

measurements of a random variable, we say:

$$\mu = \frac{\sum_i y_i}{N}$$

$$\sigma = \sqrt{\frac{\sum_i (y_i - \mu)^2}{N - 1}}$$

Other definitions that are commonly derived for distributions of random variables are:

- *Mean* (μ) = Arithmetic average of the data
- *Median* = The central value of the data (half of all values are above, half below)
- *Mode* = The most common value in the dataset
- *Range* = The difference between the highest and lowest value in the dataset
- *Standard deviation* (σ) = The square root of the average of the squared deviations from the mean
- *Variance* (σ^2) = The square of the standard deviation.

We can also define the behavior of a series of measurements with the following characteristics (Figure 8.30).

- *Accuracy* describes how close a measured value is to the "true" result.
- *Precision* gives an understanding of how much dispersion there is in a group of data.
- *Bias* is the presence of systematic error.

Figure 8.30 summarizes a few cases of these characteristics, for a series of measurements that are shown as arrows directed at a target, where the true mean value of the distribution is the target bulls-eye.

The properties of the Gaussian distribution are well known. The analytical form of the Gaussian function is:

$$P(x) = \frac{1}{\sigma\sqrt{2\pi}} \exp\left(\frac{-(x - \mu)^2}{2\sigma^2}\right).$$

$P\{x\}$ is the probability of observing a value of x in a distribution with mean μ and standard deviation σ. If we integrate the Gaussian function between two values of x, the area under the curve represents the probability of observing values in this range. The area between the mean and one standard deviation

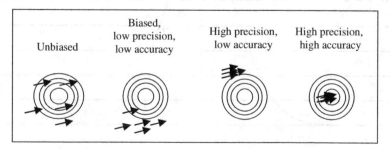

Figure 8.30 Characteristics of sample groups.

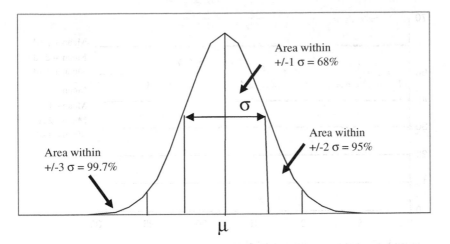

Figure 8.31 Important regions of the Gaussian distribution.

above the mean contains 34% of the total area under the curve, for example. Figure 8.31 shows three very important properties of the Gaussian curve:

- 68% of measured values will be within one standard deviation of the mean.
- 95% of measured values will be within two standard deviations of the mean.
- 99.7% of measured values will be within three standard deviations of the mean.

So, if 100 measurements were made of a random variable whose mean value is 50 and whose standard deviation is 10, we would expect that we would see 68 measurements with values between 40 and 60, 95 values between 30 and 70, and almost all of the observed values between 20 and 80. An odd value outside those ranges may always be observed; this is the nature of statistical chance. A value of, for example, 7 would be very surprising, and might indicate that some error occurred with the measuring apparatus, or it might just be a true fluctuation of the phenomenon that we are studying. This concept is often used with radiation detection quality assurance methods. After establishing the average behavior of a detector, a daily check is made of a repeatable measurement, and the data are plotted on a running chart (Figure 8.32).

The daily values should routinely fall within the mean plus or minus t standard deviations, and certainly within three standard deviations. If a measurement falls outside this range, it should be immediately repeated, before the instrument is trusted for measuring unknown samples. If the new value is back within the expected population, this value can be recorded and the instrument used as desired. If a second measurement falls outside the acceptable range, it would indicate some malfunction or drift of the detector behavior, and corrective action should be taken before any samples are measured. These charts are useful also in watching for such upward or downward drift of counts over time (a downward drift might be reasonable, if the source has a short enough half-life to show a decrease in counts over the plotting period). Note in the graph in Figure 8.32 that way more than one-half of the counts are above the supposed mean. The overall behavior looks steady and has an expected spread of the data. The mean, however, appears to be a bit low, and a new

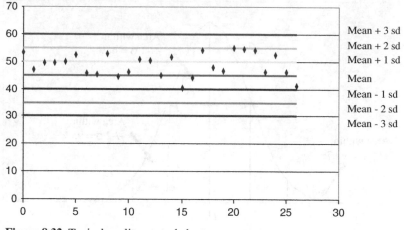

Figure 8.32 Typical quality control chart.

mean should be calculated (perhaps from the last 10–20 points taken) and the plot limits should be adjusted.

8.8.2 Poisson Distribution

A special case of the binomial distribution occurs when there are a large number of trials, but a very small probability of success per trial ($n \gg 1$, $p \ll 1$). This is called a Poisson distribution. The probability of x successes is:

$$\frac{\lambda^x e^{-\lambda}}{x!} \qquad \lambda = n \cdot p$$

In this distribution, a unique and very helpful property emerges:

$$\boxed{\begin{aligned} \mu &= \lambda \\ \sigma^2 &= \lambda \\ \sigma &= \sqrt{\mu} \end{aligned}}$$

Thus, once we know the mean of the distribution, we automatically know the standard deviation; it is just the square root of the mean. In a Poisson distribution, if there are only a few events, the distribution is skewed to the left. If there are many events, the distribution looks just like a Gaussian, but you know the standard deviation as soon as you know the mean, because of the relationship shown above. The following processes are described by Poisson distributions.

- Decay rates of radioactive sources
- Production of scintillation photons
- Conversion of scintillation photons to photoelectrons
- Production of secondary electrons at a photomultiplier tube dynode
- Production of charge carriers at the PM tube anode

When radiation of a particular energy creates pulses in a radiation detector, the pulse is made up of many individual charge carriers, whose number varies according to Poisson statistics. Thus, the distribution of pulses has a mean value and a variance. An important characteristic of radiation detector pulses

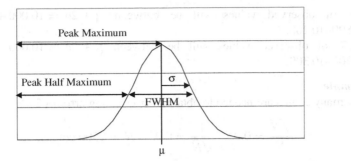

Figure 8.33 The full width at half maximum.

is the full width at half maximum. As the name implies, it describes the width of the distribution at one half of the peak height (Figure 8.33):

Mathematically, the FWHM occurs at 2.35 σ. A small FWHM is good, because it allows pulses of similar energy to be distinguished (a higher energy pulse will have a higher μ, not a higher peak size). Figure 8.20, from Knoll[4] shows the clearly superior resolution of germanium semiconductor detectors over sodium iodide scintillation detectors, making germanium detectors attractive when spectroscopy is performed with sources with complex decay schemes or unknown sources. When the absolute value of a Poisson variable is larger, of course the standard deviation is larger, but the relative standard deviation is lower:

$$\frac{\sigma}{\mu} = \frac{\sqrt{\mu}}{\mu} = \frac{1}{\sqrt{\mu}}.$$

Thus, the accuracy of a measurement is better when there are a large number of counts, and the FWHM is smaller when there are a large number of charge carriers.

Example
What are the error and the percentage error in experiments that collect 100, 1000, and 10,000 counts?

$$\mu = 100 \quad \sigma = \sqrt{100} = 10 \text{ counts}$$

$$\% \text{ error} = \frac{\sigma}{\mu} = \frac{10}{100} = 0.10 = 10\%$$

$$\text{for } \mu = 1000 \text{ counts, } \% \text{ error} = 3.2\%$$

$$\text{for } \mu = 10,000 \text{ counts, } \% \text{ error} = 1.0\%$$

For 100 counts, we expect that:

68% of observed values will be between $\mu + 1\sigma = 100 + 10 = 90\text{--}110$.
95% of observed values will be between $\mu + 2\sigma = 100 + 20 = 80\text{--}120$.
99.7% of observed values will be between $\mu + 3\sigma = 100 + 30 = 70\text{--}130$.

For the case of 10,000 counts, we expect that:

68% of observed values will be between $\mu + 1\sigma = 10,000 + 100 = 9900\text{--}10,100$.

95% of observed values will be between $\mu + 2\sigma = 10,000 + 200 = 9800$–$10,200$.

99.7% of observed values will be between $\mu + 3\sigma = 10,000 + 300 = 9700$–$10,300$.

Example

How many counts are needed to obtain a 1σ counting error of 2%?

$$\frac{\sqrt{N}}{N} = 0.02 \qquad \frac{1}{\sqrt{N}} = 0.02 \qquad N = 2500 \text{ counts.}$$

Counting Rate

If N counts are recorded during time t, the average counting rate R is N/t.
The standard deviation of the counting rate is:

$$\sigma_R = \frac{\sqrt{N}}{t} = \sqrt{\frac{N}{t^2}} = \sqrt{\frac{R}{t}}.$$

If we collected 10,000 counts in 10 min:

$$N \pm \sqrt{N} = 10,000 \pm 100 \text{ counts}$$

$$R \pm \sqrt{\frac{R}{t}} = 1000 \; \frac{\text{cts}}{\text{min}} \pm \sqrt{\frac{1000\frac{\text{cts}}{\text{min}}}{10 \text{ min}}}$$

$$= 1000 \; \frac{\text{cts}}{\text{min}} \pm 10 \; \frac{\text{cts}}{\text{min}} \quad (1\% \text{ uncertainty})$$

If we counted the source for only 1 min instead:

$$N \pm \sqrt{N} = 1000 \pm 32 \text{ counts}$$

$$R \pm \sqrt{\frac{R}{t}} = 1000 \; \frac{\text{cts}}{\text{min}} \pm \sqrt{\frac{1000\frac{\text{cts}}{\text{min}}}{1 \text{ min}}}$$

$$= 1000 \; \frac{\text{cts}}{\text{min}} \pm 32 \; \frac{\text{cts}}{\text{min}} \quad (3.2\% \text{ uncertainty})$$

8.8.3 Propagation of Errors

When we perform mathematical operations on variables, each of which has its own uncertainty, the final result of the calculation has an overall uncertainty that contains contributions from all the values in the equation. Constants, such as pi, Avogadro's number, and others, have no associated uncertainty, but all of the true variables in the equation will contribute some uncertainty to the final result. Some variables may contribute only negligible uncertainty, others more significant levels. For example, in most cases the energies of decay for characteristic radiations from a radionuclide are measured to within very small uncertainties. On the other hand, any time that a measurement of radioactivity is introduced into a calculation, the uncertainty associated with this value (the square root of the mean value) is always an important contributor.

Uncertainties in random variables do not add, subtract, multiply, or divide following the calculations. Rather, the contributions add in quadrature, according to the following rules, so that the overall uncertainty is always greater than the largest uncertainty of any of the individual variables.

1. Addition or subtraction of variables:

$$\sigma(N_1 + N_2) = \sqrt{(\sigma_1)^2 + (\sigma_2)^2}$$

$$\sigma(N_1 - N_2) = \sqrt{(\sigma_1)^2 + (\sigma_2)^2}$$

$$\text{if } \sigma_1 = \sqrt{N_1} \text{ and } \sigma_2 = \sqrt{N_2}$$

$$\sigma(N_1 - N_2) = \sqrt{N_1 + N_2}$$

A very common application of this expression is in calculating the uncertainty in a count of a radioactive sample after correcting for background.

Example
A source is counted for 10 min, and 1600 counts are observed. During the same measurement period, 900 background counts are observed. What is the value and uncertainty of the number of net counts?

$$N_{\text{net}} = N_{\text{gross}} - N_{\text{bkgd}} = 1600 - 900 = 700 \text{ counts}$$

$$\sigma_{\text{net}} = \sqrt{N_{\text{gross}} + N_{\text{bkgd}}} = \sqrt{1600 + 900} = \sqrt{2500} = 50 \text{ counts}$$

$$N_{\text{net}} \pm \sigma_{\text{net}} = 700 \pm 50 \text{ counts} \quad (7\% \text{ uncertainty})$$

Note that : $\dfrac{\sqrt{1600}}{1600} = 2.5\%$ uncertainty $\dfrac{\sqrt{900}}{900} = 3.3\%$ uncertainty

Example
In a 4 min count, 6000 gross counts and 4000 background counts are observed. What is the net count rate and its uncertainty?

$$R_{\text{net}} = R_{\text{gross}} - R_{\text{bkgd}} = 6000/4 - 4000/4 = 500 \text{ counts/min (cpm)}$$

$$\sigma_{\text{net}} = \sqrt{\frac{6000}{16} + \frac{4000}{16}} = 25 \text{ cpm}$$

2. Multiplication or division of variables:

The standard deviation of a product or quotient of two measurements is the square root of the sum of the squares of the relative individual standard deviations. If $u = x * y$ or $u = x/y$:

$$\left(\frac{\sigma_u}{u}\right)^2 = \left(\frac{\sigma_x}{x}\right)^2 + \left(\frac{\sigma_y}{y}\right)^2$$

For example, for a ratio of counts, $R = N_1/N_2$,

$$\left(\frac{\sigma_R}{R}\right)^2 = \left(\frac{\sigma_{N_1}}{N_1}\right)^2 + \left(\frac{\sigma_{N_2}}{N_2}\right)^2 = \frac{N_1}{N_1^2} + \frac{N_2}{N_2^2}$$

Example

$$N_1 = 16,265 \quad N_2 = 8192$$

$$R = N_1/N_2 = 1.985$$

$$(\sigma_R/R)^2 = (1/16{,}265) + (1/8192) = 0.0001835$$

$$\sigma_R/R = 0.0135$$

$$\sigma_R = 0.027$$

$$R \pm \sigma_R = 1.985 \pm 0.027.$$

If we only multiply or divide by a constant, the uncertainty is just multiplied by or divided by this constant:

$$u = A^*x \quad \text{or} \quad u = x/B$$

$$\sigma_u = A^*\sigma_x \quad \text{and} \quad \sigma_u = \sigma_x/B.$$

Example $x = 1120$ counts in $t = 5$ sec

$$\sigma_x = (1120)^{1/2} = 33.47$$

$$\sigma_R = 33.47/5 = 6.7 \text{ counts/sec.}$$

8.8.4 Mean Value of Multiple Independent Counts

If a source is counted multiple times, $x_1, x_2, \ldots, x_N$:

$$\Sigma = x_1 + x_2 + \cdots x_N$$

$$\sigma_\Sigma^2 = x_1 + x_2 + \cdots x_N = \Sigma$$

$$\sigma_\Sigma = \sqrt{\Sigma}$$

$$\bar{x} = \frac{\Sigma}{N} \qquad \sigma_{\bar{x}} = \frac{\sigma_\Sigma}{N} = \frac{\sqrt{\Sigma}}{N} = \frac{\sqrt{N\,\bar{x}}}{N} = \sqrt{\frac{\bar{x}}{N}}$$

Example

x	$\sqrt{x}$	%
5	2.24	44.7
7	2.65	37.8
4	2.00	50.0
5	2.24	44.7
5	2.24	44.7
6	2.45	40.8
4	2.00	50.0
7	2.65	37.8
5	2.24	44.7
6	2.45	40.8

$$\sum = 54 \quad \sqrt{54} = 7.35 \ (13.6\%)$$

$$\bar{x} = 5.4 \quad \sqrt{5.4} = 2.32 \ (43\%)$$

$$\sigma_{\bar{x}} = \sqrt{\frac{5.4}{10}} = 0.735 \ (13.6\%)$$

8.8.5 Minimum Detectable Activity

A very important concept in radiation detection is that of Minimum Detectable Activity (MDA), which is the smallest amount of activity that a radiation can reliably report, given the background count rate that the detector routinely experiences. Any time that you turn on a radiation detector, with or without a sample in the detector, a finite number of counts will be observed, due to background radiation that the detector sees. Any count that is taken will have an associated uncertainty; the exact same background count rate will not be seen with each count due to the inherent variability of the process of radioactive decay as well as some variability in the level of background. If we are counting samples that have levels of activity that are much higher than the background, we will have no trouble trusting our count and saying that we had a positive result. We can report the result with its associated uncertainty, after subtracting the background and propagating its uncertainty through the calculation, as shown above. However, much radiation counting involves study of samples with very little or no real activity in them, so the count from the detector may appear to be very similar to a simple background reading, and it is important to be able to distinguish a true result from a fluctuation of background. Two important situations in which this is applied are:

- Environmental sample analysis
- Internal exposure of workers (urine samples, whole-body counts for suspected intakes)

In the general use of statistics to make decisions, we define the variable:

$$\frac{\bar{x} - \mu}{\sigma / \sqrt{n}}$$

Here μ and σ are the true mean and standard deviation of a randomly distributed variable, and $\bar{x}$ is the observed mean value of n measurements. It can be shown that this variable has a normal distribution. We can say that there is a probability of $(1 - \alpha)$ that:

$$-z_{\alpha/2} < \frac{\bar{x} - \mu}{\sigma / \sqrt{n}} < z_{\alpha/2}$$

The area under the normal curve to the right of $z_{\alpha/2}$ is $\alpha/2$. Lookup tables have the values of the area under the Gaussian curve between any limits in which we are interested. To test a particular stated hypothesis, we perform a series of measurements, calculate the mean and standard deviation, and assume that these measured values are representative of the true mean and standard deviation. Then we devise the test criterion, calculate the test statistic, go to the lookup table, and find the z or t value. Finally, we evaluate our question in terms of the likelihood that the answer is no, which is the fraction of the area under the curve.

Example

A manufacturer makes a claim that their light bulbs have a mean life expectancy of 2000 hours. We make a series of measurements on 40 sample bulbs and find a mean life of 1970 hours with a standard deviation of 65 hours.

$$\mu = 1970 \text{ h}$$
$$\sigma = 65 \text{ h}$$
$$n = 40$$

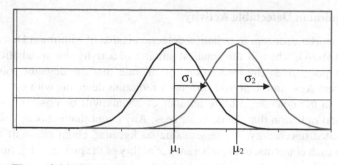

Figure 8.34 Two Gaussian distributions.

Calculation: $z = (2000 - 1970)/(65/\sqrt{40}) = 2.92$.
Conclusion: $z = 2.92$ occurs for $(1 - p) = 0.996$, so there is only a chance of 4 in 1000 ($p = 0.004$) that the mean is really this high.

We can also compare two populations of measurements and make statistically based decisions about whether they are different (see Figure 8.34). The statistic for this test is:

$$z = \frac{\bar{x}_1 - \bar{x}_2}{\sqrt{\frac{\sigma_1^2}{n_1} + \frac{\sigma_2^2}{n_2}}}$$

In Figure 8.34, the first distribution would be that of the background count rate for our detector and the second distribution is that of a sample that may have some activity, with its mean level and associated uncertainty. If μ_2 is truly zero, the measured value will simply be a value from the background distribution. If μ_2 is positive, but very small, it will be difficult to say with confidence whether the measured value is a real positive result or just a high value from the background distribution. Following this line of reasoning, Currie[8] derived a Lower Limit of Detection (LLD), which can be defined as the number of counts above background that can be seen with a 5% probability of mistakenly calling a fluctuation of background a true positive count (called a Type I error) or, conversely, mistakenly identifying a small, but truly positive reading just a fluctuation of background (Type II error; see Figure 8.35). Mathematically, Currie showed that the LLD $= 4.66 \sigma + 3$ (actually the derived value is 2.71, but it is generally shown as rounded to 3). This has units of counts.

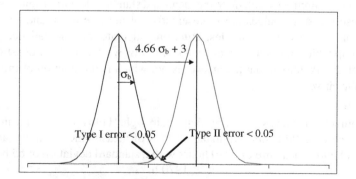

Figure 8.35 Type I and Type II errors.

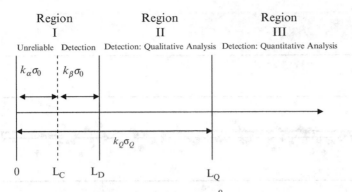

Figure 8.36 Currie's regions of detection.[8]

The minimum detectable activity is just the LLD expressed in units of activity, after dividing by the counting time, detector efficiency, possibly sample recovery, and other variables.

$$\text{MDA} = \frac{4.66\,\sigma_b + 3}{Kt}.$$

Currie defined several regions around the sample background in which the analyst's confidence can be defined (see Figure 8.36).

Currie gave mathematical values for the following variables.

- L_C *(Critical Level)*: The activity is different from zero with 95% confidence $= 2.33\,\sigma_b$.
- $L_D = $ MDA is the smallest "true" signal that can be detected, keeping Type I and Type II errors to 5%.
- L_Q *(Determination Limit)*: This is the true value of the net signal that has a relative standard deviation of 10%.

$$L_Q = 50\left\{1 + \left[1 + \frac{\sigma_B}{12.5}\right]^{1/2}\right\}$$

In practice, most health physicists will report values above the MDA, and cite values below the MDA as "less than detectable." You may note from Figure 8.36 that Currie recommended only reporting of values greater than L_Q as quantitative, however. Treatment of "less than" values is important to do correctly. The most correct method is to show all measured values (and their associated uncertainty), with the MDA shown for comparison. The two temptations in reporting these "less than" values is to (1) assign them all to be equal to the MDA or (2) assign them all zero activity. The first assumption biases these results to be too high and the second biases them to be too low. If these values are averaged with true positive values, the averages will be biased in these directions if the data are all set to a fixed value. If the true values are used, the average and the distribution will be more correctly shown. Recall that if we have a number of results that are all very low, near the background count rate, when we subtract the background (which will fluctuate also), we should get some positive and some negative values. "Negative" activity of course does not exist, but if the average of a number of samples is taken, that includes some positive and some negative values, if the samples truly have no activity, the average should be near zero. If they do contain

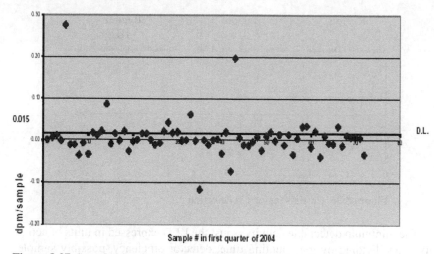

Figure 8.37 Sample plot showing many values near the detection limit.

some small amount of activity, the average may be slightly positive, but for individual samples, a reliable value may not be assumed if the value is less than the MDA. Consider the plot shown in Figure 8.37 of observed ^{239}Pu and ^{240}Pu activity in worker fecal samples over a particular sampling period (http://www.sharepoint.eqo.anl.gov/eshac/eshac_files/Bioassay/QAQC2.pdf).

Example
Calculate the MDA for a Cs-137 water sample:

Sample size $= 1$ L
Sample count time $= 100$ min
Blank: 169 cts in 100 min
Chemical yield 75%
Counting efficiency 25% (0.25)
Number of particles emitted per decay of ^{137}Cs 85% (0.85)

$$\text{MDA} = \frac{4.66\sqrt{169} + 3}{0.75 \times 0.25\frac{\text{cts}}{\text{particle}} \times 0.85\frac{\text{particles}}{\text{transition}} \times 60\frac{\text{trans}}{\text{min}-Bq} \times 1\ L \times 100\ \text{min}}$$

$\text{MDA} = 6.65 \times 10^{-2}\ Bq/L$

The formula shown above for the MDA is the most commonly used form. It is, however, a simplified form, which assumes that the counting time for the sample and the background are equal. If these times are different, the following formula should be used instead.

$$\text{MDA} = \frac{3.29\sqrt{r_b t_g \left(1 + \frac{t_g}{t_b}\right)} + 3}{Kt}.$$

8.8.6 Optimization of Limited Counting Time

If sample counting times are long, and/or if a large number of samples must be counted under deadline, we may wish to optimize the use of a total counting time, splitting the time between the background count and the sample count. We know that:

$$\sigma_n = \sqrt{\frac{r_g}{r_g} + \frac{r_b}{t_b}}.$$

If we differentiate with respect to t, and minimize by setting $d\sigma_n = 0$, we obtain:

$$\frac{t_g}{t - t_g} = \frac{t_g}{t_b} = \sqrt{\frac{r_g}{r_b}}.$$

Endnotes

1. http://en.wikipedia.org/wiki/Bomb_calorimeter#X-ray_microcalorimeter.
2. H. Fricke and E. J. Hart, Chemical dosimetry, in *Radiation Dosimetry*, edited by F. W. Attix, W. C. Roesch, and E. Tochilin (Academic Press, New York, 1966), Vol. 2, 2nd ed., pp. 167–239.
3. C. Audet and L. J. Schreiner, Multiple-site fast exchange model for spin-lattice relaxation in the Fricke-gelatin dosimeter, *Medical Physics* **24** (2, Feb.), 201–209 (1997).
4. G. Knoll, *Radiation Detection and Measurement*, 3rd ed. (Wiley, New York, 1999).
5. J. W. Luetzelschwab, C. Storey, K. Zraly, and D. Dussinger, Self absorption of alpha and beta particles in a fiberglass filter. *Health Physics* **79** (4), 425–430 (2000).
6. R. L. Bramblett, R. I. Ewing, and T. W. Bonner, A new type of neutron spectrometer, *Nuclear Instruments and Methods* **9** (1) (1960).
7. The phrase "it can be shown" indicates either that the speaker doesn't know how, or doesn't want to take the time to show how, a particular relevant result is derived. This is actually a pretty easy derivation to show, but going through the steps is not thought to be particularly helpful to the student in this context.
8. L. A. Currie, Limits for qualitative detection and quantitative determination. Application to radiochemistry, *Analytical Chem.* **40**, 586–593 (1968).

9

External Dose Assessment

We now begin two chapters on the measurement or calculation of radiation dose to humans, external and internal dose assessment. *Dose assessment* is the correct formal name for this process. In day-to-day use, however, most people will refer to this as external and internal *dosimetry*. This is the classic historical term, in use since at least the Manhattan Project in the 1940s. The term "dosimetry" contains the suffix "metry", which relates to metrology, which implies the measurement of physical quantities. In the last chapter, we looked at the issue of personnel dose-measuring devices. Much of external dose assessment does have to do with measurements, so the term "dosimetry" is mostly accurate. In this chapter, however, we show that a lot of work done in external dose assessment involves theoretical calculations of dose, generally with later verification using a survey meter or personnel monitoring devices.

In Chapter 10, we study the science of internal dose assessment, which is almost entirely founded in theoretical calculations and models. This field is generally called *internal dosimetry*, to be a correlate to external dosimetry. There is nothing profoundly wrong with the use of the term "dosimetry" to describe any of these practices; it is just slightly inaccurate. In your routine practice of health physics, you will probably use these terms in discussing this work with others. In these chapters, however, we use the more correct term "dose assessment".

External doses may be received from photon, beta, or neutron sources. There are many types of important radiation sources with which the health physicist must be familiar. Medical centers may have diagnostic X-ray machines, radiation therapy machines, discrete sources used for brachytherapy (in which radioactive sources are placed directly in or near tumor tissue), and other sources. In nuclear power plants, there are many large sources of significant neutron and gamma radiation. Small, intense radiation sources are used for industrial radiography. Particle accelerators in hospitals, research laboratories, and universities may produce intense beams of photons, electrons, or other particles, and neutrons are often produced by secondary reactions with structural and other materials.

The first step in the process of analyzing a source of external radiation is to identify the radionuclide(s) involved and their intensity. Then, relationships regarding the change in intensity with distance and intervening materials and the absorption characteristics of human tissue for the radiation involved will determine the dose rate received. Understanding of the time pattern of the dose rate (time exposed in various geometries) will allow a calculation of the cumulative dose ultimately received by different important tissues in the body.

9.1 Dose from Discrete Photon Sources

The Bragg–Gray Principle

Radiation detectors typically measure currents caused by ionization of gases within their chambers. To be useful, the readings from the meter must be tied to an absolute calibration, as discussed in Chapter 8. The readings are most often expressed in terms of exposure rate. We can, however, with careful design, produce a reading of absorbed dose itself. Consider a gas-filled cavity within a solid medium whose thickness is such that the rate of production of electrons within the wall will introduce secondary electrons into the gas cavity at a rate equal to that with which electrons formed within the cavity are escaping into the wall. This condition is known as *electronic equilibrium*. Under these conditions, the dose in the wall can be directly related to the dose measured in the gas by the equation:

$$D_W = D_G \frac{S_W}{S_G} = \frac{S_W}{S_G} wJ$$

where D is the dose in the wall (W) or gas (G), S is the mass stopping power of the wall or gas, w is the energy needed to form an ion pair in the gas, and J is the number of ion pairs produced per unit mass of the gas. For this to work correctly, the detector wall must be at least as thick as the maximum range of the secondary charged particles formed, but not so thick that it notably reduces the fluence of the measured particle field. The cavity size also should not be such that it modifies the secondary charged particle spectrum produced by the interactions of the primary radiation field with the material of the cavity wall. "Tissue equivalent" detectors exploit the Bragg–Gray principle to actually produce a readout that can be in units of equivalent dose, if desired.

In the study of external dose quantities, a special quantity is sometimes employed. This quantity is actually an acronym: Kerma is the Kinetic Energy Released in MAtter. Kerma has the same units as absorbed dose (e.g., J/kg, Gy), but measures energy released rather than energy absorbed. In radiation therapy, subjects are exposed to beams of external radiation to treat internal structures (tumors). Figure 9.1 shows how the absorbed dose builds up to a maximum level, at the depth that corresponds to the range of the secondary particles (electrons) created by the beam.[1] In the region immediately below the body surface, charged particle equilibrium is not maintained, and the absorbed dose is much smaller than the collision kerma. As we go farther into the body, electronic equilibrium will be reached at $z = z_{max}$, where z is approximately equal to the range of the secondary charged particles. Here the absorbed dose is similar to the kerma. Beyond z_{max} both dose and kerma decrease because of photon attenuation.

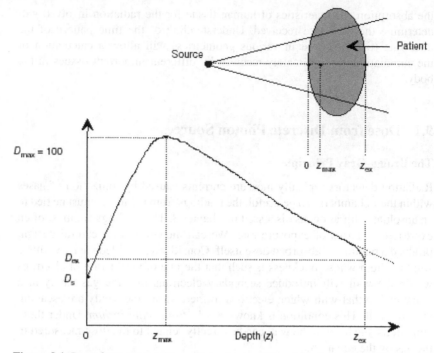

Figure 9.1 Dose from megavoltage photon beam. (From http://www-naweb.iaea.org/ nahu/dmrp/pdf_files/Chapter6.pdf, with permission from the International Atomic Energy Agency.)

9.2   Specific Gamma Ray Emission Factor

9.2.1 Point Source

An important factor used in external dose assessment gives the strength of a photon emitting point source per unit source activity at any distance from the source. This quantity is sometimes called the *specific gamma ray emission factor* (there are a number of other names that may refer to this, such as *specific gamma ray constant*, *gamma constant*, etc.) To derive this quantity for a given nuclide, we consider a point source of a gamma-ray emitter, and assume that the source is emitting gamma rays in all directions uniformly (isotropic source). If we know the source activity in Bq, which is disintegrations/sec, and if we can find the characteristic energy of the photons emitted and their relative abundance (which is easy, given access to any radionuclide decay data source), we can calculate the rate at which photon energy is being emitted:

$$\frac{\text{dis}}{s} \frac{\text{MeV}}{\gamma} \frac{\gamma}{\text{dis}} = \frac{\text{MeV}}{s} \text{ emitted.}$$

At a distance r from the source, the energy fluence rate will be this value, divided by the surface area of a sphere of radius r: $4\ \pi r^2$. The units on the energy fluence rate will be, for example, MeV cm^{-2} s^{-1}. This is the rate at which energy is being emitted; the rate of energy absorption by the medium at that point will be the fluence rate times the linear absorption coefficient (μ_a) for the medium (almost always air) for photons of that energy. Therefore the exposure rate will be:

$$\dot{X} = \frac{A\left(\frac{dis}{s}\right) E\left(\frac{MeV}{\gamma}\right) f\left(\frac{\gamma}{dis}\right) \mu_a (m^{-1})}{4\pi r^2} \times \text{unit conversions}$$

For a given nuclide, we know the values of E, f, and μ (most nuclides will not have a single value, but multiple values of E and f), and we need of course to apply the correct value of μ for each one and then sum up the contributions:

$$\dot{X} = \frac{A\left(\frac{dis}{s}\right) \sum_i E_i\left(\frac{MeV}{\gamma}\right) f_i\left(\frac{\gamma}{dis}\right) \mu_i (m^{-1})}{4\pi r^2} \times \text{unit conversions}$$

Finally, we can define a factor Γ, our specific gamma ray constant , such that

$$\boxed{\dot{X} = \frac{A\Gamma}{r^2}}$$

Traditionally, the units of Γ have been R-cm^2/mCi-hr; now the SI unit equivalent is (C/kg)-m^2/MBq-h. It should be obvious that the factor Γ is:

$$\Gamma = \frac{\sum_i E_i\left(\frac{MeV}{\gamma}\right) f_i\left(\frac{\gamma}{dis}\right) \mu_i (m^{-1})}{4\pi} \times \text{unit conversions}$$

We have not developed how to calculate the unit conversions; this is best demonstrated through an example.

Example

Calculate the specific gamma ray constant for ^{60}Co, considering the 1.17 and 1.33 McV photon cmissions.

^{60}Co has an interesting decay scheme. The two photons at 1.17 and 1.33 MeV are emitted in cascade, one after the other, both at about 100% abundance. For many nuclides, the various emissions have partial abundances that add up to 100%, but in this case, you get both photons 100% of the time. Linear absorption coefficients for these energies are (http://physics.nist.gov):

$$1.17 \, \text{MeV}: 3.5 \times 10^{-5} \text{cm}^{-1}$$

$$1.33 \, \text{MeV}: 3.4 \times 10^{-5} \text{cm}^{-1}$$

$$\Gamma = \left\{ \left(1.17\frac{MeV}{\gamma}\right)\left(1.0\frac{\gamma}{dis}\right)(0.0035 \text{ m}^{-1}) + \left(1.33\frac{MeV}{\gamma}\right)\left(1.0\frac{\gamma}{dis}\right) \right.$$

$$\left. \times (0.0034 \text{ m}^{-1}) \right\} \frac{1}{4\pi} \frac{1.6 \times 10^{-13} J}{MeV} \frac{i.p.}{34 \, eV} \frac{eV}{1.6 \times 10^{-19} J} \frac{1.6 \times 10^{-19} C}{i.p.}$$

$$\times \frac{m^3}{1.293 \text{ kg}} \frac{10^6 \, dis}{s - MBq} = 2.5 \times 10^{-12} \frac{(C/kg) \, m^2}{MBq \, s},$$

$$2.5 \times 10^{-12} \frac{(C/kg) \, m^2}{MBqs} \frac{37 \text{ MBq}}{mCi} \frac{3600 \text{ s}}{h} \frac{R}{2.58 \times 10^{-4} \text{ C/kg}} \frac{10^4 \text{ cm}^2}{m^2} = 12.9 \frac{R \text{ cm}^2}{mCi \, h}$$

Example

Calculate the exposure rate for a person standing 1.7 m from a small cylindrical ^{60}Co source of strength 0.7 Ci. The source is 1 cm in diameter by 3 cm in length.

The hard part is getting the Γ constant. Once we have that, these calculations are very easy.

$$\dot{X} = \frac{700 \text{ mCi} \times 12.9 \, R \, \text{cm}^2}{(170 \, \text{cm})^2 \, \text{mCi h}} = 0.312 \frac{R}{h} = 312 \frac{mR}{h}.$$

A general rule of thumb is that once one is more than ten times the largest dimension of a source, the source acts like a point source, for purposes of flux and dose calculations. We develop formulas shortly for how to calculate dose from extended (nonpoint) sources of gamma rays. In this case, the largest dimension of the source is 3 cm, and the individual is more than 30 cm away, so treating the source as a point is reasonable.

When using the old units of Ci and R, two mathematical rules of thumb have been used to estimate the exposure rates, in case one does not have the true Γ value available:

Exposure rate per Ci, R/hr, at 1 m $= 0.5 \times E \times f$ (E is in MeV)

Exposure rate per Ci, R/hr, at 1 ft $= 6 \times E \times f$

For example, again for this 0.7 Ci source, ^{60}Co, this would suggest an exposure rate of $0.7 \times 0.5 \times 2.5 = 0.875$ R/hr at 1 meter and $0.7 \times 6 \times 2.5 = 10.5$ R/hr at one foot. Calculated with the true Γ value, we would obtain:

$$\dot{X}_{1m} = \frac{700 \text{ m Ci} \times 12.9 \, R \, \text{cm}^2}{(100 \, \text{cm})^2 \, \text{mCi h}} = 0.903 \frac{R}{h} = 903 \frac{mR}{h}$$

$$\dot{X}_{1ft} = \frac{700 \text{ m Ci} \times 12.9 \, R \, \text{cm}^2}{(30.54 \, \text{cm})^2 \, \text{mCi h}} = 9.68 \frac{R}{h} = 9680 \frac{mR}{h}$$

The rule of thumb approximation in this case is pretty good at 1 m and reasonably accurate at 1 ft. Of course, one notes that:

$$\frac{100^2}{30.54^2} = 10.7 \quad \text{but} \quad \frac{6}{0.5} = 12.$$

So the rule itself is only approximate, as all rules of thumb must be assumed to be.

9.2.2 Line Source

All problems cannot be reduced to treatment with a point source geometry. The next level of complexity that might be encountered would be for a one-dimensional source of finite length. Examples of situations that might be reasonably treated by this geometry would be radioactive water flowing in a pipe or rod sources of photon-emitting radionuclides. If the distance to the point of interest is large compared to the diameter of the cylindrical source, then the source may be well treated as being a line source. The usual rule of $10\times$ may be used for evaluation of this situation (i.e., if the distance is more than $10\times$ the diameter of the source). Note that if the distance is also more than $10\times$ the length of the source, the source may be well treated as a point source again! If one is closer than $10\times$ the diameter of the source, the source may have to be treated as a volume source.

For a line source of gamma radiation, the expression for dose rate at a distance h from a source of strength C_l (MBq/m) and length L, divided into sublengths l_1 and l_2 (see figure) is:

$$\dot{X} = \frac{\Gamma C_l}{h}\left(\tan^{-1}\frac{l_1}{h} + \tan^{-1}\frac{l_2}{h}\right)$$

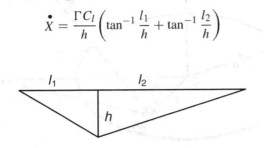

This formula, as with others in this chapter, neglects any self-shielding by the source or attenuation between the source and receptor (which is studied in detail shortly).

Example

A source of liquid ^{131}I is dumped into a sink in an upstairs laboratory. On the ground floor, the activity, at a level of 500 MBq/m, flows through a pipe of diameter 5 cm near the ceiling of another laboratory. The pipe is 6 m long, and an individual is standing on the ground, at an average distance of 3 m from the center of the pipe. Calculate the exposure rate to this individual while the source is present. $\Gamma_{I-131} = 1.53 \times 10^{-9}$ C-m^2/kg-hr-MBq.

$$\dot{X} = \frac{1.53 \times 10^{-9} \times 500}{3} \, 2 \times \left(\tan^{-1}\left(\frac{3}{3}\right)\right)$$

$$\dot{X} = 1.02 \times 10^{-7} \times \frac{\pi}{4} = 8.0 \times 10^{-8} \frac{C}{\text{kg h}}$$

9.2.3 Plane Source

The next level of complexity that we might encounter logically is that of a thin two-dimensional source. The easiest case to treat analytically is that of a plane source; this case is not uncommon to encounter as well. A spill of liquid material on a surface may indeed spread out in very much a circular pattern and be well treated by this geometry. For a thin source of radius r meters, surface concentration C_a MBq/m^2 with gamma constant Γ, the exposure rate at a distance h directly over the center of the source is given by

$$\dot{X} = \pi \times \Gamma \times C_a \times \ln\left(\frac{r^2 + h^2}{h^2}\right)\frac{C}{\text{kg h}}$$

You may have noted from the derivation of the Γ for ^{60}Co above the relationship for photons in air:

$$\frac{i.p.}{34 \text{ eV}}\frac{\text{eV}}{1.6 \times 10^{-19} \text{ J}}\frac{1.6 \times 10^{-19} \text{ C}}{i.p.} = \frac{C}{34 \text{ J}} = \frac{C/\text{kg}}{34 \text{ J/kg}}.$$

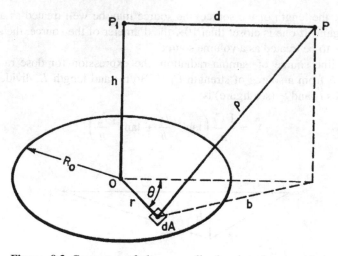

Figure 9.2 Geometry of the generalized point above a plane source[2].

Thus, we can also describe the equivalent dose rate:

$$\dot{H} = 34\,\pi \times \Gamma \times C_a \times \ln\left(\frac{r^2 + h^2}{h^2}\right)\frac{Sv}{h}$$

For a general point P above the source at any location (Figure 9.2), Fitzgerald[2] gives the formula for exposure rate:

$$\dot{X} = \pi\Gamma C_a\,\ln\left[\frac{(R_0^2 + h^2 - d^2) + \left[(R_0^2 + h^2 - d^2)^2 + 4\,d^2 h^2\right]^{1/2}}{2\,h^2}\right].$$

9.2.4 Volume Source

Some situations must be treated as truly three-dimensional. One method that can be used to estimate the exposure rate from a volume source is to reduce the volume source to a planar source by accounting for attenuation, and then apply the appropriate formula for a plane source of those dimensions. For example, consider a source of thickness t meters, and of concentration C_v MBq/m³ of uniformly distributed activity. If the linear absorption coefficient of the material is μ m^{-1}, the equivalent areal concentration C_a (MBq/m²) corresponding to the true volumetric concentration C_v (MBq/m³) is:

$$d(C_a) = C_v \cdot dx \cdot e^{-\mu x}$$

$$C_a = \int_0^t C_v e^{-\mu x}\,dx = \frac{C_v}{\mu}(1 - e^{-\mu t})$$

Using this, we can then calculate either the exposure rate or equivalent dose rate from this cylindrical source as

$$\dot{X} = \pi\,\Gamma\frac{C_v}{\mu}(1 - e^{-\mu t})\ln\frac{r^2 + h^2}{h^2}\frac{C}{kg\,h}$$

$$\dot{H} = 34\,\pi\,\Gamma\frac{C_v}{\mu}(1 - e^{-\mu t})\ln\frac{r^2 + h^2}{h^2}\frac{Sv}{h}$$

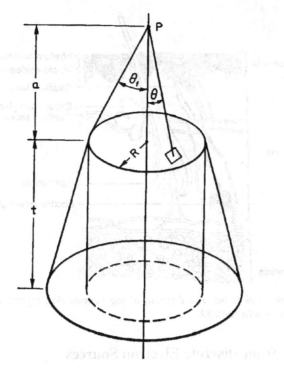

Figure 9.3 Geometry of the truncated circular cone source[2].

Theoretical formulas can be developed to explicitly treat regular volume sources, such as spheres, cylinders, cones, and the like. Again, Fitzgerald[2] is an excellent source for such formulas. The expression for a sphere of radius r containing a concentration C of a radionuclide is:

$$\dot{X} = \frac{C \Gamma}{2} \frac{4\pi}{\mu} (1 - e^{-\mu r}).$$

The exposure rate above the center of a truncated circular cone (Figure 9.3) is given as

$$\dot{X} = \Gamma \, C_V (2\pi t)[-\ln(\cos\theta_1)].$$

Obviously theoretical formulas can only go so far in treating complicated geometries. Even in fairly well-defined geometries such as those shown above, the calculation of exposure and dose rates can be quite tedious. These geometries do not treat the presence of other objects in the vicinity of the source and/or target that may be significant in causing radiation scattering; indeed theoretical formulas cannot treat such complexities. As shown later in the chapter, the use of radiation transport simulation codes is popular in the treatment of such problems. Such approaches are used to treat complex geometries in the workplace, as well as to calculate dose conversion factors for photon sources for people standing on contaminated areas, immersed in clouds of contaminated material, swimming in contaminated water, and other applications.

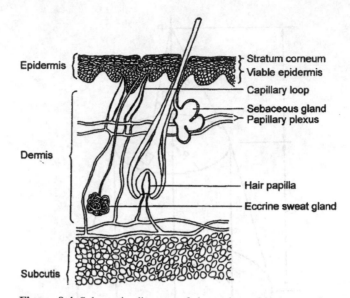

Figure 9.4 Schematic diagram of the various skin layers, from ICRP Publication 89.

9.3 Dose from Discrete Electron Sources

For an external electron source to give a significant dose to the body, the first consideration is whether the electrons have enough energy to penetrate the skin's external "dead layer" (formally called the stratum corneum) (Figure 9.4[3]). This layer of inert cells is about 70 μm thick on average (0.007 g/cm^2 = 7 mg/cm^2). On the palms of the hands and the bottoms of the feet, external skin layers are considerably thicker (up to 40 mg/cm^2), and thinner on the trunk and head (near 5 mg/cm^2). If an electron can penetrate the skin dead layer and deposit dose in the dividing cells, then a potential radiation safety concern is present.

Electron dosimetry may be approximated using measurements or calculational models. Measurements are made with specialized personal dosimeters designed for measuring beta doses. As noted in Chapter 8, monitoring badges may be fit with filters to block low-energy photons, electrons, or both, and have multiple dosimeters so that the results may be used to determine the different components. If workers will be handling small, intense radiation sources (such as nuclear medicine syringes) or working with their hands in locations where high-intensity sources may be present, other badges can be fitted to wear as finger rings.

Theoretical calculations can be performed as well. Models exist for calculating skin doses from surface contamination, contamination of the skin, submersion in a radioactive cloud, and dose from volume sources. They are somewhat difficult to apply accurately, however, due to uncertainties about the amount of backscattered radiation, angular distribution of the source activity, tissue attenuation, and other variables. Dose factors are available for important nuclides in some tabulated sources[4,5] and computer programs such as Varskin.[6] The lookup tables were, of course, originally generated using some computer program.

Table 9.1 Electron doses for a point or plane source of ^{32}P in water, from Cross et al.[5].

R (cm)	$R^2 J (R)$	Plane Source
0.0000	108.2	*****
.0060	105.4	2496
.0120	105.6	2054
.0180	106.1	1788
.0240	106.6	1597
.0300	107.0	1448
.0600	107.3	999
.0900	105.0	732
.1200	100.3	548
.1500	93.7	412
.1800	85.7	310
.2100	76.9	231
.2400	67.7	171
.2700	58.4	124
.3000	49.3	88
.3300	40.6	62
.3600	32.4	42
.3900	25.0	27
.4200	18.6	17
.4500	13.2	10
.4800	8.8	6
.5100	5.5	3
.5400	3.1	1
.5700	1.6	1
.6000	.7	
.6300	.2	
.6600	.1	

Table 9.1 is taken from Reference 5 and shows electron doses, in $mGy - cm^2/MBq - h$ for a point or plane source of ^{32}P in water (which is a good surrogate for soft tissue). The first column shows distances from the source in cm. The second column shows normalized doses at these distances from a point source and the third column shows doses as a function of distance from a plane source of activity. To calculate dose below the dead layer, we might interpolate linearly or logarithmically between the values of 0.006 and 0.012 cm. For the point source case, one can easily "eyeball" the value at around 105.5 $mGy-cm^2/MBq-h$; extensive interpolation to refine this particular value is probably not warranted. So, for a 1 MBq point source on the surface of the skin, the dose rate would be

$$105.5 \, \frac{mGy - cm^2}{MBq - h} \times \frac{1 \, MBq}{(0.007 \, cm)^2} = 2.15 \times 10^6 \frac{mGy}{h}.$$

For a plane source of ^{32}P, with the activity specified in MBq/cm^2, the dose is obtained directly from the value from column 3. For this value, a true interpolation is needed, and due to the rapidly changing values of dose, a logarithmic interpolation is the best to apply:

$$\ln(D) = \ln(2496) - \frac{0.007 - 0.006}{0.012 - 0.006}(\ln(2496) - \ln(2054)) = 7.79$$

$$D = 2420 \frac{\text{mGy} - \text{cm}^2}{\text{MBq}-\text{h}}$$

For comparison, Kocher and Eckerman[4] give a value of 2400 mGy-cm^2/MBq-h, and given a source strength of 1 μCi/cm^2, the Varskin code[6] gives a dose at 0.007 cm depth of 8.24 rad/h:

$$\frac{8.24 \text{ rad} - \text{cm}^2}{\mu\text{Ci}-\text{h}} \frac{10 \text{ mGy}}{\text{rad}} \frac{\mu\text{Ci}}{0.037 \text{ MBq}} = 2230 \frac{\text{mGy} - \text{cm}^2}{\text{MBq} - \text{h}}.$$

9.4   Hot Particles

In the mid-1980s a new problem suddenly emerged in radiation protection. Several reactor licensees reported possibly excessive skin exposures from contamination incidents involving single *hot particles* of radioactive material. Hot particles were found to be small (often microscopic) particles of radioactive material with very high specific activities. These particles may be tiny fragments of reactor (metallic) cladding material that has been activated by capturing neutrons, fragments of reactor fuel, or other particles of high specific activity that are also extremely small, and thus difficult to detect with conventional skin or body scanning methods.

In addition, the particles tend to carry electrostatic charge, and, because of their light weight, can be quite mobile. They earned the nickname "fuel fleas" because of their tendency to "jump" as their electrostatic charge causes attraction or repulsion from other objects. They have been found on people's skin, in laundered clothing at power plants, and in abundance around the failed Chernobyl reactor (see later chapter on criticality). They also create difficulties in calculating radiation dose, because the dose from the electrons emitted from the particles may be very high in a very small region and then drop off quickly (the example, in Figure 9.5, shows a three-dimensional plot of dose as a function of x and y position for a small ^{60}Co hot particle).

The problem here is that skin dose, as traditionally defined by the NRC, is the dose averaged over 1 cm^2 of skin. The peak dose rate from this type of source may be orders of magnitude higher than the average dose over an area of 1 cm^2. In addition, the radiobiological effects will be very different from those produced by more uniform doses delivered to larger areas of skin. Although very high peak doses may be estimated, there may be no observed biological effects, as only a small number of skin cells actually received a lethal dose. So how does one average the dose correctly, to assess the true hazard of the event and for reporting purposes? After much study, and input from the NCRP, the NRC[7] decided that:

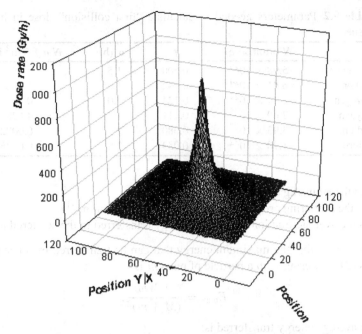

Figure 9.5 Plot of dose rate near a Co-60 "hot particle. (From http://www.ph.bham.ac.uk/research/radiation/dosimetry.htm.)

"Under the final rule, the dose to the skin will be averaged over the most highly exposed, 10 square centimeters instead of being averaged over one square centimeter, as is currently required. This change is based on scientific studies that demonstrate that risks from doses to small areas of the skin are less than risks to larger areas from the same dose".

Also, Xu[8] calculated the effective dose from hot particles assumed to be at any of 74 different positions on the surface of the body, calculating only the photon dose to the organs of the body and then estimating the effective dose. Solutions were given for discrete photon energies, so that they could be applied to any photon emitter, and a sample solution was given for ^{60}Co particles.

9.5 Dose from Discrete Neutron Sources

Dose calculations for neutron sources are best considered in the two categories of fast and slow neutrons. The former have reactions that are more oriented towards their identity as scattering particles, whereas the latter are more involved in absorption reactions.

For fast neutrons, up to about 20 MeV calculate the "first collision dose" as

$$\dot{D}_n(E) = \phi(E) \, E \sum_i N_i \sigma_i f$$

where:

$\phi(E)$ is the neutron fluence rate (n cm^{-2} s^{-1}) for neutrons of energy group E

E is the mean neutron energy of the group (MeV)

Table 9.2 Parameters needed to calculate "first collision" dose in human tissue.

	N (atoms/kg)	f	σ (b)	$N\,\sigma\,f$ (cm^2 kg^{-1})
Hydrogen	5.98×10^{25}	0.500	1.5	44.85
Carbon	6.41×10^{24}	0.142	1.65	1.502
Nitrogen	1.49×10^{24}	0.124	1.0	0.185
Oxygen	2.69×10^{25}	0.111	1.55	4.628
Sodium	3.93×10^{22}	0.080	2.3	0.00723
Chlorine	1.70×10^{22}	0.053	2.8	0.00252

N_i is the atoms/kg of the ith element

σ_i is the scattering cross-section for element i (cm^2 atom^{-1})

f is the average fraction of neutron energy transferred to the scattered atom

We recall that the maximum energy that can be transferred by a particle of mass M scattering with a particle of mass m is:

$$E_{\max} = \frac{4\,Mm}{(M+m)^2}.$$

The average energy transferred is:

$$E_{\text{avg}} = \frac{2\,Mm}{(M+m)^2} = \frac{2\,M}{(M+1)^2},$$

if the small particle is a neutron of mass 1 amu. The scattering cross-sections for human tissue needed to complete the calculation are shown in Table 9.2.

Example

Calculate the dose rate from a field of 2700 n cm^{-2} s^{-1} beam of 5 MeV neutrons to soft tissue.

The sum of the values in the right-hand column of Table 9.2 is 51.17 cm^2/kg

$$\dot{D} = \frac{2700\,n}{\text{cm}^2\,s}\,\frac{5\text{ MeV}}{n}\,\frac{1.6 \times 10^{-13} J}{\text{MeV}}\,\frac{51.17\text{ cm}^2}{\text{kg}}\,\frac{\text{kg} - Gy}{J} = 1.11 \times 10^{-7}\frac{Gy}{s}$$

For thermal neutrons, we need to add the effects from two important reactions: the dose delivered by the ^{14}N$(n, p)^{14}$C reaction and that from the ^{1}H$(n, \gamma)^2$H reaction. The two equations for calculating the dose from these reactions are shown below.

$$\dot{D}_{N-14} = \phi\,N_N\,\sigma_N\,Q$$

In this equation,

N_N = number of nitrogen atoms/kg

σ_N = nitrogen absorption cross-section for thermal neutrons

Q = energy released per reaction (0.63 MeV)

$$\dot{D}_H = \phi\,N_H\sigma_H\,\varphi\Delta$$

where:

φ = fraction of 2.23 MeV γ energy absorbed in the body (lookup table, next section)

Δ = dose rate in an infinitely large medium at a unit activity concentration

In the first reaction, the emitted particle is a proton, which will be totally absorbed by the tissues of the body. In the second reaction, a gamma ray is emitted. The gamma ray energy will only be partially absorbed by the body; some will escape. We explore the meaning of the term Δ when we study internal dose calculations in the next chapter. For now, the definition should be fairly obvious. Assume for a moment that the gamma energy is being released into an infinitely large medium made of soft tissue. Based on the energy of the gamma ray, we could look up the linear or mass absorption coefficient for gamma rays of that energy in tissue, and calculate the dose rate being delivered by the photons being emitted. Note that this is like having a radioactive source in the medium: the flux times the number of atoms times the cross-section gives the number of photons being created per unit of time. The absorption coefficient, when multiplied by the photon flux, will give the rate at which energy is being absorbed. Dividing by the mass we then have an absorbed dose rate, but only in an infinite medium. In a finite medium, only part of the emitted energy will be absorbed, and this fraction is given in the factor φ.

9.6 Dose from Extended Sources

Another category of dose calculations involves large sources of radiation in which the dose to individuals near the source cannot be characterized by fixed geometrical relationships. For example, a person may be immersed in a fluid such as air or water that is contaminated with radioactive material. When radioactive material is released from fixed sources, it will be transported downwind from the source. During the transport, the material will diffuse from areas of higher to lower concentration. At any point downwind from the source, one may calculate the concentration based on the position and modeling of these processes. We treat this situation mathematically in a later chapter. For now, we can just think of the dosimetric aspects of standing in such a cloud of radioactive material.

Workers in nuclear installations as well may be exposed to airborne radioactive material. We need to be concerned about their intakes of such nuclides and their resulting internal doses (treated in the next chapter), but we must also think about how much dose they get from exposure to the airborne material to their skin and eye lenses from photon and electron sources, and to all of their internal organs from photons. Similarly, people may have part or all of their bodies exposed to electrons or photons when immersed in radioactive liquids such as contaminated lake or river water, or if divers should need to enter spent fuel pools near nuclear reactors. Another example of an important extended source comes from standing on a ground surface that is uniformly contaminated with radionuclides.

Sources of radiation dose will include:

1. External irradiation of the skin and outer surfaces of the body (e.g., lens of the eye) by both penetrating and nonpenetrating radiations in the cloud
2. External irradiation of the inner organs of the body that are close to the body surfaces (e.g., testes, breasts, thyroid) by penetrating and nonpenetrating radiations in the cloud
3. External irradiation of "deep" internal organs of the body by penetrating radiations in the cloud

4. Irradiation of the surfaces of the lung from gas contained in the lung spaces by penetrating and nonpenetrating radiations in the air in the lung spaces
5. Irradiation of the internal organs of the body from activity absorbed into the body tissues by penetrating and nonpenetrating radiations and from activity in the lung spaces by penetrating radiations
6. From the production of radioactive progeny and irradiation through any of the above pathways

In theory, the calculation of submersion doses is not very difficult. For external irradiation, the dose rate will be directly proportional to the concentration of the radionuclide and the energy of its emissions. The more difficult task will usually be the estimation of the concentration, probably through a dispersion modeling technique. Once the concentration is known, the equation for irradiation of the outer body surfaces is:

$$H = \frac{Cskg}{\rho}$$

where:

H = the dose equivalent rate for external tissues (Sv/hr)
C = the concentration of the radionuclide (Bq/m^3)
s = the dose equivalent rate in a small element of the medium if uniformly contaminated at a concentration of 1 Bq/g (Sv$-$g/Bq$-$hr)
k = the ratio of the stopping power of tissue over the stopping power of the medium
g = some geometrical factor that accounts for the fraction of the emitted radiation which is absorbed in the target
ρ = density of the medium (g/m^3)

The factor g is 0 for alpha particles and ^{3}H beta particles (or any low-energy beta that cannot penetrate the skin dead layer), 0.5 for other betas or low-energy photons, and some value less that 1.0 but greater than 0 for high-energy photons.

The equation for irradiation of the lung would be:

$$H = \frac{CVsg}{m}$$

where:

H = the dose equivalent rate to the lung (Sv/hr)
C = the concentration of the radionuclide (Bq/m^3)
V = the lung volume (m^3)
s = the dose equivalent rate in the lungs if uniformly contaminated at a concentration of 1 Bq/g (Sv $-$ g/Bq $-$ hr)
g = some geometrical factor that accounts for the fraction of emitted radiation which is absorbed in the target
m = mass of the lung (g)

Here, the geometry factor would be 1.0 for alphas, betas, and low-energy photons, and some value between 0 and 1.0 for other photons.

Consider a simple example. Estimate the dose equivalent rate per unit concentration of ^{81}Kr to the lens of the eye considering only nonpenetrating emissions, assuming the following data.

Electron Energy	Fraction/Decay
0.0014	1.1
0.0102	0.31

Stopping power ratio tissue/air $= 1.14$.
Air density $= 1293$ g/m^3.

$$\left(1.1\frac{\text{dis}}{\text{Bq s}}\right)\left(0.0014\frac{\text{MeV}}{\text{dis}}\right) + \left(0.31\frac{\text{dis}}{\text{Bq s}}\right)\left(0.0102\frac{\text{MeV}}{\text{dis}}\right) = 0.0047\frac{\text{MeV}}{\text{Bq s}}$$

$$s = 0.0047\frac{\text{MeV}}{\text{Bq s}}\left(1.6 \times 10^{-13}\frac{\text{J}}{\text{MeV}}\right)\left(1000\frac{\text{g}}{\text{kg}}\right)\left(1\frac{\text{kg Gy}}{\text{J}}\right)\left(1\frac{\text{Sv}}{\text{Gy}}\right)$$

$$\times \left(3600\frac{s}{h}\right) = 2.71 \times 10^{-9}\frac{\text{Sv g}}{\text{Bq h}}$$

$$\frac{H}{C} = \frac{\text{sk g}}{\rho} = \frac{2.71 \times 10^{-9}\frac{\text{Sv g}}{\text{Bq h}} 1.14 \times 0.5}{1293\frac{\text{g}}{\text{m}^3}} = 1.19 \times 10^{-12}\frac{\text{Sv m}^3}{\text{Bq h}}$$

So, given a steady-state room concentration in Bq/m^3 we can calculate the equivalent dose rate in Sv/h. Or, we could use the integrated concentration (Bq-h/m^3) to get a committed dose in Sv.

Although this example demonstrates the basic principles on which submersion dose calculations are based, it does not consider the photon dose and it oversimplifies the electron dosimetry. The use of an average energy and average absorbed fraction (or geometry factor) for a submersion dose situation is really not adequate because the photons and electrons that are emitted in the cloud may interact with the medium before reaching the target organ, thereby degrading their energy. Also, if the radionuclide is a beta emitter, it will emit betas over a spectrum of energies to begin with. Furthermore, there is really no analytical way to estimate the absorbed fractions for photons of most energies.

Therefore, we need to get some help from computer simulations in order to do a reasonably accurate job of the dosimetry. Just as Monte Carlo techniques are employed to estimate the absorbed fractions for radiation originating within the body, they have been employed to estimate the radiation dose per emission to the various organs of the MIRD phantom when placed in a cloud of radioactive material. Dillman simulated the scattered energy spectrum from an infinite cloud containing a uniform distribution of a monoenergetic photon emitter.[9] Poston and Snyder[10] used Monte Carlo techniques with Dillman's results to estimate the radiation dose per emission to the various organs of the MIRD phantom when placed in a cloud of radioactive material. Although these values were derived for several discrete photon energies, they may be easily used with specific photon energies by interpolation. Poston and Synder expressed their results both as rad/photon and as rad/day per μCi/m^3. From the table, if an infinite cloud source had a uniform concentration of a 1.0 MeV photon emitter, the dose to the kidneys for immersion in a concentration of 1 μCi/m^3 would be 0.014 rad/day.

Calculating the dose from betas and electrons is often thought of as being easier than for photons, however, this is mostly because of the simplifying assumptions normally applied. Electrons are attenuated in tissue in approximately an exponential fashion, and the dose is very dependent on the distance from the air/tissue or water/tissue interface. As mentioned above, to treat the situation rigorously, one must account for the continuum of electron energies.

Martin Berger has studied electron dosimetry for many years and has contributed greatly to this field. In 1974, he published a paper[11] that defined the depth dose in tissue immersed in a cloud of air contaminated with radioactive material for a variety of radionuclides of interest in the nuclear fuel cycle. Kocher and Eckerman[12] expanded this work to provide estimates of electron doses to the skin from submersion in a radioactive air medium, water medium, or from standing on a contaminated surface (see the sample Table 9.3). David Kocher of Oak Ridge National Laboratory published a paper that gave dose rate conversion factors for immersion in contaminated air and water and from standing on contaminated ground for 240 radionuclides of importance to the nuclear fuel cycle.[13] He interpolated between the values given by Poston and Snyder to give the photon doses to the skin and internal organs from immersion in media contaminated with these nuclides. He also calculated dose rates at the body surface from betas, electrons, and photons using the relationship described in the first equation in this handout. He used the single average energy for betas using the assumption that the ratio of absorbed energy in tissue to that in the medium is independent of electron energy within a tolerance of 10% for most energies.

ICRP Publication 30, used as the basis for a revision of the Code of Federal Regulations regarding radiation exposures, has pulled together the best pieces of data from the above works to provide a comprehensive dosimetry model for submersion in radioactive clouds. Like Kocher, they employed the results of Poston and Snyder for photons to give the organ doses and average doses to the skin and lens (the latter two assumed to be the same). For electrons and betas, a rigorous treatment of the electron attenuation in tissue using some general results of Berger's was derived for the skin and lens, assumed to be at depths of 0.007 and 0.3 μm from the body surface, respectively.

Because these results were meant to apply to occupational exposures, a further correction to the photon dose equivalent rates was applied for situations in which the cloud is within a room of finite dimensions. In order to make this correction, one must consider the following expression. The uncollided flux at the center of a sphere of radius R due to photons emanating from a spherical shell of thickness dr located at a distance r from the center is[14]

$$d\phi = S \, (4 \, \pi \, r^2) \, dr \, \frac{e^{-\mu \, r}}{4 \, \pi \, r^2}.$$

Here S is the source strength (photons/cm^3-s) and μ is the medium attenuation coefficient (cm^{-1}). Integrating this expression for values of r from

Table 9.3 Electron dose-rate factors for skin for immersion in contaminated air (from Kocher and Eckerman[12])*

Nuclide	Half-life	Value at 70 μm depth	Value averaged over dermis
^{3}H	12.28 Y	0.0	0.0
^{14}C	5.73E3 Y	5.9E-03	1.1E-03
^{85}Kr	10.72 Y	4.1E-01	1.1E-01
^{85m}Kr	4.48 H	4.2E-01	1.2E-01
^{88}Kr	2.84 H	6.6E-01	3.7E-01
^{88}Kr	17.8 M	4.6E 00	3.8E 00
^{90}Sr	28.6 Y	2.9E-01	6.2E-02
^{90}Y	64.1 H	2.0E 00	1.3E 00
^{95}Zr	64.02 D	1.1E-01	2.4E-02
^{95}Nb	35.06 D	6.9E-03	2.8E-03
^{99}Tc	2.13E5 Y	5.4E-02	1.8E-02
^{106}Ru	368.2 D	0.0	0.0
^{106}Rh	29.92 S	3.1E 00	2.3E 00
^{129}I	1.57E7 Y	3.5E-03	6.6E-04
^{131}I	8.040 D	2.6E-01	5.8E-02
^{133}I	20.8 H	7.6E-01	3.2E-01
^{131m}Xe	11.84 D	1.1E-01	1.5E-02
^{133}Xe	5.245 D	8.2E-02	2.6E-02
^{133m}Xe	2.19 D	2.5E-01	5.3E-02
^{135}Xe	9.11 H	5.4E-01	1.7E-01
^{134}Cs	2.062 Y	2.3E-01	5.7E-02
^{137}Cs	30.17 Y	2.3E-01	5.5E-02
^{137m}Ba	2.552 M	1.3E-01	5.6E-02
^{154}Eu	8.8 Y	3.7E-01	1.5E-01
^{210}Pb	22.26 Y	0.0	0.0
^{214}Pb	26.8 M	4.2E-01	1.0E-01
^{210}Bi	5.013 D	7.1E-01	2.8E-01
^{214}Bi	19.9 M	1.3E 00	7.5E-01
^{210}Po	138.378 D	0.0	0.0
^{222}Rn	3.8235 D	0.0	0.0
^{226}Ra	1600 Y	3.2E-03	4.9E-04
^{230}Th	7.7E4 Y	0.0	2.2E-07
^{231}Th	25.52 H	4.6E-02	1.6E-02
^{234}Th	24.10 D	7.9E-03	1.9E-03
^{234m}Pa	1.17 M	1.7E 00	1.1E 00
^{234}U	2.445E5 Y	3.4E-05	2.6E-06
^{235}U	7.038E8 Y	5.0E-03	1.0E-03
^{236}U	2.3415E7 Y	1.9E-05	1.5E-06
^{238}U	4.468E9 Y	1.3E-05	1.1E-06
^{238}Pu	87.75 Y	0.0	0.0
^{240}Pu	6537 Y	0.0	0.0
^{241}Pu	14.4 Y	0.0	0.0
^{241}Am	432.2 Y	1.5E-05	1.8E-06

*Values are in units of Sv/yr per Bq/cm^3.

0 to R, we obtain:

$$\phi = \left(\frac{S}{\mu}\right)(1 - e^{-\mu R}).$$

The absorbed fraction for a sphere of radius R for radiations originating within the volume, for a target at the center of the sphere, is the $(1 - e^{-\mu R})$ term. So, for a cloud source in a room whose dimensions are smaller than the maximum range of the radiations originating within the cloud, a correction to the standard equation is made, using R as the radius of the "equivalent sphere", that is, the sphere that has the same volume as the room in question.

ICRP 30 gives dose equivalent rates for radioactive inert gases and elemental tritium for the semi-infinite case and for room sizes of 100, 500, and 1000 m^3. Using the ICRP system of dose equivalent limitation, they also give Derived Air Concentration (DACs) for these situations as well, based on the following criteria.

$$\sum_T w_T \dot{H}_T \int C(t)\, dt \leq 0.05 \text{ Sv}$$

$$\sum_T \dot{H}_T \int C(t)\, dt \leq 0.5 \text{ Sv}$$

$$\sum_T \dot{H}_{\text{lens}} \int C(t)\, dt \leq 0.15 \text{ Sv}$$

The DACs are based on uniform conditions throughout the year (see ICRP 30, p. 50, for a definition of terms). As always, the DACs are only applicable if (1) conditions are uniform and (2) there are no other significant sources of exposure. As with MPCs, the DACs may be exceeded if the exposure is of short duration; the real limit is on the total equivalent dose commitment.

9.7 Tritium and Noble Gases

ICRP 30 explains that for submersion in a cloud of tritiated water (^{3}H$_2$O) and/or tritium gas (^{3}H$_2$), the concern is for internal exposures from absorption of the tritiated water, and emphasizes that in most cases the exposure to tritiated water will be of overriding concern, as the concentration limit is much lower and very few environments are found in which concentrations of tritium gas are significant compared to that of tritiated water. For submersion in clouds containing other radioactive species, the internal dosimetry concerns will probably dominate, but some of the above documents may need to be consulted if there is a need to estimate the contributions from the submersion. The following quote[15] explains the NRC system.

"When air concentration is limited by submersion dose, the DAC for a particular radionuclide is the maximum concentration of that radionuclide in air that, for a 2,000-hour exposure, will result in a dose that is equal to or less than each of the applicable limits (5 rem effective dose equivalent, 15-rem eye dose equivalent, 50-rem dose equivalent to other organs and tissues, shallow dose equivalent of 50 rem to the skin). That is, the DAC for a particular radionuclide depends on which of the applicable dose limits is the most restrictive with respect to the concentration of that particular radionuclide. The dosimetric model used to calculate the DACs considers shielding of organs by overlying tissues and the degradation of the photon spectrum through scatter and attenuation by air. The dose from beta particles is evaluated at a depth of 7 mg/square cm for skin, and at a depth of 3 mm for the lens of the eye. The worker is assumed to be immersed in pure parent radionuclide, and no radiation from airborne progeny is considered. In most cases, the concentration limit for submersion is based on external irradiation of the body; it does not take into account either absorbed gas within the body or the inhalation of radioactive decay products. An exception to the preceding statement is Ar-37, for which direct exposure of the lungs by inhaled activity limits (stochastically) the concentration in air. The skin dose is limiting for Ar-39, Kr-85, and Xe-131m; the eye dose is limiting for Kr-83m."

Dose from Standing on a Contaminated Surface

Kocher and Eckerman's 1981 paper[12] included electron dose conversion factors both for immersion in a radioactive cloud and for standing on a ground surface uniformly contaminated with a radioactive substance. Kocher and Sjoreen published a similar paper regarding exposure to photon sources.[16] Clouvas et al.[17] gave similar dose conversion factors (nGy kg Bq^{-1} h^{-1}) for photon emitters in soil, using several Monte Carlo codes, and compared their work to that of Kocher and Sjoreen, as well as a few other authors. This would apply to situations involving contaminated soil (from plume fallout, runoff from a site that ended up in a soil matrix, etc.) or other contaminated extended material.

Dose from Radioactive Patients Released After Nuclear Medicine Therapy

Patients who receive therapeutic amounts of radiopharmaceuticals are a potentially significant source of radiation to their family members, members of the public whom they pass by on their way from the hospital to their homes, and others. For many decades, the release criterion for such patients was primarily activity-based. As almost all radiation therapy that involved radioactive material with a significant gamma component was ^{131}I sodium iodide, used to treat hyperthyroidism (Graves' disease) or thyroid cancer, the release limit (which no one knows how it was originally derived) was that patients could be let go when their activity level was 1100 MBq (30 mCi), or the dose rate at 1 m from the patient was 50 μSv/hr (5 mrem/hr).

In a new version of 10CFR35.75, issued in 1997, the NRC changed the system to be more objectively based on a purely dose-based criterion and to cover the many more therapeutic radiopharmaceuticals in use. Licensees are now able to release patients regardless of how much administered activity they received, if the radiation dose to any individual from exposure to the released patient can be shown to be less than 5 mSv (0.5 rem), integrated over all time after patient release. The rule states that the "licensee shall provide the released individual, or the individual's parent or guardian, with instructions, including written instruction, on actions recommended to maintain doses to other individuals as low as is reasonably achievable...." The NRC did not intend to enforce patient compliance with the instructions nor is it the licensee's responsibility to do so. But hospitals do need to keep records showing that they have ascertained that the doses to the maximally exposed individual is "not likely" to be above the stated dose limit of 5 mSv.

The NRC published a "Regulatory Guide," NRC Regulatory Guide 8.39. These documents do not carry any force of law, unless the licensee formally adopts them in his or her license as part of the facilities official procedures. The NRC has formally noted that other good methods can be used for these calculations and may be accepted by the commission if they can be shown to be sound. The method used in the Regulatory Guide was quite conservative in a number of aspects. First, the patient was treated as a point source in calculation of external exposure rates. As can be noted by solving the equations in the early parts of this chapter, lower doses will be delivered from line or volume sources than from point sources. Patients will have activity distributed throughout their entire bodies, and some self-attenuation will occur, thus the use of a point source is quite conservative. Then, the decay of activity was assumed to be only by physical decay of the radionuclide; biological elimination by the patient (explicitly treated in the next chapter) was not considered. Actual measurements on patients' family members by one group of authors indeed showed that the real doses received by people are significantly less than that assumed by the calculation methods in the Regulatory Guide.[18]

The equation used was:

$$D(\infty) = \frac{34.6\Gamma Q_0 T_p OF}{r^2}.$$

Here, $D(\infty)$ is the dose integrated over all time, Γ is the radionuclide specific gamma constant, Q_0 is patient activity at time of release, T_p is the radionuclide half-life, OF is the assumed occupancy factor, and r is the assumed average distance from a subject over the time of irradiation. For short-lived nuclides ($T_p \leq 1$ d), an OF of 1.0 was used, and for others an OF of 0.25 was used. The default average distance from a subject was assumed to be 1 m. The NRC provided a default table of activity levels and dose rates for various radionuclides at which they deem the dose criterion will be met. Table 9.4 shows a sample portion of that table.

9.8 Computer Modeling in External Dose Assessment

It is very difficult in a textbook to document adequately computer-related resources. Knowledge of available resources is limited by the author's experience and recent contact with such resources in a particular area. The availability and capabilities of the resources also change frequently, so what is put in print at one point in time may be completely unrepresentative of what is going on in the field whenever a reader may study the material. I give a brief overview here of some widely used resources, as a general guide to the state of the art at the time of this printing. Even by the time that the book leaves the printer and reaches readers, the situation may have changed significantly, so there is no guarantee that the list here is particularly accurate or adequate.

There are a number of widely supported, general-purpose radiation transport codes that are in widespread use. These codes typically have very steep learning curves, as they contain many features and possible applications. Users must pay close attention to the description of the problem geometry, characterization of the source and target regions in the problem, format of the data input, and interpretation of the output. Once a code is learned, however, it yields great benefits in that the technical basis is maintained by a consistent team effort,

Table 9.4 Activities and Dose Rates for Authorizing Patient Release.

| | COLUMN I | | COLUMN 2 | |
| | Activity at or Below Which Patients May Be Released | | Dose Rate at 1 Meter, at or Below Which Patients May Be Released* | |
Radionuclide	(GBq)	(mCi)	(mSv/hr)	(mrern/hr)
^{111}Ag	19	520	0.08	8
^{198}Au	3.5	93	0.21	21
^{51}Cr	4.8	130	0.02	2
^{64}Cu	8.4	230	0.27	27
^{97}Cu	14	390	0.22	22
^{67}Ga	8.7	240	0.18	18
^{123}I	6.0	160	0.26	26
^{125}I	0.25	7	0.01	1
^{125}I implant	0.33	9	0.01	1
^{131}I	1.2	33	0.07	7
^{111}In	2.4	64	0.2	20
^{192}tr implant	0.074	2	0.008	0.8
^{32}P	**	**	**	**
^{103}Pd implant	1.5	40	0.03	3
^{186}Re	28	770	0.15	15
^{188}Re	29	790	0.20	20
^{47}Sc	11	310	0.17	17
^{75}Se	0.089	2	0.005	0.5
^{153}Sm	26	700	0.3	30
^{117m}Sn	1.1	29	0.04	4
^{89}Sr	**	**	**	**
^{99m}Tc	28	760	0.58	58
^{201}Ti	16	430	0.19	19
^{90}Y	**	**	**	**
^{169}Yu	0.37	10	0.02	2

as opposed to individual smaller codes developed by specific individuals or groups, whose robustness and longevity vary considerably. Three important codes at present are as follows.

The Electron Gamma Shower (EGS) Code Series

The electron gamma shower (EGS4) code is a general-purpose Monte Carlo simulation package for coupled transport of electrons and photons in many geometries for particles with energies from a few keV up to several TeV. The code series is copyrighted by Stanford University in California and has been continually improved ever since its first operation[19] by a diverse team of professionals based at various institutions. It was originally designed with a focus on modeling radiation beams and calculating doses in clinical radiation therapy. It is used, however, in a wide variety of applications in external and internal dose calculations, detector modeling, and other applications.

The Monte Carlo N–Particle Transport (MCNP) Code Series[20]

The general-purpose code, which was developed at the Los Alamos National Laboratory, can be used for neutron, photon, electron, or coupled neutron/photon/electron transport, in complicated three-dimensional geometries. This code is owned and maintained by a large group at Los Alamos National Laboratory (New Mexico). MCNP was originally designed principally to study neutron and photon transport and dosimetry in reactor environments. With the addition of robust electron transport capabilities, the ability for model repeating, regular geometries, and other features such as EGS, this code has found wide application in internal and external dosimetry, detector modeling, and in solving many other general radiation transport and dose problems.

GEANT[21]

GEANT is a toolkit that simulates the passage of various kinds of particles over a wide energy range (from 250 eV to the TeV energy range) through matter. It also models complex geometries, and may be adapted for use in many different applications.

ICRP Publication 74[22] provides dose coefficients for a number of external sources of radiation, and also gives an overview of some other computer codes available (in 1997).

- DEEP[23] is a computer code that incorporates the Monte Carlo code MORSE-CG[24] to calculate effective dose and the equivalent dose to 60 tissues of an adult modified-MIRD-phantom for photons in the energy range 6.2 keV to 12 MeV.
- ETRAN (electron transport) is a "condensed-history" type Monte Carlo code that treats the transport of electron–photon cascades with energies from the GeV region down to 1 keV.
- FANEUT[25] is a Monte Carlo code specially developed for calculation of the characteristics of the absorbed dose from neutrons and secondary photons in tissue, equivalent materials using a one-dimensional geometry that treats slab and sphere geometries.
- Gesellschaft für Strahlenforschung (GSF) Code[26,27] is an adaptation of the Oak Ridge National Laboratory code ALGAM,[28] which computes the dose deposited by photons from an external or internal source in various sections of a different media phantom approximating the human body. The phantom could be either MIRD-type, voxel type, or a simple sphere or slab.
- HADRON is a Monte Carlo code designed to calculate hadron-cascade transport in complex hydrogenous media, is based on the cascade-exciton model of inelastic nuclear interactions.[29]
- HL-PH is a Monte Carlo neutron transport code[30] that treats neutrons and photons, and uses standard Monte Carlo methods using combinatorial geometry methods.
- LAHET is the Los Alamos High-Energy Transport code that transports neutrons, photons, and light nuclei up to ^{4}He.
- The MORSE-CG Code. The MORSE code[24] is similar to MCNP in performing neutron/photon transport using combinatorial geometry capacity.

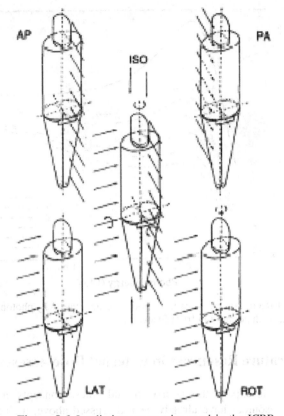

Figure 9.6 Irradiation geometries used in the ICRP 74. (Reprinted from Ref. 22, with permission from Elsevier).

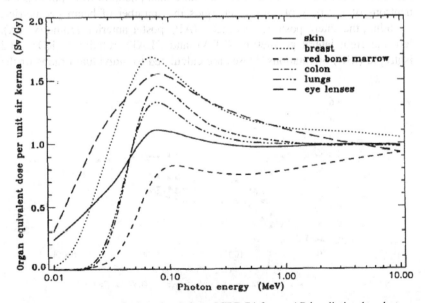

Figure 9.7 Organ equivalent doses from ICRP 74 for an AP irradiation by photons of different energies. (Reprinted from Ref. 22, with permission from Elsevier.)

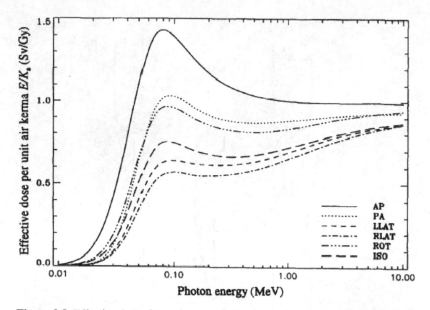

Figure 9.8 Effective doses for various irradiation geometries by photons. (Reprinted from Ref. 22 with permission from Elsevier.)

9.9   Literature Resources in External Dose Assessment

If one wishes to expand this course to study these concepts in more depth, many good resources have already been discussed above, including Fitzgerald's book, the references on beta skin dose, the NRC Regulatory Guide on patient release, and ICRP 74. ICRP 74, as the title shows, gives dose conversion factors for irradiation of the stylized standard man phantom from photons and neutrons of a variety of starting energies in a number of beam geometries, including the antero-posterior geometry (AP), poster-anterior geometry (PA), left and right lateral geometries (LLAT and RLAT), rotational (ROT) and isotropic (ISO) orientations. Doses are calculated to individual organs of the

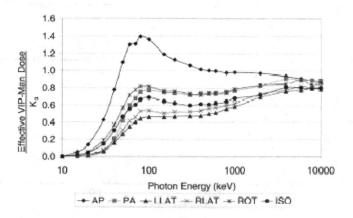

Figure 9.9 Kerman conversion coefficients for a realistic phantom irradiated by external photons. (From Ref. 31, reprinted with permission from the Health Physics Society.)

phantom, then expressed as effective doses. Some sample figures from this reference are shown in Figures 9.6 to 9.8.

Similar data have been developed by colleagues at Rennselaer Polytechnic Institute for external irradiation of a realistic phantom with photons, neutrons, protons and electrons.[31−32] Figure 9.9 shows a sample of their results (from Reference 31).

Endnotes

1. E. B. Podgorsak, External photon beams: Physical aspects. Chapter 6, **http://www.naweb.iaea.org/nahu/dmrp/pdf_files/Chapter6.pdf**.

2. J. J. Fitzgerald, *Applied Radiation Protection and Control* (Gordon and Breach, New York, 1970) pp. 822, 838.

3. International Commission on Radiological Protection. ICRP Publication 89: *Basic Anatomical and Physiological Data for Use in Radiological Protection: Reference Values*. (Elsevier, St. Louis, MO, 2003).

4. D. C. Kocher and K. F. Eckerman, Electron dose-rate conversion factors for external exposure of the skin from uniformly deposited activity on the body surface. *Health Physics* **53**(2, August):135–141 (1987).

5. W. G. Cross, G. Williams, H. Ing, N. O. Freedman, and J. Mainville, Tables of beta ray dose distributions in water, air and other media (Atomic Energy of Canada, AECL-7617, 1982).

6. J. S. Durham, W. D. Reece, and S. E. Merwin, Modelling three-dimensional beta sources for skin dose calculations using VARSKIN Mod 2, *Radiat. Prot. Dosim.* **37**(2), 89–94 (1991).

7. http://www.nrc.gov/reading-rm/doc-collections/news/2002/02-039.html.

8. X. G. Xu, The effective dose equivalent and effective dose for hot particles on the skin. *Health Physics* **89**(1, July), 53–70 (2005).

9. L. Dillman, Scattered energy spectrum for monoenergetic gamma emitter uniformly distributed in an infinite cloud (ORNL 4584, Health Physics Division Annual Progress Report for Period Ending July 1970, October 1980).

10. J. Poston and W. Synder, Model for exposure to a semiinfinite cloud of a photon emitter, *Health Physics* **26**, 287–293 (1974).

11. M. Berger, Beta-ray dose in a tissue-equivalent material immersed in a radioactive cloud, *Health Physics* **26**, 1–12 (1974).

12. D. Kocher and K. Eckerman, Electron dose-rate conversion factors for external exposure of the skin, *Health Physics* **40**, 467–475 (1981).

13. D. Kocher, Dose-rate conversion factors for external exposure to photon and electron radiation from radionuclides occurring in routine releases from fuel cycle facilities, *Health Physics* **38**, 543–621 (1980).

14. K. Morgan and J. Turner, *Principles of Radiation Protection* (Wiley, New York, 1967), p. 285.

15. http://www.nrc.gov/what-we-do/radiation/ hppos/qa426.html.

16. D. Kocher and A. L. Sjoreen, Dose-rate conversion factors for external exposure to photon emitters in soil, *Health Physics* **48**, 193–205 (1985).

17. A. Clouvas, S. Xanthos, M. Antonopoulos-Domis, and J. Silva, Monte Carlo calculation of dose rate conversion factors for external exposure to photon emitters in soil. *Health Physics* **78**(3), 295–302 (2000).

18. F. J. Rutar, S. C. Augustine, D. Colcher, et al., Outpatient treatment with 131I-anti-B1 antibody: radiation exposure to family members, *J. Nucl. Med.* **42**,907–915 (2001).

19. R. L. Ford and W. R. Nelson, The EGS code system: Computer programmes for the Monte Carlo simulation of electromagnetic cascade showers. Report No.

SLAC-210, Version 3 (Stanford Linear Accelerator Center, Stanford University, Stanford, CA, 1978).

20. J. Briesmeister, MCNP - A general Monte Carlo n-particle transport code, version 4B. Los Alamos National Laboratory, report LA-12625-M (1997).

21. S. Agostinelliae, J. Allisonas, K. Amako, et al., Geant4—A simulation toolkit, *Nuclear Instruments Methods Physics Research A* **506,** 250–303 (2003).

22. International Commission on Radiological Protection, ICRP Publication 74: *Conversion Coefficients For Use In Radiological Protection Against External Radiation*, 74 (Elsevier, St. Louis, MO, 1997).

23. Y. Yamaguchi, DEEP code to calculate dose eyuivulents in human phantom for external photon exposure by Monte Carlo method. Report No. JAERI-M 90-235 (Japan Atomic Energy Research Institute, Tokai-mura, Ibaraki- Ken, 1991).

24. M. B. Emmett, The MORSE Monte Carlo radiation transport system. Report No. ORNL-4972 (Oak Ridge National Laboratory, Oak Ridge, TN, 1975).

25. I. Kurochkin, Institute for High Energy Physics, Serpukhov, private communication to the ICRP (1994).

26. R. Kramer, Ermittlung von Konversionsjaktoren zwischen Korperdosen und relevanten Struhlungskenngriissen bei externer Riintgen- und Gammabestrahlung, GSF-Bericht S-556. Gesellschaft fiir Strahlen- und Umweltforschung mbH, Munich (1979).

27. R. Veit, M. Zankl, N. Petoussi, E. Mannweiler, G. Williams, and G. Drexler, Tomographic anthropomorphic models. Part I: Construction technique and description of models of an S-week-old baby and a 7-year-old child, GFS-Bericht No. 3/89. Gesellschaft fir Strahlen und Umweltforschung mbH, Munich (1989).

28. G. G. Warner and A. M. Craig, Jr., ALGAM, a computer program for estimating internal dose for gamma-ruy sources in a human phantom. Report No. ORNL-TM-2250 (Oak Ridge National Laboratory, Oak Ridge, TN, 1968).

29. V. T. Golovachik, V. N. Kustrarjov, E. N. Savitskaya, and A. V. Sannikov, Absorbed dose and dose equivalent depth distributions for protons with energies from 2 to 600 MeV. *Rudiut. Prot. Dosim.* **28**, 189–199 (1989).

30. R. A. Hollnagel, Effective dose equivalent and organ doses for neutrons from thermal to 14 MeV. *Radiat. Prot. Dosim.* **30**, 149–159 (1990).

31. Chao TC, Bozkurt A, Xu XG. Conversion coefficients based on the VIP-Man anatomical model and EGS4 Code for external monoenergetic photons from 10 keV to 10 MeV.. Health Phys. 2001 Aug; **81**(2):163–83

32. Bozkurt A, Chao TC, Xu XG. Fluence-to-dose conversion coefficients based on the VIP-Man anatomical model and MCNPX code for monoenergetic neutrons above 20 MeV. Health Phys. 2001 Aug; **81**(2):184–202.

10

Internal Dose Assessment

10.1 Basic Concepts in Internal Dose Calculations

Internal dose concepts are employed when radioactive material enters the body through any pathway, either in a workplace situation, in the nuclear medicine clinic (where the intakes are quite intentional), from eating or drinking contaminated food and water, and other such situations. In the workplace, the most common modes of intake are:

- Inhalation (the material is breathed in, as a dust, gas, or vapor)
- Ingestion (the material is swallowed in food or liquids)
- Injection (the material is directly introduced into the bloodstream, by a puncture wound, adsorption across the skin barrier, or other means)

These modes are in order of likelihood for an occupational setting. In the nuclear medicine clinic, they are typically in the reverse order: injection is the most common, then infrequently some materials are inhaled as gases (to study lung function) or eaten as radioactively labeled meals (to study, e.g., gastric motility). Radioactive material that enters the body will be very mobile. Most external sources are typically stationary and easy to characterize.

Radioactive material that enters the lung or gastrointestinal (GI) tract, moves within these regions, and then may be absorbed into blood, where it is carried around in the body, deposited in different regions, and then cleared from these regions into excretory pathways at different rates. As one cannot measure internal doses (as people tend to object when you try to place dosimeters into their organs), internal doses are instead calculated through the use of mathematical models. The mobile nature of the radioactive source introduces another complexity to the situation that must be dealt with. This chapter provides an overview of models and methods for internal dose assessment (also called *internal dosimetry* by some, as explained in the introduction to Chapter 9).

To estimate absorbed dose for all significant tissues, one must determine for each tissue the quantity of energy absorbed per unit mass. This yields the quantity of absorbed dose, if expressed in proper units, and can be extended to calculation of dose equivalent if desired. What quantities are then needed to calculate the two key parameters, energy and mass? To facilitate this analysis, imagine an object that is uniformly contaminated with radioactive material. Depending on the identity of the radionuclide, particles or rays of characteristic energy and abundance will be given off at a rate dependent upon the amount of activity present. Our object must have some mass. Already we have almost all of the quantities needed for our equation: energy per decay (and number per decay), activity, and mass of the target region. One other factor needed is the fraction of emitted energy that is absorbed within the target. This quantity is most often called the *absorbed fraction* and is represented by the symbol ϕ.

For photons (gamma rays and X-rays) some of the emitted energy will escape objects of the size and composition of interest to internal dosimetry (mostly soft tissue organs with diameters of the order of centimeters). For electrons and beta particles, most energy is usually considered to be absorbed, so we usually set the absorbed fraction to 1.0. Electrons, beta particles, and the like are usually grouped into a class of radiations referred to as 'nonpenetrating emissions', whereas X and gamma rays are called 'penetrating radiations'. We can show a generic equation for the absorbed dose rate in our object as

$$\dot{D} = \frac{kA \sum_i n_i \, E_i \, \phi_i}{m}$$

where:

$\dot{D}$ = absorbed dose rate (rad/hr or Gy/sec)
A = activity (μCi or MBq)
n = number of radiations with energy E emitted per nuclear transition
E = energy per radiation (MeV)
ϕ = fraction of energy emitted that is absorbed in the target
m = mass of target region (g or kg)
k = proportionality constant (rad-g/μCi-hr-MeV or Gy-kg/MBq-sec-MeV)

It is extremely important that the proportionality constant be properly calculated and applied. The results of our calculation will be useless unless the units within are consistent and they correctly express the quantity desired. The application of radiation weighting factors to this equation to calculate the dose equivalent rate is a trivial matter; for most of this chapter, we consider only absorbed doses for discussion purposes.

The investigator is not usually interested only in the absorbed dose rate; more likely an estimate of total absorbed dose from an administration is desired. In the above equation the quantity activity (nuclear transitions per unit time) causes the outcome of the equation to have a time-dependence. To calculate cumulative dose, the time integral of the dose equation must be calculated. In most cases, the only term that has a time-dependence is activity, so the integral is just the product of all the factors in the above equation and the integral of the time–activity curve.

Regardless of the shape of the time–activity curve, its integral, however obtained, will have units of the number of nuclear transitions (activity, which is transitions per unit time, multiplied by time; Figure 10.1). Therefore, the equation for cumulative dose would be

$$D = \frac{k \, \tilde{A} \sum_i n_i \, E_i \, \phi_i}{m}$$

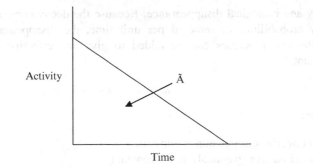

Figure 10.1 Generalized internal source time–activity curve.

where D is the absorbed dose (rad or Gy) and $\tilde{A}$ is the cumulated activity (μCi-hr or MBq-s). Details and examples of the use of these equations is developed later in the chapter.

10.2 Effective Half-Time

We know that radioactive materials decay according to first order kinetics, that is, a certain fraction of the remaining activity is removed during a specific time interval:

$$\frac{dN}{dt} = -\lambda N.$$

The well-known solution to this equation is:

$$N = N_0 \, e^{-\lambda t} \qquad A = A_0 \, e^{-\lambda t}.$$

In these equations, N is the number of atoms, N_0 is the initial number of atoms, A is the amount of activity, and A_0 is the initial activity ($A = \lambda N$). Many materials are also cleared from the body or certain organs by first-order processes. If we develop an equation for the reduction in the amount of a nonradioactive substance by a first-order system, it would look much like the equations above:

$$X(t) = X_0 \, e^{-\lambda_b \, t}$$

where:

$X(t)$ = the amount of the substance at time t
X_0 = the initial amount of substance
λ_b = the biological disappearance constant = $0.693/T_b$
T_b = the biological halftime for removal

A biological halftime for removal is exactly analogous to a radioactive (or physical) halflife; that is, it is the time in which half of the remaining material is removed.

If we now consider a certain amount of radioactive material in the body that is being cleared from the body by a first-order process, two first-order processes are involved in removing activity from the body: radioactive decay and biological disappearance. Because the decay constants are essentially probabilities of removal per unit time, the disappearance constants for the two processes can be added to give an "effective disappearance constant:"

$$\lambda_e = \lambda_b + \lambda_p,$$

where:

λ_e = effective disappearance constant
λ_p = radioactive (physical) decay constant
λ_b = biological disappearance constant

We can also define an "effective halftime" equal to $0.693/\lambda_e$, which is the actual time for half of the activity to be removed from the body or organ. The effective halftime is related to the other two halftimes by the relationship:

$$T_e = \frac{T_b \times T_p}{T_b + T_p}$$

For materials that can be described by this type of relationship, the integral of the time–activity curve may be easily evaluated:

$$\tilde{A} = \int_0^\infty A(t)dt = \int_0^\infty (f\ A_0)\ e^{-\lambda_e t}\ dt = \frac{(f\ A_0)}{\lambda_e} = 1.443\ f\ A_0\ T_e,$$

where A_0 is the administered activity, and f is the fraction of administered activity in a region at time zero. So, effective halftime is a critical parameter in the determination of cumulated activity and cumulative dose.

Note that the effective half-time for a compound will always be less than or equal to the shorter of either the biological or radiological half-time. As two processes are contributing to the removal of the element, the action of the two together must be faster than that of either acting alone. Note also that to solve the equation for effective half-time, the units for the biological and physical half-times must be the same.

Examples

$$T_b = 7 \text{ days} \quad T_p = 10 \text{ days} \quad T_{\text{eff}} = \frac{10 \times 7}{10 + 7} = 4.12 \text{ days}$$

$$T_b = 7 \text{ days} \quad T_p = 7 \text{ days} \quad T_{\text{eff}} = \frac{7 \times 7}{7 + 7} = 3.5 \text{ days}.$$

(Note: This is not a coincidence. Every time that the biological and physical half-times are the same, the effective half-time is exactly half of either value, because the expression is $(x \bullet x)/2x = x/2$.)

$$T_b = 7 \text{ days} \quad T_p = 100 \text{ days} \quad T_{\text{eff}} = \frac{100 \times 7}{100 + 7} = 6.54 \text{ days}$$

$$T_b = 7 \text{ days} \quad T_p = 10^6 \text{ days} \quad T_{\text{eff}} = \frac{10^6 \times 7}{10^6 + 7} = 7.00 \text{ days}.$$

So, as one half-time gets very long relative to the other, the effective half-time approaches the shorter of the two.

$$D = \int \dot{D}\, dt = \frac{k \sum_i n_i E_i \phi_i}{m} \int A\, dt$$

$$\int A\, dt = \int f\, A_0\, e^{-\lambda_e t}\, dt = \frac{f\, A_0}{\lambda_e}(1 - e^{-\lambda_e t})$$

If we integrate from 0 to ∞, this turns out to be just A_0/λ_e. Life is good.

$$D = \frac{k \int A\, dt \sum_i n_i E_i \phi_i}{m}$$

$$D = \frac{k \tilde{A} \sum_i n_i E_i \phi_i}{m}$$

$$D = \frac{k\, 1.443\, f\, A_0\, T_e \sum_i n_i E_i \phi_i}{m}$$

Now consider that we have two objects that are contaminated with radioactive material, and are able to irradiate themselves, each other, and possibly other objects in the system:

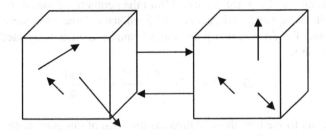

$$D_1 = \frac{k \tilde{A}_1 \sum_i n_i E_i \phi_i (1 \leftarrow 1)}{m_1} + \frac{k \tilde{A}_2 \sum_i n_i E_i \phi_i (1 \leftarrow 2)}{m_1} + \cdots$$

$$D_2 = \frac{k \tilde{A}_1 \sum_i n_i E_i \phi_i (2 \leftarrow 1)}{m_2} + \frac{k \tilde{A}_2 \sum_i n_i E_i \phi_i (2 \leftarrow 2)}{m2} + \cdots$$

10.3 Dosimetry Systems

The equations above are generic cumulative dose equations. Many authors have developed this equation in one form or another to apply to different situations. Usually many of the factors in the equations are grouped together to simplify calculations, particularly for radionuclides with complex emission spectra. Some of the physical quantities such as absorbed fraction and mass can also be combined into single values. However these quantities may be grouped, hidden, or otherwise moved around in different systems, all of them incorporate the concepts from these equations, and all are based on the same principles. Given the same input data and assumptions, the same results will be obtained. Sometimes, the apparent differences between the systems and their complicated-appearing equations may confuse and intimidate the user who may be frustrated in trying to make any two of them agree for a given problem. Careful investigation to discern these grouped factors can help to resolve apparent differences. Let us try to understand each of the systems and see how they are equivalent.

10.3.1 Marinelli–Quimby Method

Publications by Marinelli et al. and Quimby and Feitelberg[1,2] gave the dose from a beta emitter that decays completely in a tissue as

$$D_\beta = 73.8 \, C E_\beta T,$$

where D_β is the dose in rad, C is the concentration of the nuclide in microcuries per gram, E_β is the mean energy emitted per decay of the nuclide, and T is the half-life of the nuclide in the tissue. We have seen that the cumulated activity is given as 1.443 times the half-life times the initial activity in the tissue. The other terms in the equation are: $k = (73.8/1.443) = 51.1$; C is activity per mass; and for beta emitters, we assume that ϕ is 1.0. For gamma emitters, values of ϕ were estimated from the geometrical factors of Hine and Brownell[3] for spheres and cylinders of fixed sizes. Dose rates were based on expressions for dose near a point-source gamma emitter integrated over the source volume:

$$D_\gamma = 10^{-3} \, \Gamma C \int_v \frac{e^{-\mu r}}{r^2} \, dV \quad \frac{\text{rad}}{h}$$

It is difficult to see how this equation fits the form of our general equation, but it does. The factor C is still the activity per unit mass. The specific gamma rate constant Γ essentially gives the exposure rate per disintegration into an infinite medium from a point-source (equivalent to $k \times \Sigma n_i \times E_i$ in our generic equation). Finally, the factor $[\int \exp(-\mu r)/r^2 \, dV]$ acts as an absorbed fraction (μ is an absorption coefficient and $1/r^2$ is essentially a geometrical absorbed fraction). The integral in this expression can be obtained analytically only for simple geometries. Solutions for several standard objects (spheres, cylinders, etc.) were provided in the geometrical factors in Hine and Brownell's text.

10.3.2 International Commission on Radiological Protection

The ICRP has developed two comprehensive internal dosimetry systems intended for use in occupational settings (mainly the nuclear fuel cycle). ICRP publication II[4] became part of the basis for the first set of complete radiation protection regulations in this country (Code of Federal Regulations (CFR), Title 10, Chapter 20, or 10 CFR 20). These regulations were only replaced (completely) in 1994 when a revision of 10 CFR 20 incorporated the new procedures and results of the ICRP 30 series.[5] Even these two systems, published by the same body, appear on the surface to be completely different.

We have already noted, however, that they are completely identical in concept and differ only in certain internal assumptions. Both of these systems, dealing with occupational exposures, were used to calculate dose equivalent instead of just absorbed dose.

In the ICRP II system, the dose equivalent rate is given by

$$H = \frac{51.2 \; A\xi}{m}.$$

This looks somewhat like our original equation, converted to dose equivalent, but a lot seems to be missing. The missing components are included in the factor ξ:

$$\xi = \sum_i n_i \; E_i \; \phi_i \; Q_i.$$

The factor 51.2 is k, which puts the equation into units of rem per day, for activity in microcuries, mass in grams, and energy in megaelectron volts (and note that the ICRP included a quality factor, Q, to express the results in equivalent dose). The ICRP developed a system of limitation of concentrations in air and water for employees from this equation and assumptions about the kinetic behavior of radionuclides in the body. These were the well-known Maximum Permissible Concentrations (MPCs). Employees could be exposed to these concentrations on a continuous basis and not receive an annual dose rate to the so-called critical organ that would exceed established limits.

In the ICRP 30 system, the cumulative dose equivalent is given by

$$H_{50,T} = 1.6 \times 10^{-10} \sum_S U_S \; SEE(T \leftarrow S).$$

In this equation, T represents a target region and S represents a source region.

This equation looks altogether new; nothing much is similar to any of the previous equations. This is simply, however, the same old equation wearing a new disguise. The factor SEE is merely

$$SEE = \frac{\sum_i n_i \; E_i \; \phi_i(T \leftarrow S) \; Q_i}{m_T}.$$

The factor U_s is another symbol for cumulated activity, and the factor 1.6×10^{-10} is k. Note that the symbol Q (quality factor) is shown here instead of the current notation w_R (radiation weighting factor), as this is how it appeared in ICRP 30. In this system (based on the Système International unit system), this value of k produces cumulative dose equivalents in sievert, from activity in becquerels, mass in grams, energy in megaelectron volts, and appropriate quality factors. As in ICRP II, this equation was used to develop a system of dose limitation for workers, but unlike the ICRP II system, limits are placed on activity intake during a year, which would prevent cumulative doses (not continuous dose rates) from exceeding established limits. These quantities of activity were called Annual Limits on Intake (ALIs); derived air concentrations, which are directly analogous to MPCs for air, are calculated from ALIs.

The real innovation in the ICRP 30 system is the so-called Effective Dose Equivalent (H_e or EDE). As we defined in Chapter 7, certain organs or organ systems were assigned dimensionless weighting factors that are a function of their assumed relative radiosensitivity for expressing fatal cancers or genetic defects. The assumed radiosensitivities were derived from the observed rates of expression of these effects in various populations exposed to radiation. Multiplying an organ's dose equivalent by its assigned weighting factor gives a weighted dose equivalent. The sum of weighted dose equivalents for a given exposure to radiation is the effective dose equivalent. It is the dose equivalent that, if received uniformly by the whole body, would result in the same total risk as that actually incurred by a nonuniform irradiation. It is entirely different from the dose equivalent to the whole body that is calculated using values of *SEE* for the total body. Whole-body doses are often meaningless in internal dose situations because nonuniform and localized energy deposition is averaged over the mass of the whole body (70 kg).

One real difference that exists between doses calculated with the ICRP II system and the ICRP 30 (and MIRD) system is that the authors of ICRP II used a very simplistic phantom to estimate their absorbed fractions. All body organs and the whole-body were represented as spheres of uniform composition. Furthermore, organs could only irradiate themselves, not other organs. So, although contributions from all emissions were considered, an organ could only receive a dose if it contained activity, and the absorbed fractions for photons were different from those calculated from the more advanced phantoms used by ICRP 30 and MIRD (described next).

10.3.3 Medical Internal Radiation Dose (MIRD) System

The equation for absorbed dose in the MIRD system[6] is deceptively simple:

$$D_{r_k} = \sum_h \tilde{A}_h \, S(r_k \leftarrow r_h)$$

In this equation, r_k represents a target region and r_h represents a source region. No one is fooled by now, of course. The cumulated activity is there; all other terms must be lumped in the factor S, and so they are:

$$S(r_k \leftarrow r_h) = \frac{k \sum_i n_i \, E_i \, \phi_i(r_k \leftarrow r_h)}{m_{r_k}}.$$

In the MIRD equations, the factor k is 2.13, which gives doses in rad, from activity in microcuries, mass in grams, and energy in MeV. The MIRD system was developed primarily for use in estimating radiation doses received by patients from administered radiopharmaceuticals; it was not intended to be applied to a system of dose limitation for workers. In the MIRD system, one may sometimes also see the use of the term Δ_i. The factor $\Delta_i = k \times n_i \times E_i$ for a given radionuclide emission, and the equations may be represented as

$$S(r_k \leftarrow r_h) = \frac{\sum_i \Delta_i \, \phi_i(r_k \leftarrow r_h)}{m_{r_k}}$$

$$D_{r_k} = \sum_h \tilde{A}_h \, S(r_k \leftarrow r_h) = \sum_h \tilde{A}_h \frac{\sum_i \Delta_i \, \phi(r_k \leftarrow r_h)}{m_{r_k}}$$

10.3.4 RADAR

In the early 21st century, an electronic resource was established on the Internet to provide rapid worldwide dissemination of important dose quantities and data. The RAdiation Dose Assessment Resource established a Web site at www.doseinfo-radar.com, and provided a number of publications on the data and methods used in the system. The RADAR system has about the simplest manifestation of the dose equation[7]:

$$D = N \times DF$$

where N is the number of disintegrations that occur in a source organ, and DF is:

$$DF = \frac{k \sum_i n_i E_i \phi_i}{m}$$

The DF is mathematically the same as an "S value" as defined in the MIRD system. The number of disintegrations is the integral of a time–activity curve for a source region. RADAR members produced compendia of decay data, dose conversion factors, and catalogued standardized dose models for radiation workers and nuclear medicine patients, among other resources. They also produced the widely used OLINDA/EXM personal computer software code,[8] which used the equations shown here and the input data from the RADAR site.[9]

10.4 Internal Dose Calculations for Radiation Workers

Radiation workers are a very important group for which internal dose calculations are performed. Loose radioactive contamination in the workplace may be taken into the body, with the deposition of energy in many body tissues. In the majority of workplace situations, the design goal is to avoid all intakes of radioactive materials. Such intakes generally represent minor or major failures of containment and monitoring systems, accidental situations, or other avoidable scenarios. In some industries, however, such as uranium and thorium mining and milling, airborne radioactive dust is everywhere, and a certain low level of intake of radioactive material is to be expected. In the first case, monitoring for radioactive material will generally show negative results most of the time (i.e., results that are less than the detection system's MDA; see Chapter 8). In the second case, a certain low level of body retention or excretion is to be expected (above the natural levels of such elements in the environment (this difference may be difficult to characterize) and monitoring is performed to ensure that the levels of intake are in the acceptable range.

Modes of Intake

As noted in the beginning of the chapter, the most common mode of intake in occupational settings is inhalation, followed by ingestion and then injection. Airborne radioactive contaminants may be in the form of aerosols, dusts, or vapors. In particular, iodine (e.g., ^{131}I, ^{125}I) is volatile and working with liquid solutions always has the hazard of inhalation of volatilized activity. Mobile ^{3}H will most often exist as ^{3}H$_2$O vapor in any humid environments. Ingestion is rare, and represents perhaps an accidental intake (contaminated hands touching

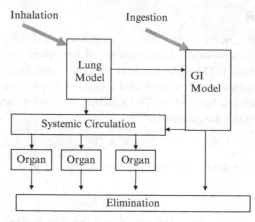

Figure 10.2 Generalized internal retention model.

the mouth directly, handling food or other substances placed in the mouth), material splashed in the face, or material deposited on the throat when mouth breathing.

Injection incidents are quite rare and almost always accidental. People working with syringes to inject radioactive material always run the risk of a puncture incident, even through protective gloves are worn. Handling glass beakers or other containers of liquid radioactive materials can lead to occasional breakage, with sharp pieces of glass cutting through protective gloves or clothing and cutting the skin. Jagged pieces of radioactive metal, for example, depleted uranium being machined for particular purposes, may also cause direct injection of radioactive material into the bloodstream. Most situations, and thus most mathematical models for treating intakes, however, assume that the intakes were from inhalation or ingestion. Thus much work has been done in developing kinetic models to simulate the movement of radioactive material in these two important organ systems. Note that the organs involved in the deposition of radioactive material and its potential transfer to the systemic circulation and other body organs (respiratory tract airways, lungs, and pulmonary lymph nodes in the case of inhalation and the stomach and intestines in the case of ingestion) are also organs for which we wish to know the radiation dose. A general systemic model may be shown as in Figure 10.2.

Kinetic Models for the GI Tract

The dosimetric model for the gastrointestinal tract (GI tract) given in ICRP 30 (Figure 10.3) is a very simple, straight-through, four-compartment model.[10] The four sections are stomach, small intestine, upper large intestine, and lower large intestine, sometimes abbreviated ST, SI, ULI, and LLI, respectively. Ingestion is a common means of intake of radioactive material, either through swallowing of material somehow introduced into the mouth or through transfer of material from the various regions of the lung system to the throat and subsequent swallowing. The sections of the GI tract are treated as separate target tissues according to the recommendations in the ICRP 30 dosimetric system. They are not assigned any specific weighting factors, however, and the weighting factors recommended for the "remainder" are assumed to apply to any significant committed dose equivalents.

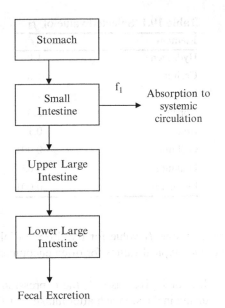

Figure 10.3 ICRP 30 GI tract model.

Translocation from one compartment to the next is assumed to be governed by first-order kinetics (exponential removal). The only way for material to leave the system is through excretion from the LLI or absorption from the SI into the transfer compartment. The removal to the transfer compartment is also assumed to be a first-order process. Most often quoted in the literature is the fraction of stable material reaching the body fluids following ingestion, f_1. The transfer coefficient from the SI to the transfer compartment is called λ_B and is defined numerically as

$$\lambda_B = \frac{f_1 \lambda_{SI}}{1 - f_1}.$$

Transfer rates are assumed to be fixed for the various compartments regardless of the chemical form of the material. The only parameter affected by the chemical form is the value f_1 (and therefore λ_B). Values of f_1 may be given for several chemical forms of the elements. When f_1 is very small, most of the activity will pass through the GI tract and be excreted in the feces. Thus, the principal radiation doses of concern will be to the segments of the GI tract itself. When f_1 is large, most of the activity ingested will be transferred to the systemic circulation and little will pass through the GI tract beyond the small intestine. When f_1 is 1.0 (i.e., 100%) the activity is assumed to pass into the systemic circulation, the transfer is assumed to occur very quickly, and the GI tract organs are not involved at all. Essentially the model assumes that the material passes directly from the stomach to the circulation, as if the material were actually just injected at time zero into the blood.

With intermediate values of f_1, both the systemic organs and the GI tract organs may receive significant radiation doses. Some representative f_1 values are shown in Table 10.1. (Note: In many cases, more than one chemical class of an element may be known and defined in the ICRP models, so

Table 10.1 Selected value of f_1.

Element	f_1
Hydrogen	1.0
Cesium	1.0
Iodine	1.0
Cobalt	0.3
Iron	0.1
Gallium	0.001
Uranium	0.05
Plutonium	0.001

there may be more than one f_1 value for the different inhalation classes, as defined below. These are typical values for often encountered classes of these elements.)

If you use the values of λ given as s^{-1}, the expression for the number of transformations occurring in the stomach after intake of 1 Bq of activity is:

$$\tilde{A}_{ST} = \frac{1}{\lambda_{ST} + \lambda_R},$$

where:

λ_{ST} = the rate constant for loss of stable material from the stomach to small intestine (24 day^1 = 1 hr^{-1} = 0.00028 s^{-1})
λ_R = the radioactive decay constant for the radionuclide (s^{-1})

The expression for the number of transformations in the small intestine after ingestion of 1 Bq of activity is:

$$\tilde{A}_{SI} = \frac{\lambda_{ST}}{(\lambda_{ST} + \lambda_R)(\lambda_{SI} + \lambda_B + \lambda_R)},$$

where:

λ_{SI} = rate constant for loss of stable material from small intestine to upper large intestine (6.0 day^{-1} = 0.25 hr^{-1} = 6.9 × 10^{-5} s^{-1})
λ_B = rate constant for loss of stable material from small intestine to transfer compartment, as defined above (s^{-1})

The expression for the number of transformations in the upper large intestine after ingestion of 1 Bq is:

$$\tilde{A}_{ULI} = \frac{\lambda_{ST}\,\lambda_{SI}}{(\lambda_{ST} + \lambda_R)(\lambda_{SI} + \lambda_B + \lambda_R)(\lambda_{ULI} + \lambda_R)},$$

where λ_{ULI} = rate constant for loss of stable material from upper large intestine to large intestine (1.8 day^1 = 0.075 hr^{-1} = 2.1 × 10^{-5} s^{-1}).

The expression for the number of transformations in the lower large intestine after ingestion of 1 Bq is:

$$\tilde{A}_{LLI} = \frac{\lambda_{ST}\,\lambda_{SI}\,\lambda_{ULI}}{(\lambda_{ST} + \lambda_R)(\lambda_{SI} + \lambda_B + \lambda_R)(\lambda_{ULI} + \lambda_R)(\lambda_{LLI} + \lambda_R)},$$

where λ_{LLI} = rate constant for loss of stable material from lower large intestine to the large intestine ($1.0 \text{ day}^1 = 0.0417 \text{ hr}^{-1} = 1.16 \times 10^{-5} \text{ s}^{-1}$).

If you watched closely as the expressions developed, you might have noticed that a pattern developed through which the number of transformations in one compartment could be predicted by the number in the previous compartment:

$$\tilde{A}_i = \frac{\tilde{A}_{i-1}\,\lambda_{i-1}}{\lambda_i + \lambda_R}.$$

Radioactive Progeny

The expressions for production of radioactive daughters within the tract are equally straightforward, but much more complicated. The expressions are shown below:

$$\tilde{A}_{ST-\text{progeny}} = \frac{\tilde{A}_{ST-\text{parent}}\,\lambda_{R-\text{progeny}}}{\lambda_{ST} + \lambda_{R-\text{progeny}}}$$

$$\tilde{A}_{SI-\text{progeny}} = \frac{\tilde{A}_{ST-\text{parent}}\,\lambda_{R-\text{progeny}}\,\lambda_{ST}}{(\lambda_{ST} + \lambda_{R-\text{progeny}})(\lambda_{SI} + \lambda_B + \lambda_{R-\text{progeny}})}$$
$$+ \frac{\tilde{A}_{SI-\text{parent}}\,\lambda_{R-\text{progeny}}}{\lambda_{SI} + \lambda_B + \lambda_{R-\text{progeny}}}$$

$$\tilde{A}_{ULI-\text{progeny}} = \frac{\tilde{A}_{ST-\text{parent}}\,\lambda_{R-\text{progeny}}\,\lambda_{ST}\lambda_{SI}}{\left(\lambda_{ST} + \lambda_{R-\text{progeny}}\right)\left(\lambda_{SI} + \lambda_B + \lambda_{R-\text{progeny}}\right)\left(\lambda_{ULI} + \lambda_{R-\text{progeny}}\right)}$$
$$+ \frac{\tilde{A}_{SI-\text{parent}}\,\lambda_{R-\text{progeny}}\,\lambda_{SI}}{\left(\lambda_{SI} + \lambda_B + \lambda_{R-\text{progeny}}\right)\left(\lambda_{ULI} + \lambda_{R-\text{progeny}}\right)}$$
$$+ \frac{\tilde{A}_{ULI-\text{parent}}\,\lambda_{R-\text{progeny}}}{\lambda_{ULI} + \lambda_{R-\text{progeny}}}$$

$$\tilde{A}_{LLI-\text{progeny}}$$
$$= \frac{\tilde{A}_{ST-\text{parent}}\,\lambda_{R-\text{progeny}}\,\lambda_{ST}\lambda_{SI}\lambda_{ULI}}{\left(\lambda_{ST} + \lambda_{R-\text{progeny}}\right)\left(\lambda_{SI} + \lambda_B + \lambda_{R-\text{progeny}}\right)\left(\lambda_{ULI} + \lambda_{R-\text{progeny}}\right)\left(\lambda_{LLI} + \lambda_{R-\text{progeny}}\right)}$$
$$+ \frac{\tilde{A}_{SI-\text{parent}}\,\lambda_{R-\text{progeny}}\,\lambda_{SI}\lambda_{ULI}}{\left(\lambda_{SI} + \lambda_B + \lambda_{R-\text{progeny}}\right)\left(\lambda_{ULI} + \lambda_{R-\text{progeny}}\right)\left(\lambda_{LLI} + \lambda_{R-\text{progeny}}\right)}$$
$$+ \frac{\tilde{A}_{ULI-\text{parent}}\,\lambda_{R-\text{progeny}}\,\lambda_{ULI}}{\left(\lambda_{ULI} + \lambda_{R-\text{progeny}}\right)\left(\lambda_{LLI} + \lambda_{R-\text{progeny}}\right)} + \frac{\tilde{A}_{LLI-\text{parent}}\,\lambda_{R-\text{progeny}}}{\left(\lambda_{LLI} + \lambda_{R-\text{progeny}}\right)}$$

Dose Factors—Absorbed Fractions

Absorbed fractions for photons are calculated by Monte Carlo methods using the Fisher–Snyder phantom, as described shortly. For beta particles, the assumption is made that the dose rate to the surface of the segment wall is half that in the contents, which is calculated by the usual theoretical expression. For alpha particles and fission fragments, this value of one-half is further modified to account for selfabsorption within the contents by multiplying by 0.01. Recoil atoms are assumed not to irradiate the GI tract walls.

Kinetic Model for the Lungs

The lungs are much more complicated than the GI tract to model. Unlike the GI tract model, which has not changed since its creation in 1966,[10] the lung model has been continually evolving to more and more complex forms. In its earliest form, given in ICRP II[4] (1960), the model had only two major compartments, the "upper respiratory tract" and the "lower respiratory tract" (Figure 10.4), with half-lives in the lower respiratory tract of <1 day for "soluble" materials, 120 days for "insoluble" materials, and 365 days and 1460 days for plutonium and thorium.

The ICRP 30 lung model[5] had eight compartments (ten if you include the lymph node compartment; Figure 10.5), with varying clearance half-times which depended on the type of aerosol assumed to be inhaled (class D, W, or Y, corresponding to clearance times of the order of days, weeks, or years, relating to material that had varying degrees of biological mobility). Material deposited in the upper or lower respiratory tract either moved into the bloodstream or into the GI tract (via mucociliary transport into the esophagus). Although tedious, this model could be solved by hand or with a small computer program or spreadsheet. The model structure is shown in the diagram in Figure 10.5. The deposition model was based on particle size.

The figure shows the deposition model, which gives the deposition probability as a function of a parameter known as the Aerosol Median Aerodynamic Diameter (AMAD). Aerosols encountered in the workplace are often not of uniform diameter, and the size of the particles may cover a large range of values (polydisperse aerosol). Such aerosols will typically have lognormal distributions of particle sizes, and the best parameter to characterize the distribution is the median aerodynamic diameter. This diameter may be characterized according to the particle mass, activity, or other criteria; if we use activity, we define the AMAD. Figure 10.6, from ICRP 30, shows that from particles between 0.2 and 10 μm, the deposition in the tracheobronchial region (TB) is about the same, about 0.08 (or 8% of the inhaled particles), deposition increases dramatically in the nasopharyngeal (NP) region with increasing particle size but decreases in the pulmonary (P), or deep lung, region with increasing

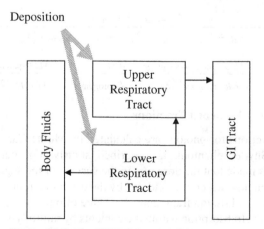

Figure 10.4 The ICRP II lung model.

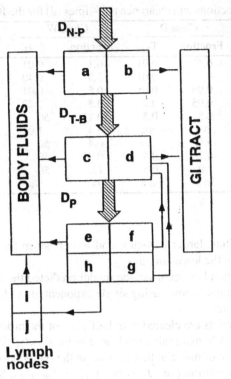

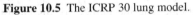

Figure 10.5 The ICRP 30 lung model.

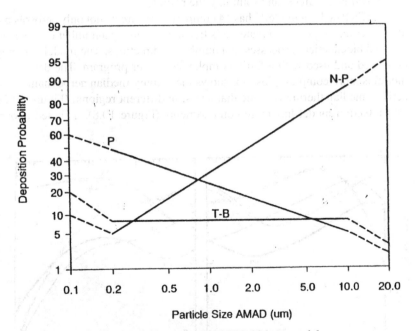

Figure 10.6 Aerosol deposition model in ICRP 30 lung model.

Table 10.2 Fractions and clearance half-times (d) for the ICRP 30 lung model.

Region		Class D		Class W		Class Y	
		Fraction	$T_{1/2}$	Fraction	$T_{1/2}$	Fraction	$T_{1/2}$
NP	a	0.5	0.01	0.1	0.01	0.01	0.01
	b	0.5	0.01	0.9	0.40	0.99	0.40
TB	c	0.95	0.01	0.5	0.01	0.01	0.01
	d	0.05	0.2	0.5	0.20	0.99	0.20
P	e	0.8	0.5	0.15	50	0.05	500
	f	n/a	n/a	0.4	1.0	0.4	1.0
	g	n/a	n/a	0.4	50	0.4	500
	h	0.2	0.5	0.05	50	0.15	500
L	i	1.0	0.5	1.0	50	0.9	1000
	j	n/a	n/a	n/a	n/a	0.1	∞

particle size. Thus, larger particles tend to deposit in the upper airways and finer particles in the lower airways.

Once deposition has occurred, the model predicted the clearance from each of the compartments above, using single exponential models. Table 10.2 summarizes the model.

Class D materials are cleared with half-lives of the order of days or shorter, and class W and Y materials are cleared more slowly. For class Y material, there is a fraction of material that ends up in the pulmonary lymph nodes with infinitely long retention (i.e., $T_b = \infty$, $T_{eff} = T_{physical}$). So it would be quite tedious, but one could calculate the number of disintegrations in the lung by adding up all of the components of the deposition and retention, calculating the individual $\tilde{A}$ values, and summing the results.

The ICRP 66 lung model[11] has 14 compartments, treats not only aerosols but also gases and vapors, and material is translocated using a number of mechanical and biochemical processes in a number of directions. The model can only be solved and used with a fairly complex computer program. The deposition model is more complex (this plot shows the activity median aerodynamic and activity median thermodynamic diameters, in different regions) (Figure 10.7). The retention model has more compartments (Figure 10.8), as noted above,

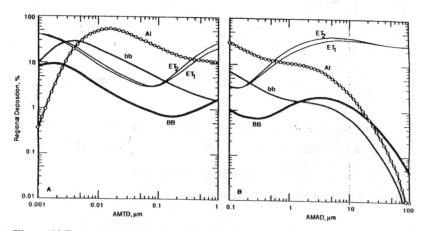

Figure 10.7 Aerosol deposition model in ICRP 66 lung model.

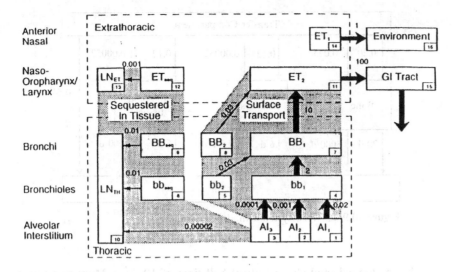

Figure 10.8 ICRP 66 lung model.

but also has far more highly complex translocation kinetics, which can only be solved using a computer program.

Systemic Models for Individual Elements

After material enters the body through the lung or GI tracts, it may deposit energy in the compartments of these models, which are organs at risk for radiation exposure, but the material may as well be absorbed into the systemic circulation and deposited in various organs of the body and then eliminated. Thus the organs in which the activity is deposited, as well as the excretory organs will be exposed to radiation dose from particulate radiations, and all organs and tissues may be exposed to a significant degree from photons originating in the organs where the activity is concentrated (the lung, GI, and other important organs with radionuclide uptake). As discussed above, the characterization of the kinetics within any organ is often made with a quantity called the *biological half-time*, as the clearance of materials from organs of the body tends to be first-order, just like radioactive decay, and thus treatable with an exponential model. When first-order biological clearance (biological half-time) is combined with the radioactive decay of the radionuclide (physical half-time), the effective half-time (as defined above) is obtained.

Biological Models for the Elements

Biological models for the different elements of concern to occupational dose assessment are derived based on whatever available literature values may be found, from experiments involving animals or measurements in human subjects. Some of the simplest models involve an element being uniformly distributed in the whole-body and being eliminated with one or more exponential terms. For example, tritium (^{3}H) was characterized for years by a uniform whole-body distribution with a single exponential clearance component

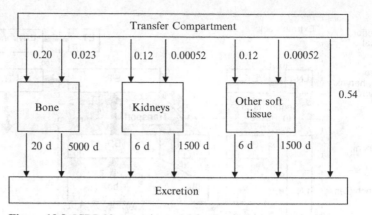

Figure 10.9 ICRP 30 systemic model for uranium.

with a biological (and also effective) half-time of 10 days. More recently, a longer-lived component, called Organically Bound Tritium (OBT) has been defined with one or more longer half-times.

The model for cesium is treated as a uniform whole-body model with two effective half-times. Ten percent of the material is seen as clearing the body with a 2 day biological half-time and the other ninety percent with a 110 day biological half-time. It is not uncommon, in the whole-body or organs, to have material cleared with more than one half-time. Sometimes there is a reasonable and understandable reason for this (e.g., some material is cleared by systemic processes, having to do with blood flow and other material with clearance processes within the organ, after uptake and release by the cells). In other cases, it may not be entirely clear why there are multiple phases of clearance, but it is simply observed during the collection of biokinetic data.

Most of the biokinetic models in ICRP 30, which form the basis for the current U.S. regulatory structure for protecting workers from intakes of radionuclides, are of the form in which material moves from the "transfer" compartment" (basically, the systemic circulation) to different organs, where it is eliminated with one or more biological half-times. For example, Figure 10.9 shows the ICRP 30 model for uranium. In this model, 54% of the activity that reaches the circulation is excreted quickly (the ICRP 30 model uses a biological half-time of 0.25 d), then about 22% goes to bone, where 20% is excreted with a 20 d biological half-time and the rest with a 5000 d biological half-time.

Kidneys and other tissues receive a little over 12% of the activity, with most of this being excreted with a 6 d biological half-time and a small fraction being retained with a 1500 d half-time. One might be tempted to ignore this small fraction in the dose calculations, but recall that the number of disintegrations is given by the product of the initial activity in a region and the effective half-time. So here, for example, $0.12 \times 6 = 0.72$ and $0.00052 \times 1500 = 0.78$, so these two components will contribute almost equally to the total number of disintegrations occurring in these regions. This kind of model, also sometimes called a "once through" model, is not terribly realistic, although it probably gives a reasonably good estimate of the number of disintegrations occurring

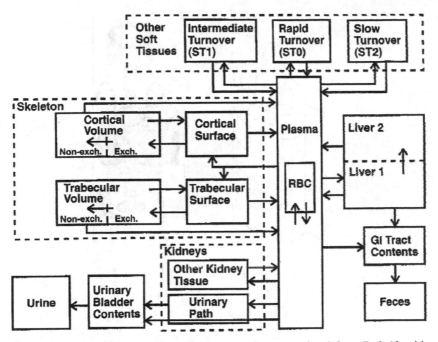

Figure 10.10 ICRP 69 systemic model for uranium. (Reprinted from Ref. 12 with permission from Elsevier.)

in each of the important source regions. This kind of model, at least for the more complicated cases, is being replaced by more realistic models in which material recirculation occurs, as in the example below, from ICRP 69[12] for uranium (Figure 10.10).

Calculation of Organ Doses

Once we have completely characterized the biokinetic model for an element (including the kinetics of the intake component, lung or GI), we have the U_S values for all of the important source regions, and just need to multiply them by the appropriate dose conversion factors (*SEE*s), as shown above. To calculate an *SEE*, one needs decay data, data on standard organ masses, and the "absorbed fractions" for photons. Decay data are not difficult to find. For electrons, we assume that the absorbed fraction is 1.0 for an organ irradiating itself and 0.0 for irradiation of other organs. Electrons have short ranges in human tissue relative to the dimensions of the organs, whereas photons will travel and scatter throughout all tissues of the body, and some photon energy will likely escape the body entirely. Absorbed fractions for photons are calculated using Monte Carlo radiation transport simulation methods in anthropomorphic phantoms, that is, mathematical representations of the human body.

The sizes of the organs are based on values given by the ICRP for standard reference persons, so we now have all of the pieces needed to calculate our *SEE* values. The first complete descriptions of a phantom representing the reference adult were given in MIRD Pamphlets 5 and 5, Revised[13,14] (Figure 10.11). These absorbed fractions were used to develop the *SEE* values in ICRP 30[5]

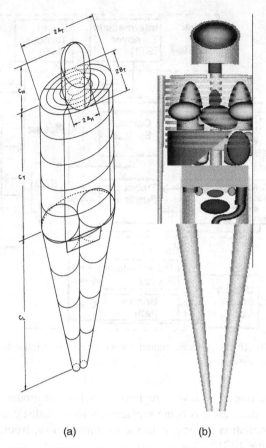

(a)   (b)

Figure 10.11 Stylized adult male Fisher–Snyder model showing (a) exterior view, and (b) the skeleton and internal organs.

as well as the S values in MIRD Pamphlet No. 11[15] (we study the MIRD system next). An improved set of absorbed fractions for a slightly different adult phantom and for five other individuals representing children of different ages (newborns, 1-yr-olds, 5-yr-olds, 10-yr-olds, 15-yr-olds) was published by Cristy and Eckerman in 1987.[16] Then, in 1995, four phantoms representing the adult female, both nonpregnant and at three stages of pregnancy were published by Stabin et al.[17] Before 1995, the Cristy and Eckerman 15-yr-old phantom was often used to represent the adult female. The Stabin et al. adult female phantom is somewhat different from the Cristy–Eckerman model. Others have as well proposed more detailed models of some organs, including the brain,[18,19] eye,[20] peritoneal cavity,[21] prostate gland,[22] and others.

Example
Ignoring organically bound tritium for the moment, we can calculate the number of transformations occurring from a 1 Bq intake of ^{3}H, using the standard model that has a uniform, whole-body distribution with a biological retention half-time of 10 days. As the physical half-life of ^{3}H is 12.3 years, the biological half-time is equal to the effective half-time. Thus, the number of

transformations is easily calculated as

$$U_{\text{whole body}} = \frac{1 \frac{\text{transformation}}{s}}{\left[\frac{0.693}{10\ d \times 86400\frac{s}{d}}\right]} = 1.2x10^6 \text{ transformations}$$

The *SEE* for tritium in the whole body may be found to be 9×10^{-8} MeV/g-transf. Thus, the committed dose from a 1 Bq intake is:

$$H_{50} = 1.6 \times 10^{-10} \frac{\text{Sv g}}{\text{MeV}}\ 1.2\times 10^6\ \text{transf}\ 9 \times 10^{-8} \frac{\text{MeV}}{\text{g transf}} = 1.7\times 10^{-11}\text{Sv}$$

Thus our dose coefficient for ^{3}H intakes is 1.7×10^{-11} Sv/Bq intake. Given an intake of 1 kBq, we could immediately calculate an estimated committed dose of 1.7×10^{-8} Sv.

For ^{137}Cs, as noted above, there are two components to the retention curve. 10% of the cesium in the body is cleared with a 2 day biological half-time and the rest with a 110 day biological half-time. To calculate the number of disintegrations resulting from the intake of 1 Bq of ^{137}Cs, we would need to calculate the effective half-times and apply the two fractions for the distribution:

$$T_{eff-1} = \frac{2 \times 10960}{2 + 10960} \approx 2\ d \qquad T_{eff-2} = \frac{110 \times 10960}{110 + 10960} - 109\ d$$

$$\lambda_{eff-1} = \frac{0.693}{2\ d} = 0.346\ d^{-1} \qquad \lambda_{eff-2} = \frac{0.693}{109\ d} = 0.0064\ d^{-1}$$

$$U_{\text{whole body}} = \frac{0.1}{0.346\ d^{-1}} + \frac{0.9}{0.0064\ d^{-1}} = 0.29 + 141$$

$$= 141\,Bq - d = 1.22 \times 10^7 \text{ transformations}$$

$$U_{\text{whole body}} = 1.443\,[0.1 \times 2\ d + 0.9 \times 109\ d] = 0.29 + 141 = 141\,Bq - d$$

$$= 1.22 \times 10^7 \text{ transformations}$$

Calculation of Permissible Intake Limits

In the ICRP system, one is not content to simply calculate doses to individuals. In this system, we wish to take the calculation a step further and calculate the amount of activity that would give a certain dose (the permissible annual dose) from the dose conversion coefficients for a given nuclide. The dose conversion factors are usually calculated as 50-year committed dose (Sv) per Bq of intake (by inhalation or ingestion). Knowing our annual dose limits in Sv, we can thus calculate the number of Bq that are permissible to take in during one year of work. The only complication is that we really have two dose limits that must be satisfied at the same time: the stochastic limit (50 mSv effective whole-body) and the nonstochastic limit (500 mSv to any organ). We resolve this by calculating the permissible amount of activity that will satisfy both limits and choose the smaller of the two values as our controlling limit.

Example

Calculate the ALI and the DAC for inhalation of class D ^{32}P. Solution of the biokinetic model gives the following values for the number of disintegrations in various source organs for a 1 Bq intake.

Lungs:	1.8×10^4
ULI contents:	1.4×10^3
LLI contents:	2.5×10^3
Cortical bone:	1.5×10^5
Trabecular bone:	1.5×10^5
Other tissues:	2.8×10^5

We may find the dose conversion factors (*SEE*s) in ICRP 30, and solve for the H$_{50}$ values as follows.

$$H_{50,\text{gonads}} = 1.6x10^{-10} \frac{\text{Sv g}}{\text{MeV}} \times 2.8x10^5 \frac{\text{transf}}{\text{Bq intake}} \times 9.9x10^{-6} \frac{\text{MeV}}{\text{g transf}}$$

$$= 4.4x10^{-10} \frac{\text{Sv}}{\text{Bq}}$$

$$H_{50,\text{breasts}} = 1.6x10^{-10} \frac{\text{Sv g}}{\text{MeV}} \times 2.8x10^5 \frac{\text{transf}}{\text{Bq intake}} \times 9.9x10^{-6} \frac{\text{MeV}}{\text{g transf}}$$

$$= 4.4x10^{-10} \frac{\text{Sv}}{\text{Bq}}$$

$$H_{50,\text{marrow}} = 1.6x10^{-10} \frac{\text{Sv g}}{\text{MeV}} \times 2.8x10^5 \frac{\text{transf}}{\text{Bq intake}} \times 9.9x10^{-6} \frac{\text{MeV}}{\text{g transf}}$$

$$+ 1.6x10^{-10} \frac{\text{Sv g}}{\text{MeV}} \times 1.5x10^5 \frac{\text{transf}}{\text{Bq intake}} \times 2.3x10^{-4} \frac{\text{MeV}}{\text{g transf}}$$

$$= 6.0x10^{-9} \frac{\text{Sv}}{\text{Bq}}$$

$$H_{50,\text{lungs}} = 1.6x10^{-10} \frac{\text{Sv g}}{\text{MeV}} \times 2.8x10^5 \frac{\text{transf}}{\text{Bq intake}} \times 9.9x10^{-6} \frac{\text{MeV}}{\text{g transf}}$$

$$+ 1.6x10^{-10} \frac{\text{Sv g}}{\text{MeV}} \times 1.8x10^4 \frac{\text{transf}}{\text{Bq intake}} \times 6.9x10^{-4} \frac{\text{MeV}}{\text{g transf}}$$

$$= 2.4x10^{-9} \frac{\text{Sv}}{\text{Bq}}$$

$$H_{50,ULI} = 1.6x10^{-10} \frac{\text{Sv g}}{\text{MeV}} \times 2.8x10^5 \frac{\text{transf}}{\text{Bq intake}} \times 9.9x10^{-6} \frac{\text{MeV}}{\text{g transf}}$$

$$+ 1.6x10^{-10} \frac{\text{Sv g}}{\text{MeV}} \times 1.4x10^3 \frac{\text{transf}}{\text{Bq intake}} \times 1.6x10^{-3} \frac{\text{MeV}}{\text{g transf}}$$

$$= 8.0x10^{-10} \frac{\text{Sv}}{\text{Bq}}$$

$$H_{50,LLI} = 1.6x10^{-10} \frac{\text{Sv g}}{\text{MeV}} \times 2.8x10^5 \frac{\text{transf}}{\text{Bq intake}} \times 9.9x10^{-6} \frac{\text{MeV}}{\text{g transf}}$$

$$+ 1.6x10^{-10} \frac{\text{Sv g}}{\text{MeV}} \times 2.5x10^3 \frac{\text{transf}}{\text{Bq intake}} \times 2.6x10^{-3} \frac{\text{MeV}}{\text{g transf}}$$

$$= 1.5x10^{-9} \frac{\text{Sv}}{\text{Bq}}$$

$$H_{50,\text{bone surfaces}} = 1.6x10^{-10}\frac{\text{Sv g}}{\text{MeV}} \times 2.8x10^5\frac{\text{transf}}{\text{Bq intake}} \times 9.9x10^{-6}\frac{\text{MeV}}{\text{g transf}}$$

$$+1.6x10^{-10}\frac{\text{Sv g}}{\text{MeV}} \times 1.5x10^5\frac{\text{transf}}{\text{Bq intake}} \times 8.7x10^{-5}\frac{\text{MeV}}{\text{g transf}}$$

$$+1.6x10^{-10}\frac{\text{Sv g}}{\text{MeV}} \times 1.5x10^5\frac{\text{transf}}{\text{Bq intake}} \times 1.4x10^{-4}\frac{\text{MeV}}{\text{g transf}}$$

$$= 5.9x10^{-9}\frac{\text{Sv}}{\text{Bq}}$$

Having the individual H_{50} values, we can choose the appropriate tissue weighting factors for each organ and also calculate the effective dose equivalent:

Organ	H_{50}	w_T	$w_T \times H_{50,T}$
Gonads	4.4×10^{-10}	0.25	1.1×10^{-10} Sv/Bq
Breast	4.4×10^{-10}	0.15	6.6×10^{-11} Sv/Bq
Red marrow	6.0×10^{-9}	0.12	7.2×10^{-10} Sv/Bq
Lungs	2.4×10^{-9}	0.12	2.9×10^{-10} Sv/Bq
Bone surface cells	5.9×10^{-9}	0.03	1.8×10^{-10} Sv/Bq
ULI	8.0×10^{-10}	0.06	4.8×10^{-11} Sv/Bq
LLI	1.5×10^{-9}	0.06	9.0×10^{-11} Sv/Bq
			$\Sigma = 1.5 \times 10^{-9}$ Sv/Bq

Now we need to calculate our two possible intake values (called Annual Limits on Intake, or ALIs). For the stochastic *ALI*, we divide the stochastic dose limit into the effective dose (the sum of the right hand column). For the nonstochastic *ALI*, we divide the nonstochastic limit into the highest of the dose/intake values for the individual organs:

$$ALI_{\text{stochastic}} = \frac{0.05 \frac{\text{Sv}}{y}}{1.5x10^{-9} \frac{\text{Sv}}{\text{Bq}}} = 3.3x10^7 \frac{\text{Bq}}{y}$$

$$ALI_{\text{non-stochastic}} = \frac{0.5 \frac{\text{Sv}}{y}}{6.0x10^{-9} \frac{\text{Sv}}{\text{Bq}}} = 8.3x10^7 \frac{\text{Bq}}{y}$$

As the stochastic limit is less than the nonstochastic limit, this becomes the limiting value, and is our actual *ALI* (3.3×10^7 Bq).

Calculation of Permissible Air Concentrations

Now, for the calculation of an air concentration that may be present all year that a worker may breathe without exceeding the dose limit, we just divide the chosen *ALI* by the amount of air breathed in a year (2400 m^3, based on a breathing rate of 0.02 m^3/min):

$$\text{DAC} = \frac{3.3x10^7 \text{ Bq}}{2400 \text{ m}^3} = 1.4x10^4 \frac{\text{Bq}}{\text{m}^3}$$

Important notes:

- If one takes in exactly one *ALI* of any nuclide, one is exposed exactly at the dose limit, and may have no other sources of exposure during that year. Thus, the true compliance equation is:

$$\sum \frac{\text{Intake}_i}{ALI_i} + \frac{H_{\text{ext}}}{50 \text{ mSv}} \leq 1.0$$

- The DAC gives the concentration that may be present continuously throughout a (2000 hr) working year. Thus, the true limit on air concentrations is based on the idea of DAC-hours: if one is exposed to 2000 DAC-hours in a year, one takes in exactly 1 *ALI* by inhalation, and thus is exposed exactly at the dose limit. Another form of the compliance equation thus may be:

$$\sum \frac{\text{Intake}_i}{ALI_i} + \sum \frac{\text{DAC} - \text{hours}}{2000} + \frac{H_{\text{ext}}}{50 \text{ mSv}} \leq 1.0$$

One thus may be exposed to a level of 2000 DACs for 1 hour and still be within permissible dose limits. This again assumes that the individual had no external radiation exposures during the year, and had no other intakes, either by inhalation or ingestion, during that year.

10.5   Internal Dose Calculations for Nuclear Medicine Patients

Just as for radiation workers, to calculate radiation dose for nuclear medicine patients, one needs a kinetic model, based on measurements made in animal or human studies, and dose conversion factors. The kinetic data are obtained from animal or human studies, as for radiation workers. However, radionuclides in nuclear medicine are bound to a very wide variety of compounds, and a separate kinetic model must be developed for each compound. To obtain approval from the Food and Drug Administration to distribute a new radiopharmaceutical, a company must show that the drug is both safe and efficacious. Safety concerns include, but are not limited to, radiation doses expected to be received by patients who receive the radiopharmaceutical. Dose conversion factors are available for all of the standard models, as they are for radiation workers; in fact, the same anthropomorphic models are used. Kinetic models must be derived for each new compound, which involves the proper design and execution of experiments in either animals or humans to obtain the necessary data to build a kinetic model.

Kinetic Data

To design and execute a good kinetic study, one needs to collect the correct data, enough data, and express the data in the proper units. The basic data needed are the fraction (or percentage) of administered activity in important source organs and excreta samples. We discuss later how these data are gathered from an animal or human study. It is very important, in either type of study, to take enough samples to characterize both the distribution and retention of the radiopharmaceutical over the course of the study. The following criteria are essential.

- Catch the early peak uptake and rapid washout phase.
- Cover at least three effective half-times of the radiopharmaceutical.
- Collect at least two time points per clearance phase.
- Account for 100% of the activity at all times.
- Account for all major paths of excretion (urine, feces, exhalation, etc.).

Some knowledge of the expected kinetics of the pharmaceutical are needed for a good study design. For example, the spacing of the measurements and the time of the initial measurement will be greatly different if we are studying a ^{99m}Tc labeled renal agent that is 95% cleared from the body in 180 minutes or an ^{131}I labeled antibody that clears about 80% in the first day and the remaining 20% over the next two weeks. A key point that researchers can overlook is the characterization of excretion. Very often, the excretory organs (intestines, urinary bladder) are the organs that receive the highest absorbed doses, as 100% of the activity (minus decay) will eventually pass through one or both of these pathways at different rates. If excretion is not quantified, the modeler must make the assumption that the compound is retained in the body and removed only by radioactive decay. For very short-lived radionuclides, this may not be a problem and in fact may be quite accurate. For moderately long-lived nuclides, this can cause an overestimate of the dose to most body organs and an underestimate of the dose to the excretory organs, perhaps significantly.

Development of Kinetic Data

Data for kinetic studies are generally derived from one of two sources:

- Animal studies, usually performed for submission of an application for approval to the Food and Drug Administration for use of a so-called Investigational New Drug (IND)
- Human studies, usually performed in Phase I, II, or III of approval of a New Drug Application (NDA)

Animal Studies

In an animal study, the radiopharmaceutical is administered to a number of animals which are then sacrificed at different times, and the organs harvested and counted for activity (or perhaps imaged). The extrapolation of animal data to humans is not an exact science. One method of extrapolating animal data is the % kg/g method.[23] In this method, the animal organ data need to be reported as percent of injected activity per gram of tissue, and the animal whole-body weight must be known. The extrapolation to humans then uses the human organ and whole-body weight, as follows.

$$\left[\left(\frac{\%}{g_{\text{organ}}}\right)_{\text{animal}} x\,(\text{kg}_{TB \text{ weight}})_{\text{animal}}\right] x \left(\frac{g_{\text{organ}}}{\text{kg}_{TB \text{ weight}}}\right)_{\text{human}} = \left(\frac{\%}{\text{organ}}\right)_{\text{human}}$$

Human Studies

In human studies, data are collected with a nuclear medicine gamma camera. Quantification of data gathered with these cameras is rather involved. The gamma camera, also called an "Anger" camera, relies on the basic principles and design of inventor Hal Anger.[24] Briefly, the camera employs a single large sheet of scintillation material, coupled (usually via a light guide) to a group of photomultiplier tubes, with a collimator between the patient and scintillator. Collimators are typically made of lead, are about 4 to 5 cm thick and 20 by 40 cm on a side. The collimator contains thousands of square, round, or hexagonal parallel holes through which gamma rays are allowed to pass, supposedly photons coming from angles other than parallel to the camera are attenuated (but this is not strictly true: septal penetration is an important source of noise in gamma camera images). There are many designs of collimators, to handle low-, medium-, or high-energy nuclides on different systems.

Some solid-state gamma cameras are under development, but the majority in use today use scintillation technology. Because the photomultiplier tubes are large compared to the desired final image resolution, a signal-weighted average of at least several of the photomultipliers is used to estimate the actual interaction location in the scintillator. A projection image of the object under study (organs within a patient) is thus developed after many thousands or millions of events strike the crystal and are processed. As this is a projection image, the actual depth of the object within the patient is not known. Also, the image contains events caused by Compton scattered photons, photons from energies other than that of the photopeak of interest, and possibly other interferences.

The number of counts at individual points across the Field Of View (FOV) is provided directly by the gamma camera computer. The FOV is usually a square field, with 128×128, 256×256, or some other number of points (pixels) in the view. One may draw Regions Of Interest (ROIs) around images of objects that will be recognizable as internal organs or structures; the number of counts in a ROI, however, cannot be used directly to calculate how much activity is in the organ. A number of corrections are needed to the observed number of counts to obtain a reliable estimate of activity in this object.[25]

Depth Dependence

To remove uncertainties about the depth of the object, usually images are taken in front of and behind the patient, and a geometric mean of the two values is taken. This geometric mean has been shown to be relatively independent of depth for most radionuclides of interest, and thus this quantity is thought to be the most reliable for use in quantification.

Attenuation

Photons leaving the object and striking either detector will have been attenuated within the patient to some degree. Each setup will have slightly different characteristics, depending on the camera itself, the collimator employed, and the nuclide of interest. A study to establish the attenuation coefficient for each combination must be done before collecting data. Patient thickness in a region where activity is to be quantified is estimated by using a sheet, or flood, source of a particular nuclide (often ^{57}Co), with knowledge of this nuclide's attenuation characteristics in the same setup.

Scatter

Gamma cameras can collect counts in each pixel within a user-defined energy "window," which is generally set to be $\pm 15\%$ or 20% of the photopeak energy (e.g., 140 keV $\pm$ 28 keV). If the photopeak is the highest energy in the radionuclide's emission spectrum (as is true for ^{99m}Tc, but not for ^{131}I), scatter events in the photopeak window will be due to small angle scattering, which can be approximately corrected by subtracting counts from an equal width window below the photopeak area. If there are other high-energy events that may contribute to photopeak counts, another window is needed above the photopeak window to correct for these events. Other, mathematical, techniques (convolution-based) have been suggested (e.g., Floyd et al.[26]) in addition to double- or triple-energy window techniques; other authors developed a method to correct for scatter by evaluating depth-dependent buildup factors for objects of different size. After scatter correction has been applied, the activity of the source within the ROI is thus given by

$$A_{\text{ROI}} = \sqrt{\frac{I_A I_P}{e^{-\mu_e t}} \frac{f_j}{C}}$$

where I_A and I_P are the anterior and posterior counts in the region, μ_e is the effective attenuation coefficient, t is the average patient thickness over the ROI, f_j is the source self-attenuation coefficient (given as $[(\mu_e t/2)/\sinh(\mu_e t/2)]$, but which is rarely of much impact in the calculation and so is usually neglected), and C is a source calibration factor (cts/s per Bq), obtained by counting a source of known activity in air.

Thus, activity in identifiable regions of the body, such as the liver, spleen, kidneys, and so on may be determined at individual times. ROIs may also be drawn over the entire body, to track the retention and excretion of the compound in the body. Excreta samples may also be taken to study excretion pathways. If only a single excretion pathway is important, knowledge of whole-body clearance may be used to explain excretion.

Tomography

In the derivation given above, the gamma camera is assumed to stay in a fixed position with respect to the patient, with image data gathered in the anterior and posterior camera heads and then processed. The projection image thus obtained gives a two-dimensional (2D), or planar, image of the patient. This may be perfectly adequate for gathering the diagnostic information needed or for calculating dose estimates. If the camera heads are rotated around the body, with projection images taken at multiple angles around the body, 3D image information may be reconstructed from these multiple images, using techniques common to tomographic imaging in other modalities (such as CT). This 3D information may thus be displayed in slices or 3D renderings, yielding a more detailed understanding of the activity distribution.

Single Photon Emission Computed Tomography (SPECT) is applied to most radionuclides, whereas Positron Emission Tomography (PET) applies to positron-emitting nuclides. Projection images are taken at many angles around the body, typically every 3 or 6° in a 360° range. Each projection is then "back-projected" across the entire field; as areas intersect from different projections in an individual "slice" of the body in space they reinforce, giving a 2D image in that slice of the object (Figure 10.12). Direct back-projection introduces artifacts into the reconstructed images, so back-projection is usually performed by applying a filter to the data as they are back-projected.

It is convenient in dealing with the volumes of data involved in these calculations to convert information from the spatial domain into the frequency domain by taking the Fourier transform of the data. Small objects in the spatial

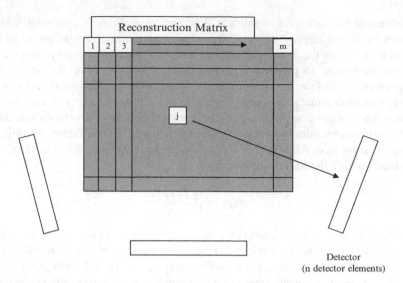

Figure 10.12 Reconstruction of image data from tomography.

domain correspond to high frequencies in the frequency domain. Data filters are used to try to eliminate as much noise and background from the images as possible without removing true information from the object. Many types of filters have been developed, and most have adjustable parameters that can be manipulated to obtain the best estimates possible of the 3D image for use in diagnosis or activity quantification.

With a filtered reconstructed image, one has still not corrected for attenuation. Every "voxel" (3D correlate of a 2D "pixel") in a given slice will have been back-projected through a different amount of tissue (of variable attenuation characteristics, perhaps including soft tissue, bone, or lung) from the original projection image pixel. One may assume an approximately uniform attenuation field for the body, or, if possible, an attenuation "map" may be defined by using a CT image of the same patient prior to SPECT imaging. By using multiple projections and studying the detected radiation at detectors on the opposite side of the patient, attenuation characteristics of each voxel in the to-be reconstructed image field can be defined. Then, by successive iterative approximations, one may attempt to obtain an optimized 3D image that has been corrected for attenuation (scatter may also have been removed by methods similar to that used in the planar approach above). This is the approach taken with the Maximum Likelihood Expectation Maximization (MLEM) or Ordered Sets Expectation Maximization (OSEM) reconstruction algorithm,[27] which has a coupled CT/SPECT camera system so that all the needed data can be obtained without moving the patient and having to deal with issues of image registration.

Figure 10.12 illustrates the idea of use of an iterative reconstruction algorithm for attenuation correction. Every voxel j in the reconstruction matrix ($m \times m$) has an attenuation factor for a given projection angle and detector element (1 through n), depending on the number of elements between them, the cross-section of each element in the projection, and the material composition of each element.

Even with the best possible images, however, the edges of the object image will not perfectly correspond to those of the true object, because of the inherent resolution limits of the gamma camera system (typically of the order of 10s of mm currently for most nuclides and collimators). Studies must be undertaken with objects of known mass and activity, in scattering media with and without background activity, to evaluate volume and activity recovery coefficients for a range of object sizes and activity contrast values ((source − bkgd)/source) anticipated in practice. Patient motion can also interfere with the quality of image data.

Analysis of Kinetic Data

Let us now assume that we have gathered a series of measurements that represent either retention and/or excretion. Our task is to interpret these measurements in such a way as to derive a workable kinetic model that can be used to estimate $\tilde{A}$ for each significant source region in our system. In general, there are three levels of complexity that our analysis can take.

Direct Integration

One can directly integrate under the actual measured values by a number of methods. This does not give very much information about your system, but it does allow you to calculate τ rather easily. The most common method used is the trapezoidal method, simply approximating the area by a series of trapezoids.

Least Squares Analysis

An alternative to brute-force integration of the data is to attempt to fit curves of a given shape to the data. The curves are represented by mathematical expressions that can be directly integrated. The most common approach is to attempt to characterize a set of data by a series of exponential terms, as many systems are well represented by this form, and exponential terms are easy to integrate. In general, the approach is to minimize the sum of the squared distance of the data points from the fitted curve. The curve will have the form:

$$A(t) = a_1 e^{-b_1 t} + a_2 e^{-b_2 t} + \cdots$$

The method looks at the squared difference between each point and the solution of the fitted curve at that point, and minimizes this quantity by taking the partial derivative of this expression with respect to each of the unknowns a_i and b_i and setting it equal to zero. Once the ideal estimates of a_i and b_i are obtained, the integral of $A(t)$ from zero to infinity is simply:

$$\int_0^\infty A(t)dt = \frac{a_1}{b_1} + \frac{a_2}{b_2} + \cdots$$

If the coefficients a_i are in units of activity, this integral represents cumulated activity (the units of the b_i are time^{-1}).

Compartmental Models

The situation frequently arises wherein you either know quite a bit about the biological system under investigation or you would like to know in greater detail how this system is working. In this case, you can describe the system as a group of compartments linked through transfer rate coefficients. Solving for $\tilde{A}$ of the various compartments involves solving a system of coupled differential equations describing transfer of the tracer between compartments and elimination from the system. The solution to the time activity curve for each compartment will usually be a sum of exponentials, but not obtained by least squares fitting each compartment separately, but by varying the transfer rate coefficients between compartments until the data are well fit by the model.

Dose Calculations

Dose calculations in the MIRD system are completely identical to those in the ICRP 30 system, we just use different symbols:

$$U_S \Rightarrow \tilde{A}$$

$$SEE \Rightarrow S$$

$$D_1 = \tilde{A}_1 S(1 \leftarrow 1) + \tilde{A}_2 S(1 \leftarrow 2)$$

$$D_2 = \tilde{A}_1 S(2 \leftarrow 1) + \tilde{A}_2 S(2 \leftarrow 2)$$

$$D_3 = \tilde{A}_1 S(3 \leftarrow 1) + \tilde{A}_2 S(3 \leftarrow 2)$$

Looking up these values in tables can make the calculation of internal doses a not-too-painful exercise. Software programs have shortcut the process even further.

Example: First Principles

ICRP Report #23 describes Reference Man as containing 140 grams of potassium. Estimate the average beta dose rate in rad/week to the whole body, given the following information.

Reference Man	70,000 g
^{40}K in potassium	0.012%
^{40}K beta decay probability	90%
^{40}K maximum beta energy	1.3 MeV
^{40}K half-life	1.2×10^9 years
Avogadro's number	6.025×10^{23}
Energy conversion	1.6021×10^{-6} erg/MeV
Dose conversion	100 erg/g = 1 rad

We use no particular system to evaluate this dose rate, just the principle that absorbed dose is energy absorbed per unit mass. Here we have a uniform distribution of radioactivity in 70,000 g of material, and we just want to estimate the beta dose rate. First, we need to find how much activity we have, which is calculated from the number of grams, from the well-known relationship:

$$A = \lambda N$$

In this problem:

$$A = \left(\frac{0.693}{1.2x \ 10^9 \ \text{yr}} \right) \left(\frac{1\text{yr}}{\pi x \ 10^7 \ \text{sec}} \right) (140 \ \text{g K}) \left(\frac{1.2x \ 10^{-4} \ \text{g K} - 40}{\text{g K}} \right)$$

$$\times \left(\frac{\text{mol K} - 40}{40 \ \text{g K} - 40} \right) \left(\frac{6.025x \ 10^{23} \ \text{atoms}}{\text{mol}} \right)$$

$$A = 4.65x \ 10^3 \ \text{atoms/sec} = 4.65x \ 10^3 \ \text{dis/sec}$$

We can directly calculate the dose rate in the object of 70,000 g now, because we know the activity. Dose rate is the product of the activity, the average energy of decay, the abundance of decay, the absorbed fraction, and the factor k (which just converts units), divided by the mass. There is a small trick in this problem, in that I gave you the maximum beta energy for ^{40}K; you must use the average energy to get the average dose rate. If you can look this up in a decay data book it would be better; here we just use the rule of thumb that the average beta energy is one-third the maximum.

$$D = \left(\frac{4.65x10^3 \text{ dis}}{\text{sec}}\right)\left(\frac{6.048x10^5 \text{ sec}}{\text{week}}\right)\left(\frac{1.3 \text{ MeV}}{3\beta}\right)\left(\frac{0.9\beta}{\text{dis}}\right)$$

$$\times \left(\frac{1.602x10^{-6} \text{ erg}}{\text{MeV}}\right)\left(\frac{\text{g} - \text{rad}}{100 \text{ erg}}\right)\left(\frac{1}{70000 \text{ g}}\right)$$

$$D = 2.5x \, 10^{-4} \text{ rad/week}$$

In this problem, k is the product of the number of seconds in a week, the erg/MeV over the factor of 100 erg/g-rad. The first term in the dose rate equation is the activity we calculated above, the third term is the average energy per decay (1.3/3), and the next term (0.9) is the abundance. The absorbed fraction is 1.0, and is given in the same term as the mass (70,000 g). This is a pretty accurate analysis of the beta dose rate that you get every week from the natural ^{40}K in your body. Of course there might be a bit less or a bit more activity in your body, you might weigh a bit less or more than 70 kg, and so on. And if you wanted to study the total dose rate, you would need to evaluate the gamma component, too, the methods for which we look at in other examples.

Example: Contributions to Organ Dose

Calculate the dose to liver, spleen, and lungs from ^{90}Y activity in the liver and spleen as described below. To understand the contributions to the total dose to any organ, consider separately:

1. Dose from nonpenetrating radiations (nonpenetrating self-dose)
2. Dose from penetrating radiations when source and target are the same organ (penetrating self-dose)
3. Dose from penetrating radiations when source and target are not the same organ (cross-irradiation dose)

Organ	Mass (kg)	$\tilde{A}$ (μCi·hr)
Liver	1.91	2000
Spleen	0.183	2000
Lungs	1.00	0

For ^{90}Y:

$\Delta_p = 0.000$ g-rad/μCi-hr
$\Delta_{np} = 1.99$ g-rad/μCi-hr
(p = penetrating, np = nonpenetrating)

Target (r_T)	$\phi(r_T$ <- Liver)		$\phi(r_T$ <- Spleen)	
	p	np	p	np
Liver	NA	1	NA	0
Spleen	NA	0	NA	1
Lungs	NA	0	NA	0

Yttrium-90 has no penetrating component. Absorbed fractions for any organ for nonpenetrating emissions are one and for any other organ are zero. The organ doses, in rad, are

| Target Organ | Source Organ | | | | |
| | Liver | | Spleen | | |
	p	np	p	np	Total
Liver	NA	2.1	NA	0	2.1
Spleen	NA	0	NA	22	22
Lungs	NA	0	NA	0	0.0

The liver dose is (2000 μCi·hr $\times$ 1.99 g-rad/μCi·hr)/1910 g = 2.1 rad.
The spleen dose is (2000 μCi·hr $\times$ 1.99 g-rad/μCi·hr)/183 g = 22 rad.
The lung dose is of course zero, as the lung had no assigned activity, and activity in liver and spleen were assumed not to irradiate lung.

Teaching Points
1. Liver and spleen had equal cumulated activities. Because spleen is about 10× smaller, the dose is 10× higher (exactly true for beta emitters, results will vary for gamma emitters, but the trend will be the same).
2. Units, units, units. The mass was given in kg; we needed to convert it to g. If we didn't, our answers were off by a factor of 1000, and our calculations would have wrongly indicated that the patient may have received an extremely high dose.

Example: Contributions to Organ Dose
Now, repeat the same problem for ^{99m}Tc.
Complete the data tables below to calculate the doses. Consider 140 keV to be the only penetrating decay energy.

Organ	Mass (kg)	$\tilde{A}$ (μCi·hr)
Liver	1.91	2000
Spleen	0.183	2000
Lungs	1.00	0

For ^{99m}Tc:

$\Delta_p = 0.267$ g-rad/μCi-hr
$\Delta_{np} = 0.0342$ g-rad/μCi-hr
(p = penetrating, np = nonpenetrating)

	Absorbed Fractions*			
	$\phi(r_T \leftarrow \text{liver})$		$\phi(r_T \leftarrow \text{spleen})$	
Target Organ (r_T)	p	np	p	np
Liver	0.162	1	0.0071	0
Spleen	0.0062	0	0.072	1
Lungs	0.0098	0	0.0085	0

* Absorbed fractions from reference 9

	Source Organ				
	Liver		Spleen		
Target Organ	p	np	p	np	Total
Liver	0.045	0.036	0.002	0	0.083
Spleen	0.0018	0	0.21	0.37	0.59
Lungs	0.0052	0	0.0045	0	0.0097

Teaching Points

1. Again, with equal cumulated activities in liver and spleen, roughly a tenfold difference in dose, due to the mass difference.
2. Now lung gets a dose, even with no assigned activity. This is due to penetrating dose from the liver and spleen, that is, photons that originate in these organs but deposit dose in the lungs.

Example: Calculation of S-Value for Average Organ Dose

We can calculate an S-value for liver self-irradiation from ^{99m}Tc by combining the appropriate decay data with calculated absorbed fractions. Table 10.3 shows the decay scheme for ^{99m}Tc, from the RADAR Web site.[28]

At first glance there appear to be a considerable number of emissions to consider. However, for our purposes, we can consider ^{99m}Tc to have only five emissions: one γ ray, three X-rays, and a group of nonpenetrating emissions. We can group the nonpenetrating emissions together because they are all multiplied by the same absorbed fraction (1.0), and so in the sum $\Sigma n_i E_i \phi_i$, we may sum the $n_i E_i$ and multiply the whole sum by $\phi = 1.0$. To calculate the S-value for liver irradiating itself, then, we need only to look up the appropriate absorbed fractions for the penetrating emissions,[8] and sum over all emissions:

Emission	n	E	$k\Sigma n_i E_i$	ϕ	$k\Sigma n_i E_i\, \phi_i$
$\gamma 2$:	0.891	0.1405	0.267	0.162	0.0432
Kα1 x-ray	0.04	0.0184	0.0016	0.82	0.0013
Kα2 x-ray	0.021	0.0182	0.0008	0.82	0.00066
Kα1 x-ray	0.0068	0.0206	0.0003	0.78	0.00023
Nonpenetrating	—	—	0.0343	1.0	0.0343
				TOTAL	= 0.080

Table 10.3 99m-Tc-43 decay mode: IT half-life: 6.01 H.

Emission Type	Mean Energy (MeV)	Frequency
ce-M e_	0.0016	0.7460
Auger-L e_	0.0022	0.1020
Auger-K e_	0.0155	0.0207
ce-K e_	0.1195	0.0880
ce-K e_	0.1216	0.0055
ce-L e_	0.1375	0.0107
ce-L e_	0.1396	0.0017
ce-M e_	0.1400	0.0019
ce-N+ e_	0.1404	0.0004
ce-M e_	0.1421	0.0003
L X-ray	0.0024	0.0048
$K\alpha 1$ X-ray	0.0183	0.0210
$K\alpha 2$ X-ray	0.0184	0.0402
$K\beta$ X-ray	0.0206	0.0120
γ	0.1405	0.8906
γ	0.1426	0.0002

We set k equal to 2.13, which causes the units on the third and fifth columns to be g-rad/μCi-hr, given the energy in MeV. The S-value is simply the sum of the values in the fifth column divided by the mass of the liver, 1800 g:

$$S(\text{liver} \leftarrow \text{liver}) = 0.080/1800\,\text{g} = 4.4 \times 10^{-5}\,\text{rad}/\mu\text{Ci-hr}$$

Example: Dose to One Organ

Data extrapolated from an animal study yield the following parameters for a new compound tagged to ^{99m}Tc.

$$\text{Liver} \quad f_1 = 0.3 \quad T_{e1} = 0.5\,\text{hours}$$
$$f_2 = 0.1 \quad T_{e2} = 5.5\,\text{hours}$$
$$\text{Kidneys} \quad f = 0.2 \quad T_e = 1.2\,\text{hours}$$

where f is the fraction of injected activity (note that only 60% of the injected activity is accounted for by considering only these two organs). Let's calculate the dose to the liver. If this were a real problem, we would calculate dose to the liver, kidneys, gonads, red marrow, and perhaps a few other organs. We find the following S-values in MIRD 11[15].

$$S(\text{liver} \leftarrow \text{liver}) = 4.6 \times 10^5\,\text{rad}/\mu\text{Ci-hr}$$
$$S(\text{liver} \leftarrow \text{kidneys}) = 3.9 \times 10^6\,\text{rad}/\mu\text{Ci-hr}$$

(The liver to liver S-value is slightly different than we had calculated, as MIRD 11 used slightly different decay data.) Assume $A_0 = 1\,\text{mCi} = 1000\,\mu\text{Ci}$; then,

$$\tilde{A}(\text{liver}) = 1.443 \cdot 1000\,\mu\text{Ci}\,(0.3 \cdot 0.5\,\text{hr} + 0.1 \cdot 5.5\,\text{hr}) = 1010\,\mu\text{Ci-hr}$$

$$\tilde{A}(\text{kidneys}) = 1.443 \cdot 1000\,\mu\text{Ci} \cdot 0.2 \cdot 1.2\,\text{hr} = 350\,\mu\text{Ci-hr}$$

$$D(\text{liver}) = 1010\,\mu\text{Ci-hr} \cdot 4.6 \times 10^{-5}\,\text{rad}/\mu\text{Ci-hr} + 350\,\mu\text{Ci-hr} \cdot 3.9$$
$$\times 10^{-6}\,\text{rad}/\mu\text{Ci-hr}$$

$$D(\text{liver}) = 0.0465\,\text{rad} + 0.0014\,\text{rad} = 0.048\,\text{rad}.$$

Note that the liver contributes 97% of its total dose. Dividing by the injected activity, the dose, given these input assumptions, is 0.048 rad/mCi. So, if we redesigned the study to use 3 mCi, the liver absorbed dose would be 3 mCi × 0.048 rad/mCi = 0.14 rad.

Example

1 MBq of ^{99m}Tc is given to a patient: 40% goes to the liver and has a 10 hr biological half-time; the other 60% goes to the spleen and has an infinite biological half-time (never leaves).

$$A_{\text{liver}} = 4 \times 10^5 \text{Bq} \times 1.443 \times \frac{6 \times 10h}{6 + 10} x \frac{3600s}{h}$$

$$A_{\text{liver}} = 7.79 \times 10^9 \text{Bq} - s$$

$$A_{spleen} = 6 \times 10^5 \text{Bq} \times 1.443 \times 6h \times \frac{3600s}{h}$$

$$A_{spleen} = 1.87 \times 10^{10} \text{Bq} - s$$

$$S_{\text{liver}\leftarrow\text{liver}} = 4.1x10^{-5} \text{ rad}/\mu\text{Ci} - \text{hr} = 3.08x10^{-6} \text{ mGy/MBq} - s$$

$$S_{\text{spleen}\leftarrow\text{spleen}} = 3.1x10^{-4} \text{ rad}/\mu\text{Ci} - \text{hr} = 2.33x10^{-5} \text{ mGy/MBq} - s$$

$$S_{\text{spleen}\leftarrow\text{liver}} = 9.6x10^{-7} \text{ rad}/\mu\text{Ci} - \text{hr} = 7.2x10^{-8} \text{ mGy/MBq} - s$$

$$S_{\text{liver}\leftarrow\text{spleen}} = 9.6x10^{-7} \text{ rad}/\mu\text{Ci} - \text{hr} = 7.2x10^{-8} \text{ mGy/MBq} - s$$

$$D_{\text{liver}} = 7.79x10^3 \text{ MBq} - s \times 3.08x10^{-6} \text{ mGy/MBq} - s$$
$$+ 1.87x10^4 \text{ MBq} - s \times 7.2x10^{-8} \text{ mGy/MBq} - s = 0.025 \text{ mGy}$$

$$D_{spleen} = 7.79x10^3 \text{ MBq} - s \times 7.2x10^{-8} \text{ mGy/MBq} - s$$
$$+ 1.87x10^4 \text{ MBq} - s \times 2.33x10^{-5} \text{ mGy/MBq} - s = 0.43 \text{ mGy}$$

Example: Dose to the Fetus

MIRD Dose Estimate Report No. 13[29] gives the following numbers of disintegrations for intravenous administration of ^{99m}Tc MDP.

Cortical bone	1.36 μCi-hr/μCi administered
Cancellous bone	1.36 μCi-hr/μCi administered
Kidneys	0.148 μCi-hr/μCi administered
Urinary bladder	0.782 μCi-hr/μCi administered
Remainder of body	1.64 μCi-hr/μCi administered

If 17 mCi of ^{99m}Tc-MDP has been given to a woman who is two weeks pregnant, what is the likely absorbed dose to the fetus? In early pregnancy, the dose to the nongravid uterus is a reasonably good estimate of the fetal dose, because the size and shape of the uterus relative to other organs has not changed substantially. Therefore, we can use S-values for these source organs irradiating the uterus:

S(Uterus ← Cortical bone)		= 5.7 × 10^{-7} rad/μCi-hr
S(Uterus ← Cancellous bone)		= 5.7 × 10^{-7} rad/μCi-hr
S(Uterus ← Kidneys)		= 9.4 × 10^{-7} rad/μCi-hr

$S(\text{Uterus} \leftarrow \text{Urinary bladder}) \quad = 1.6 \times 10^{-5} \text{ rad}/\mu\text{Ci-hr}$
$S(\text{Uterus} \leftarrow \text{Total body}) \quad = 2.6 \times 10^{-6} \text{ rad}/\mu\text{Ci-hr}$

The last S-value is not exactly what we need. It is the S-value for an organ being irradiated by activity uniformly distributed in the whole body (i.e., including bone, kidneys, etc.). The formula for calculating the S-value for remainder of the body for a given configuration of other source organs is:[30]

$$S(r_k \leftarrow RB) = S(r_k \leftarrow TB)\left(\frac{m_{TB}}{m_{RB}}\right) - \sum_h S(r_k \leftarrow r_h)\left(\frac{m_h}{m_{RB}}\right)$$

where:

$S(r_k \leftarrow RB)$ is the S-value for the remainder of the body irradiating target region r_k
$S(r_k \leftarrow TB)$ is the S-value for the total body irradiating target region r_k
$S(r_k \leftarrow r_h)$ is the S-value for source region h irradiating target region r_k
m_{TB} is the mass of the total body
m_{RB} is the mass of the remainder of the body, that is, the total body minus all other source organs used in this problem
m_h is the mass of source region h

For this problem, the S-value for remainder of the body to uterus is 2.7×10^{-6} rad/μCi-hr (4% higher than that for the total body). The total dose to the uterus is calculated as

1.36 μCi-hr/μCi administered $\times$ 5.7 $\times$ 10^{-7} rad/μCi-hr = 7.8 $\times$ 10^{-7} rad/μCi

1.36 μCi-hr/μCi administered $\times$ 5.7 $\times$ 10^{-7} rad/μCi-hr = 7.8 $\times$ 10^{-7} rad/μCi

0.148 μCi-hr/μCi administered $\times$ 9.4 $\times$ 10^{-7} rad/μCi-hr = 1.4 $\times$ 10^{-7} rad/μCi

0.782 μCi-hr/μCi administered $\times$ 1.6 $\times$ 10^{-5} rad/μCi-hr = 1.2 $\times$ 10^{-5} rad/μCi

1.64 μCi-hr/μCi administered $\times$ 2.7 $\times$ 10^{-6} rad/μCi-hr = 4.4 $\times$ 10^{-6} rad/μCi

TOTAL = 1.8 $\times$ 10^{-5} rad/μCi

Total dose from incident = 1.8 $\times$ 10^{-5} rad/μCi $\times$ 17,000 μCi = 0.30 rad.

It would probably be more accurate to use the 57 kg model for the adult female[11] instead of the 70 kg adult male model to calculate this estimate. Using S-values for the adult female, a dose of 2.3 $\times$ 10^{-5} rad/μCi is estimated, leading to an estimate of the total dose of 0.39 rad.

Example: Dose to Several Organs
In MIRD Dose Estimate Report No. 12[31] the following residence times are found for intravenous administration of ^{99m}Tc DTPA.

Kidneys   0.092 μCi-hr/μCi administered
Urinary bladder   0.84 μCi-hr/μCi administered (2.4 hr voiding intervals)
                               1.72 μCi-hr/μCi administered (4.8 hr voiding intervals)
Remainder of body   2.84 μCi-hr/μCi administered

Let's calculate the absorbed dose to these organs and to the ovaries, testes, and red marrow. For each target organ, then, we will need all of the S-values for the three source organs. We also have two conditions to the problem: 2.4 hour and 4.8 hour voiding intervals for the urinary bladder. As in the previous example, we will have three contributions to each target organ's total dose for each group of cumulated activity values. An easy way to represent what proves to be a rather substantial amount of math for a simple problem is through the use of matrices. If the set of dose estimates we want is a 2×6 matrix (two sets of dose estimates by six target organs: kidneys, bladder, ovaries, testes, red marrow, and total body), this can be found by multiplication of a 2×3 matrix of cumulated activity values and a 3×6 matrix of S-values:

$$D = \tau S$$

$$D = \begin{bmatrix} \tilde{A}_{\text{kid}_1} & \tilde{A}_{\text{blad}_1} & \tilde{A}_{\text{RB}_1} \\ \tilde{A}_{\text{kid}_2} & \tilde{A}_{\text{blad}_2} & \tilde{A}_{\text{RB}_2} \end{bmatrix}$$

$$\times \begin{bmatrix} S(\text{kid} \leftarrow \text{kid}) & S(\text{ov} \leftarrow \text{kid}) & S(\text{mar} \leftarrow \text{kid}) & S(\text{test} \leftarrow \text{kid}) & S(\text{blad} \leftarrow \text{kid}) & S(\text{TB} \leftarrow \text{kid}) \\ S(\text{kid} \leftarrow \text{blad}) & S(\text{ov} \leftarrow \text{blad}) & S(\text{mar} \leftarrow \text{blad}) & S(\text{test} \leftarrow \text{blad}) & S(\text{blad} \leftarrow \text{blad}) & S(\text{TB} \leftarrow \text{blad}) \\ S(\text{kid} \leftarrow \text{RB}) & S(\text{ov} \leftarrow \text{RB}) & S(\text{mar} \leftarrow \text{RB}) & S(\text{test} \leftarrow \text{RB}) & S(\text{blad} \leftarrow \text{RB}) & S(\text{TB} \leftarrow \text{B}) \end{bmatrix}$$

$$D = \begin{bmatrix} 0.092 & 0.842 & 2.84 \\ 0.092 & 1.72 & 2.84 \end{bmatrix}$$

$$\times \begin{bmatrix} 1.9x10^{-4} & 1.1x10^{-6} & 3.8x10^{-6} & 8.8x10^{-8} & 2.8x10^{-7} & 2.1x10^{-6} \\ 2.6x10^{-7} & 7.3x10^{-6} & 1.6x10^{-6} & 4.7x10^{-6} & 1.6x10^{-4} & 1.9x10^{-6} \\ 1.4x10^{-6} & 2.4x10^{-6} & 2.9x10^{-6} & 1.7x10^{-6} & 1.9x10^{-6} & 2.0x10^{-6} \end{bmatrix}$$

$$D = \begin{bmatrix} D_{\text{kid}_1} & D_{\text{ov}_1} & D_{\text{mar}_1} & D_{\text{test}_1} & D_{\text{blad}_1} & D_{\text{TB}_1} \\ D_{\text{kid}_2} & D_{\text{ov}_2} & D_{\text{mar}_2} & D_{\text{test}_2} & D_{\text{blad}_2} & D_{\text{TB}_2} \end{bmatrix}$$

$$D = \begin{bmatrix} 2.2x\,10^{-5} & 1.3x\,10^{-5} & 1.0x\,10^{-5} & 8.8x\,10^{-6} & 1.4x\,10^{-4} & 7.5x\,10^{-6} \\ 2.2x\,10^{-5} & 1.9x\,10^{-5} & 1.2x\,10^{-5} & 1.3x\,10^{-5} & 2.8x\,10^{-4} & 9.1x\,10^{-6} \end{bmatrix}$$

(units are rad/mCi)

Note from the results the increase in absorbed dose to the bladder, as well as to the gonads, from the increase in the number of disintegrations occurring in the bladder.

Endnotes

1. Marinelli L, Quimby E, Hine G. (1948) Dosage determination with radioactive isotopes II, practical considerations in therapy and protection, *Am J Roent Radium Ther* 59:260–280.
2. Quimby E and Feitelberg S (1963) *Radioactive isotopes in medicine and biology*, Philadelphia, Lea and Febiger.
3. Hine G, Brownell G. (1956) *Radiation dosimetry*, New York, Academic Press.
4. International Commission on Radiological Protection. (1960) Report of committee II on permissible dose for internal radiation, *Health Phys.* vol 3.
5. International Commission on Radiological Protection. (1979) Limits for Intakes of Radionuclides by Workers. ICRP Publication 30, Pergamon Press, New York.
6. Loevinger R, Budinger T, Watson E: MIRD Primer for Absorbed Dose Calculations, Society of Nuclear Medicine, 1988.
7. M.G. Stabin and J. A. Siegel. Physical Models and Dose Factors for Use in Internal Dose Assessment. Health Physics, 85(3):294–310, 2003.

8. M. G. Stabin, R.B. Sparks, E. Crowe. OLINDA/EXM: The Second-Generation Personal Computer Software for Internal Dose Assessment in Nuclear Medicine. J Nucl Med 2005;46 1023–1027.

9. http://www.doseinfo-radar.com

10. Eve, IS. (1966) A review of the physiology of the gastrointestinal tract in relation to radiation doses from radioactive materials. Health Phys. 12:131–161.

11. International Commission on Radiological Protection. Human Respiratory Tract Model for Radiological Protection. ICRP Publication 66. International Commission on Radiological Protection, Pergamon Press, New York, 1995.

12. International Commission on Radiological Protection. (1995) Age-dependent doses to members of the public from intakes of radionuclides: Part 3. Ingestion dose coefficients. ICRP Publication 69, Pergamon Press, New York.

13. Snyder W, Ford M, Warner G, Fisher H, Jr. (1969) MIRD Pamphlet No 5 - Estimates of Absorbed Fractions for Monoenergetic Photon Sources Uniformly Distributed in Various Organs of a Heterogeneous Phantom. J Nucl Med Suppl No 3, 5.

14. Snyder W, Ford M, Warner G. (1978) MIRD Pamphlet No 5, Revised - Estimates of Specific Absorbed Fractions for Photon Sources Uniformly Distributed in Various Organs of a Heterogeneous Phantom. Society of Nuclear Medicine, New York.

15. Snyder W, Ford M, Warner G, Watson S. (1975) "S," absorbed dose per unit cumulated activity for selected radionuclides and organs, MIRD Pamphlet No. 11, Society of Nuclear Medicine, New York, NY.

16. Cristy M. and Eckerman K. (1987) Specific absorbed fractions of energy at various ages from internal photons sources. ORNL/TM-8381 V1-V7. Oak Ridge National Laboratory, Oak Ridge, TN.

17. Stabin M, Watson E, Cristy M, Ryman J, Eckerman K, Davis J, Marshall D., Gehlen K. (1995) Mathematical models and specific absorbed fractions of photon energy in the nonpregnant adult female and at the end of each trimester of pregnancy. ORNL Report ORNL/TM-12907; Oak Ridge National Laboratory, Oak Ridge, TN, USA.

18. Eckerman KF, Cristy M, and Warner GG. (1981) Dosimetric Evaluation of Brain Scanning Agents; In Third International Radiopharmaceutical Dosimetry Symposium; eds. E.E. Watson, A.T. Schlafke-Stelson, J.L. Coffey, and R.J. Cloutier; HHS Publication FDA 81–8166, U.S. Dept. of Health and Human Services, Food and Drug Administration, Rockville, MD; pp 527–540.

19. Bouchet L, Bolch W, Weber D, Atkins H, Poston J, Sr. (1999) MIRD Pamphlet No 15: Radionuclide S values in a revised dosimetric model of the adult head and brain. J Nucl Med 40:62S–101S.

20. Holman BL, Zimmerman RL, Shapiro JR, Kaplan ML, Jones AG, Hill TC. (1983) Biodistribution and dosimetry of n-isoproyl p-123I iodoamphetamine in the primate. J Nucl Med 24:922–931.

21. Watson EE, Stabin MG, Davis JL, and Eckerman KF. (1989) A Model of the Peritoneal Cavity for Use in Internal Dosimetry. J Nucl Med 30:2002–2011.

22. Stabin MG. (1994) A Model of the Prostate Gland for Use in Internal Dosimetry. J Nucl Med 35(3):516–520.

23. Kirschner, A; Ice, R.; Beierwaltes, W. Radiation dosimetry of [131]I-19-iodocholesterol: the pitfalls of using tissue concentration data, the author's reply. J Nucl Med 16(3):248–249, 1975.

24. Anger HO. Scintillation camera. (1958) Rev Sci Instrum 29:27–33.

25. Siegel J, Thomas S, Stubbs J et al. (1999) MIRD Pamphlet No 16 – Techniques for Quantitative Radiopharmaceutical Biodistribution Data Acquisition and Analysis for Use in Human Radiation Dose Estimates. J Nucl Med 40:37S–61S.

26. Floyd CE, Jr., Jaszczak RJ, Greer KL, Coleman RE. (1985) Deconvolution of Compton scatter in SPECT. J Nucl Med 26:403–408.

27. Rosenthal MS, Cullom J, Hawkins W, Moore SC, Tsui BMW, Yester M. (1995) Quantitative SPECT imaging: a review and recommendations by the focus

committee of the Society of Nuclear Medicine Computer and Instrumentation Council. J Nucl Med 36:1489–1513.

28. MG Stabin, CQPL da Luz. New Decay Data For Internal and External Dose Assessment, Health Phys. 83(4):471–475, 2002.

29. D. Weber et al. MIRD Dose Estimate Report No 13: Radiation Absorbed Dose from ^{99m}Tc Labeled Bone Agents. *J Nucl Med* 30(6):1117–1122, 1989.

30. Cloutier, R, Watson, E, Rohrer, R, Smith, E: Calculating the radiation dose to an organ, *J Nucl Med* 14(1):53–55, 1973.

31. S. Thomas et al. MIRD Dose Estimate No. 12 - Radiation Absorbed Dose from ^{99m}Tc Diethylenetriaminepentaacetic Acid (DTPA). *J Nucl Med* 25:503–505, 1984.

11

Radiation Protection Practice/Evaluation

11.1 Introduction

The radiation protection program at any institution must not be a static entity, but one that is continually being scrutinized and evaluated to introduce continuous quality improvements. The healthiest programs will have a chief Radiation Safety Officer (RSO), perhaps an assistant RSO, and a number of other professional HPs and HP technicians, and, in addition, an oversight board comprised of a number of persons with HP and other scientific expertise. This board should meet periodically (e.g., monthly or quarterly) to receive a report composed by the RSO over the period's activities, incidents, and program changes.

The program can always be improved, with reductions in personnel doses, elimination of conditions that lead to incidents, record losses, and so on, and better and more detailed procedures. The goal is not to bury people in a morass of paperwork of course, but improvements generally tend to lead to more, rather than less, formalized procedures, so the amount of material used in the radiation protection evaluation will tend to increase. On the other hand, some of the best ways to improve radiation safety in practice may involve methods of streamlining and simplifying procedures (so that workers spend less time exposed to higher amounts of radiation and radioactivity). Particularly in high potential dose situations, mock drills should be conducted frequently to optimize the efficiency of individuals and working groups so that when actually handling hot sources or working in high dose rate environments, the work goes as smoothly as possible.

Routine logging of the number of incidents (minor overexposures, spills, etc.) and understanding of the causes will be helpful in reducing the number of such incidents over time and educating the workforce in methods to avoid such incidents. Plotting of variables such as collective dose, in various forms (see Figure 11.1, from Strom et al.[1]), can be useful in the staff's vigilance of the overall function of the program. Assuming that all employees are comfortably under regulatory limits on a regular basis, the next consideration is how effective the organization's ALARA program is functioning. If groups of workers performing similar functions have very different exposure histories, discussions with the higher-dose workers may bring about changes in their

Trends in Collective Dose per Unit Work

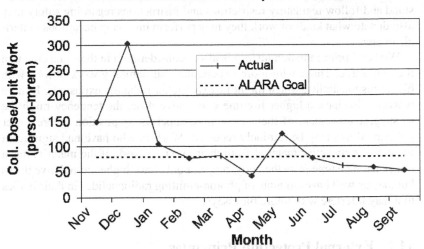

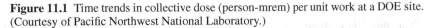

Figure 11.1 Time trends in collective dose (person-mrem) per unit work at a DOE site. (Courtesy of Pacific Northwest National Laboratory.)

work habits that can lower their doses while still getting the same work done. If these discussions bring to light real differences in approaches that have a justifiable basis, then perhaps no changes are needed or possible. But in almost all cases, doses can be reduced in many cases through continued analysis and improvement.

Worker Surveillance

All employers have a general responsibility for the surveillance of the health of their workers. This responsibility encompasses all aspects of a person's health and well being related to their employment, including all hazards, physical, chemical, biological, and radiological; their general physical health, mental and emotional well being; and so on.

Radiation protection is the specialized focus on the protection of workers from the hazards of radiation. The radiation protection professional must have a broader vision, however, regarding worker safety than just the technical aspects of radiation protection science. For example, a situation may indicate that respiratory protection is indicated for a given job. If an employee has a health condition that might interfere with her ability to safely wear a respirator (heart or lung condition, etc.), then the job might be done without the use of respiratory protection if possible or perhaps transferred to another worker. Workers that need to do hazardous jobs in hot, dusty, or otherwise stressful environments should be monitored for radiation, certainly, but also for their tolerance of the conditions of the environment. The "stay time" (time that a worker may work in a high-radiation environment, based on permissible or desired doses to be received) may not be a suitable indicator of the real time that any individual worker may be able to tolerate in a given situation.

Workers may have other conditions that prevent them from wearing respiratory protection. For example, breathing difficulties that would make such use hazardous, or if the shape of a worker's face or the presence of facial hair may lead to leakage around a partial face respirator, the use of such respiratory protection might be contraindicated. Workers' ability and willingness to understand and follow regulatory restrictions and instructions regarding safety may also dictate what kinds of work they may perform involving hazardous materials or areas.

Workers' prior radiation history is also a consideration in their placement in the workforce. Those with more experience will naturally work at jobs with higher responsibility, which might involve higher radiation exposures. If these workers also have a higher lifetime cumulative dose, the tendency might be to suggest movement of them to a more supervisory position where lower radiation doses may be routinely received. Workers who have had significant intakes of radioactive material from their previous work, if the material has a long clearance time from the body, may not only have higher cumulative doses but may as well have amounts of photon-emitting radionuclides in their bodies that may interfere with future bioassays.

11.2   External Protection Principles

The three key strategies to the protection of workers against exposure to external sources of radiation are:

- Time
- Distance
- Shielding

This is often even indicated by the acronym TDS.

Time
This a rather obvious and easily applied principle. Cumulative exposure is given by the time integral of the exposure rate. If this rate is constant, cumulative exposure is simply the exposure rate multiplied by the exposure time. If the rate varies, then the integral is more involved, but usually is calculated as just a simple sum of doses accumulated across fixed intervals. It is obvious that simply reducing the time near radiation sources will generally reduce dose in a linear fashion. Thus, the concept of "stay times" for high dose rate area jobs is implemented via calculation of the desired dose and knowledge of the expected dose rate. Workers should work quickly and efficiently in such environments, but of course not rush their work such that it is poorly done and needs to be done over at another time (with more dose being received by them or other workers).

Distance
Distance relationships are also relatively easy to understand. Referring to Chapter 7 on external dose assessment principles, we recall that for a point source, the dose drops off with the square of the distance from the source. So, doubling one's distance from a point source reduces the dose rate by a factor of four, not two. Trebling the distance changes the dose rate by a factor of nine, and so on. For extended sources, the dropoff is more linear with distance. Keeping one's distance from a high-intensity source and limiting one's time near it are instinctive measures that a worker will naturally apply. Knowledge of the exact relationships is also helpful. Shielding of sources

requires considerably more attention and development; we turn our attention to this third principle now, for photon, electron, and neutron sources.

11.3 Shielding of Photon Sources

In Chapter 4, we learned that under "good geometry" situations, the intensity of a beam of photon radiation passing through an attenuator of thickness t can be given as

$$I = I_0 \, e^{-\mu t}.$$

Here μ is the linear attenuation coefficient for photons of the specified energy in the attenuator material. The "good geometry" condition is rarely encountered. The conditions for this are that we have a well-collimated beam of photons passing through the attenuator and being detected by an also well-collimated detector. Thus no significant scattered radiation is seen by the detector, and the only effect causing a reduction in the photon intensity is the attenuation of the shielding material. In most cases encountered in real life, neither the source nor the detector is well collimated. This is called "poor geometry" or "broad beam" geometry. Figure 11.2 shows schematically a number of photon interactions in which the photons have been scattered by the shield or other materials in the system but still eventually reach the detector and contribute to the signal detected.

Under conditions of "poor geometry," the equation that describes the reduction in intensity for a given amount of shielding material is similar to that for "good geometry:"

$$I = B \times I_0 e^{-\mu t}.$$

Here the factor B is called the *buildup factor*, and is defined as the ratio of the intensity of the radiation, including both primary and scattered radiation to the intensity of the primary radiation only at a given point in space. The buildup factor is a value always greater than 1.0. Values of B are obtained empirically either by measurement or radiation transport simulation. They are dependent on:

- The energy of the primary photons
- The type of shielding material
- The thickness of the shielding material

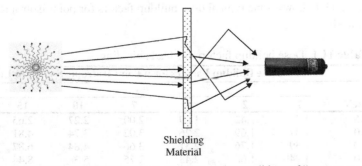

Shielding
Material

Figure 11.2 Shielding of photon sources under conditions of "poor geometry."

It is probably obvious that B would depend on the energy of the photons and the type of material; as more intervening material is added, the amount of scattered radiation also increases, so B increases also. The student might immediately note that this is going to cause a difficulty in calculating the thickness of shielding material needed for a given problem. In "good geometry" situations, once we know the reduction in intensity needed and the appropriate μ to use, calculation of t is straightforward. Now, inasmuch as B is a function of t, we cannot solve directly for t and must use an iterative approach. There may be a shortcut to the solution, however, and this will be demonstrated.

Buildup factors, however they were derived, are generally given in lookup tables for discrete photon energies and thicknesses of different standard materials used frequently for shielding photon sources. The factors may be fitted to mathematical functions, for example, of the form:

$$ B = \left[\left(1 + \frac{\beta}{\alpha} \right) e^{\alpha \gamma x} - \frac{\beta}{\alpha} \right]^{-\frac{1}{\gamma}} $$

Shielding Material	α (cm^{-1})	β (cm^{-1})	γ
Lead	1.7772	−0.5228	0.5457
Concrete	0.1539	−0.1161	2.0752
Steel	0.5704	−0.3063	0.6326

http://www.acmp.org/meetings/scottsdale_2004/diagnostic/MartinPET.ppt (Doug Simpkin, fit to Archer Eq).

Buildup factors are typically given as a function of the thickness of the shielding material, not in absolute units of length, but in terms of

- The number of mean free paths of a photon of the specified energy in that medium
- The number of relaxation lengths of a photon of the specified energy in that medium

A *mean free path* (mfp) of a photon in a medium is the average distance traveled by a photon before interacting, and is given as $1/\mu$, where μ is the medium's linear attenuation coefficient for the given photon energy. The number of *relaxation lengths* is defined as $\mu \cdot x$, where x is the thickness of the material. The relaxation length is also the thickness of a shield that will attenuate a narrow beam of radiation to 1/e times its original intensity. Table 11.1 shows some typical dose buildup factors for point isotropic sources in lead.

Table 11.1 Dose buildup factors for lead shielding.

	Dose Buildup Factors for a Point Isotropic Source in Lead						
	μx						
MeV	1	2	4	7	10	15	20
0.5	1.24	1.42	1.69	2.00	2.27	2.65	2.73
1.0	1.37	1.69	2.26	3.02	3.74	4.81	5.86
2.0	1.39	1.76	2.51	3.66	4.84	6.87	9.00
3.0	1.34	1.68	2.43	2.75	5.3	8.44	12.3

http://cted.inel.gov/cted/radpro/rp16.pdf.

Use of buildup factor tables will usually involve some interpolation, in order to apply the values to specific problems. Some computer programs exist that can facilitate the calculations. As with any computer program, however, users must exercise extreme caution in their use and must be familiar with how to do the calculations by hand before trusting them.

Example

Calculate the exposure rate at 1 m from a 50 GBq ^{137}Cs source with 3 cm of intervening lead, assuming good geometry and poor geometry conditions.

Assume a ^{137}Cs exposure rate constant of 2.3×10^{-9} (C/kg)-m^2/MBq-h and a linear attenuation coefficient for the 662 keV photons (assumed to contribute the majority of the exposure) to be 1.24 cm^{-1}.

Good geometry:

$$\dot{X} = \frac{50000 \text{ MBq} \times 2.3x10^{-9} \text{ (C/kg)-m}^2}{1 \ m^2 \qquad \text{MBq} - \text{h}} e^{-1.24 \text{ cm}^{-1} \times 3 \text{ cm}}$$

$$= 2.79x10^{-6}\frac{\text{C/kg}}{\text{h}} = 0.0108\frac{R}{h}$$

Poor geometry:
(More likely the case) $\mu = 1.24$ cm^{-1}, $x = 3$ cm, $\mu x = 3.72$. Linear interpolation in Table 11.1:

0.5 MeV:

$$B_{0.5} = 1.42 + \frac{3.72 - 2}{4 - 2}(1.69 - 1.42) = 1.65$$

1.0 MeV:

$$B_{1.0} = 1.69 + \frac{3.72 - 2}{4 - 2}(2.26 - 1.69) = 2.18$$

$$B = 1.65 + \frac{0.662 - 0.5}{1.0 - 0.5}(2.18 - 1.65) = 1.82$$

$$\dot{X} = 1.82 \ x \ 2.79x10^{-6} \ \frac{\text{C/kg}}{\text{h}} = 5.08x10^{-6}\frac{\text{C/kg}}{\text{h}} = 0.020\frac{R}{h}$$

Example

Calculate the amount of shielding needed for the source above to reduce the intensity to an exposure rate of 2.5 mR/hr.

The unshielded intensity is:

$$\frac{50000 \text{ MBq} \times 2.3x10^{-9} \text{ (C/kg)} - \text{m}^2}{1 \text{ m}^2 \qquad \text{MBq} - \text{h}} = 1.15x10^{-4} \ \frac{\text{C/kg}}{\text{h}} = 0.446\frac{R}{h}.$$

The reduction factor needed is $0.0025/0.446 = 0.0056$. With no assumption of buildup, we would estimate a shield thickness of:

$$0.0056 = e^{-1.24 \times t} \qquad t = 4.18 \text{ cm}.$$

Of course we now know that 4.2 cm will not be adequate, because of at least a factor of 2 buildup that we will encounter. We can't calculate the buildup exactly until we know the thickness, and we don't know the thickness we need until we know the buildup! So we have to iteratively play around with some guesses until we find the right answer. A shortcut, however, usually will get us there in one or two steps. The trick is this:

1. Calculate the amount of shielding that appears to be necessary, giving no consideration to buildup.
2. Add one Half Value Layer (HVL) of shielding material.

We have performed the first step. An HVL for concrete at this energy will be approximately $0.693/1.24 = 0.56$ cm of concrete. So, with an initial guess of $4.18 + 0.56 = 4.74$ cm of lead, and $\mu x = 5.88$ we find:

0.5 MeV:

$$B_{0.5} = 1.69 + \frac{5.88 - 4}{7 - 4}(2.0 - 1.69) = 1.88$$

1.0 MeV:

$$B_{1.0} = 2.26 + \frac{5.88 - 4}{7 - 4}(3.02 - 2.26) = 2.74$$

$$B = 1.88 + \frac{0.662 - 0.5}{1.0 - 0.5}(2.74 - 1.88) = 2.16$$

$$\dot{X} = 2.16 \; x \; 1.15 x 10^{-4} \; \frac{C/kg}{h} \; e^{-1.24 \, cm^{-1} \times 4.74 \, cm}$$

$$\dot{X} = 6.96 x 10^{-7} \frac{C/kg}{h} = 0.0027 \frac{R}{h}$$

So we got this calculation almost right the first time! This does not always happen, but the shortcut method usually gets you close to the right answer, and the addition or subtraction of a small amount of material will then provide the solution. Of course, one would not design a shield with exactly 4.74 cm thickness. One would typically round up a bit anyway, to be conservative, but with lead not too much, so that the shield is not excessively heavy and expensive.

11.4 Graded or Laminated Shielding

In some instances it is useful to use multiple layers of shielding against photon-emitting sources. Two examples include the following.

Graded-Z shields place materials with decreasing atomic numbers towards the detector to absorb the lead X-rays created in a primary lead shield around low-level radiation detectors. Lead has a K X-ray at 80 keV. Cadmium has an HVL for 80 keV X-rays of about 0.3 mm, so 10 HVLs will reduce the intensity of the lead X-ray by a factor of 1000. Cadmium emits X-rays of 22 keV. Copper has an HVL of 0.03 mm for 20 keV photons and 1 mm for 80 keV photons. Thus, placing thin layers of cadmium and then copper within a (for example) cylindrical lead shield around a photon detector will reduce the level of 80 keV X-rays significantly. Other materials, such as Al and Lucite, may be used as well in graded-Z shields used to study photons of lower energy.

When considering the idea of using a multiple-materials approach to a shielding problem, it is important to know that high-Z materials have a stronger

Table 11.2 Dose buildup factors in water and lead.

Material	E (MeV)	μx 1	2	4	7	10	15
Water	0.5	2.63	4.29	9.05	20.0	35.9	74.9
	1.0	2.26	3.39	6.27	11.5	18.0	30.8
	2.0	1.84	2.63	4.28	6.96	9.87	14.4
	3.0	1.69	2.31	3.57	5.51	7.48	10.8
	4.0	1.58	2.10	3.12	4.63	6.19	8.54
Lead	0.5	1.24	1.39	1.63	1.87	2.08	
	1.0	1.38	1.68	2.18	2.80	3.40	4.20
	2.0	1.40	1.76	2.41	3.36	4.35	5.94
	3.0	1.36	1.71	2.42	3.55	4.82	7.18
	4.0	1.28	1.56	2.18	3.29	4.69	7.70

tendency to stop low-energy photons than do low-Z materials via the photo-electric effect. Photons scattered by multiple Compton interactions to these low energies will thus be effectively removed by a high-Z material such as lead. If a high-Z material is placed close to a radiation source of moderate to high energy, the spectrum of photons emerging from the other side of the high-Z shield will be "hardened," or be comprised of a higher fraction of higher-energy photons than the spectrum incident on the shield. On the other hand, if a low-Z material is placed near the source, there is more tendency to scatter the photons to lower energies, which are then effectively removed by the (second) high-Z shield. Note from Table 11.2 (taken from a different source than Table 11.1, slight differences in the buildup factors for lead are evident) the larger buildup factors for water than lead for the same number of relaxation lengths, showing the higher degree of photons that are scattered (rather than being absorbed). Thus in designing a graded shield, the low-Z material should be placed nearer to the source, followed by the high-Z material, to produce a shield whose overall design is most efficient. The total buildup factor for the graded shield will be the product of the factors in each segment.[2,3]

11.5 Shielding of X-Ray Sources

X-rays are of course identical to photons in nature, so all of the same principles that we applied to general photon sources will apply to the shielding of X-ray sources. X-ray sources are generally of lower energy, so less shielding will be needed than for medium- to high-energy gamma ray sources. Techniques for X-ray shielding, however, have been developed using different approaches than for general gamma emitters. The guiding document at present that describes these techniques is NCRP Report No. 147[4] (this report is an update on a previous, very similar report, NCRP Report No. 49 from 1976).

Two categories of areas will be protected involving X-ray sources:

- Controlled areas (only radiation workers present)
- Uncontrolled areas (members of the general public may be present)

Three types of radiation from either diagnostic or therapeutic X-ray units are considered (Figure 11.3):

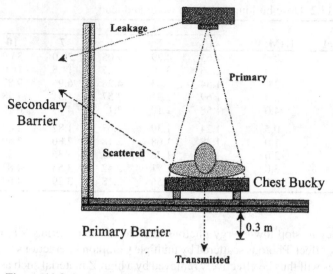

Figure 11.3 Sources of radiation from X-ray sources. (Reprinted with permission of the National Council on Radiation Protection and Measurements, NCRP Report No. 147.)

- Primary radiation (designed output of the machine)
- Scattered radiation (radiation scattered from objects in the room, including the patient)
- Leakage radiation (output from the machine that is not part of the useful beam)

Six "occupancy factors" (T) are assumed for various areas, which is the fraction of time that an area is assumed to be possibly occupied by individuals to be protected (Figure 11.4):

- Administrative offices, laboratories, waiting rooms, and other areas ($T = 1$)
- Patient examination rooms ($T = 0.5$)
- Corridors, lounges, and the like ($T = 0.2$)
- Corridor doors ($T = 0.125$)
- Rest rooms, vending areas, storage areas, patient holding areas ($T = 0.05$)
- Outdoor areas with transient pedestrian traffic, parking lots, and the like ($T = 0.025$)

The output of an X-ray machine is given as unit "workload" (W), which is the cumulative total of the milliampere-minutes per week that the unit puts out. The energy of the X-rays for any procedure is given in units of kilovolts potential (kVp); the amount of radiation at this energy that is given off depends on the tube current (mA) multiplied by the amount of time that the unit is energized (s), so has units of mAs, milliampere-seconds. The radiation coming from the unit has a spectrum of energies up to the maximum kVp used to create the beam, although the lower energies are often reduced in intensity through the use of beam filters of various materials (Al, Cu, Mo, Rh, or plastic, generally).

The next variable of interest is called the "use factor" (U), and is the fraction of the primary beam workload that is directed towards any given primary barrier to the radiation. Some X-ray units are pointed in the same direction

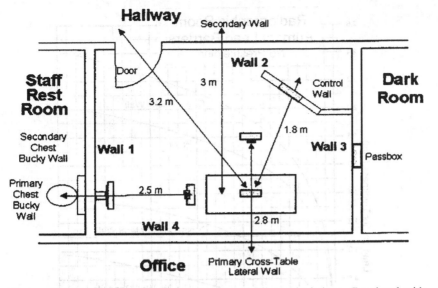

Figure 11.4 Radiographic room diagram for shielding calculations. (Reprinted with permission of the National Council on Radiation Protection and Measurements, NCRP Report No. 147.)

almost all the time, but others can be oriented against a wall, down onto a table, up towards a subject reclining on a table, or moved around at most any angle to take particular images. Fluoroscopic units in particular are frequently moved around in various orientations to help physicians view progress with various invasive procedures.

Shielding is studied for two types of barriers: the primary barrier (towards which the useful radiation beam is directed, and that may be exposed to primary, scattered, and leakage radiation), and all secondary barriers (floors, ceilings, and other walls that are exposed to scattered and leakage radiation). If P is the permissible dose rate in an area (25 μSv/hr or 2.5 mrem/hr in controlled areas, also 1 mSv/week or 100 mrem/week and 0.0005 mSv/hr or 0.05 mrem/hr, also 0.02 mSv/week or 2 mrem/week in uncontrolled areas), and we define K to be the unshielded air kerma per patient at 1 m from the source, the "barrier transmission factor" that we need to shield a given source would be:

$$B = \frac{Pd^2}{KTUN}.$$

Here, N is the number of patients imaged per week and d is the distance between the radiation source and an individual beyond the primary barrier. This is the notation used in NCRP 147. It is unfortunate that the symbol B is used for the transmission factor. This factor is really $(B \cdot e^{-\mu x})$ in our previous equations, where B is the buildup factor and μ is the linear attenuation coefficient. For a secondary barrier, the equation is the same, except that the factor U does not appear, as it is assumed to always be 1.0 for secondary barriers, as scattered and leakage radiation are assumed to be emitted isotropically. The method for solving individual shielding problems is handled graphically at this point. The authors of NCRP 147 provide plots of lead (and other) barrier thicknesses, typically in mm, against the variable $(N \cdot T/P \cdot d^2)$. Considerations will also

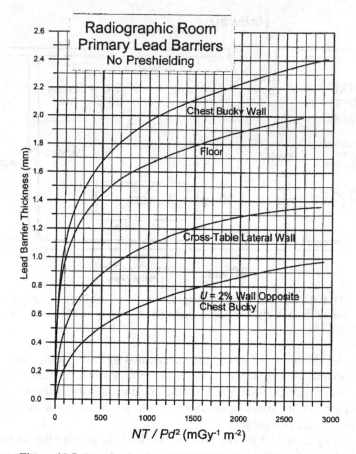

Figure 11.5 Sample plot from NCRP 147 of lead thickness versus
NT/Pd^2. (Reprinted with permission of the National Council on
Radiation Protection and Measurements, NCRP Report No. 147.)

include how much shielding is already in place, just due to the existing building
materials in a room (called "preshielding") (Figure 11.5).

Example
(From NCRP 147.) A dedicated chest X-ray unit performs images on 300
patients per week. The beam is always pointed in one direction in the room,
towards a wall-mounted chest bucky image receptor. The area behind this wall
is an uncontrolled area, assumed to be always occupied, and the distance from
the source to an individual behind this wall is about 3 m. A typical chest unit
has an output of $\sim$1.2 mGy patient^{-1} at 1 m. The barrier transmission factor is
thus:

$$B = \frac{0.02\frac{\text{mGy}}{\text{week}}(3\text{m})^2}{1.2\frac{\text{mGy-m}^2}{\text{patient}} \times 1 \times 1 \times 300\frac{\text{patients}}{\text{week}}} = 5x10^{-4}.$$

From Figure 11.6 ('Chest Room') we find that 2.2 mm of lead would be
required to meet this shielding requirement. The bucky is assumed to provide
an equivalent lead thickness of 0.85 mm, so the additional shielding required
for this room would be 2.2 − 0.85 = 1.4 mm of lead.

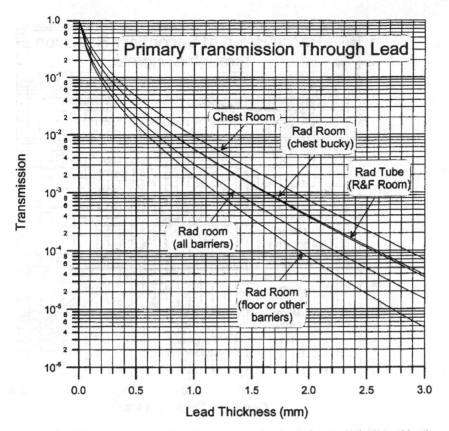

Figure 11.6 Broad beam transmission factors for lead for a clinical workload. (Reprinted with permission of the National Council on Radiation Protection and Measurements, NCRP Report No. 147.)

We can also look at attenuation in other building materials, such as concrete and gypsum wallboard (Figures 11.7 and 11.8).

Alternatively, barrier thickness values for the lead case may be provided from fits to empirical formulas:

$$x_{\text{barrier}} = \frac{1}{\alpha\gamma} \ln\left[\frac{\left(\frac{NTK}{Pd^2}\right)^{\gamma} + \frac{\beta}{\alpha}}{1 + \frac{\beta}{\alpha}} \right], \text{ here}$$

$$x_{\text{barrier}} = \frac{1}{2.283 \times 0.637} \ln\left[\frac{\left(\frac{300 \times 1 \times 1 \times 1.2}{0.02 \times 3^2}\right)^{0.637} + \frac{10.74}{2.283}}{1 + \frac{10.74}{2.283}} \right] = 2.2 \text{ mm}$$

Now consider the adjacent wall, which only receives scattered radiation from the unit. NCRP 147 provides empirical factors that give the fraction of the primary beam that is generally observed from scattered and leakage radiation. For this type of unit, the total unshielded secondary air kerma for leakage plus 90° scatter at 1 m is given as 2.7×10^{-3} mGy/patient. Persons behind this barrier are at a distance of about 2.1 m from the source. The calculation

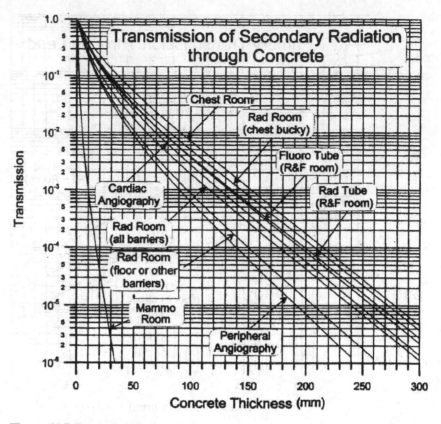

Figure 11.7 Transmission of secondary radiation through concrete. (Reprinted with permission of the National Council on Radiation Protection and Measurements, NCRP Report No. 147.)

is thus:

$$B = \frac{0.02 \frac{\text{mGy}}{\text{week}} (2.1\text{m})^2}{2.7 x 10^{-3} \frac{\text{mGy-m}^2}{\text{patient}} \times 1 \times 1 \times 300 \frac{\text{patients}}{\text{week}}} = 1.1 x 10^{-1}.$$

Use of the appropriate graph yields a required thickness of 0.42 mm of lead. The empirical equation approach can also be used, which gives the same result. In a real-life situation, the calculations above must be repeated many times, to account for all rooms on all sides of the use room, including above and below (Figure 11.9).

For units that are used in fluoroscopic mode, the total output from the continuous irradiation studies must be added to any spot films made in the room each week. Curves are provided for mammography units, which use lower kVp settings to achieve imaging of the breast. Treatment of Computed Tomography (CT) units is unique, in that the dose values reported are unusual. The CT Dose Index (CTDI) is a dose index obtained by integrating over a dose profile from a single CT axial rotation. Another quantity sometimes quoted is the Dose-Length Product (DLP), which is given as

$$\text{DLP} = \frac{L}{p} \left(\frac{1}{3} \text{CTDI}_{100,\text{center}} + \frac{2}{3} \text{CTDI}_{100,\text{peripheral}} \right).$$

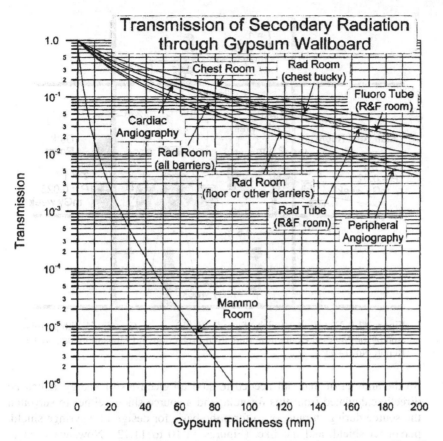

Figure 11.8 Transmission of secondary radiation through gypsum wallboard. (Reprinted with permission of the National Council on Radiation Protection and Measurements, NCRP Report No. 147.)

The quantities $CTDI_{100,center}$ and $CTDI_{100,center}$ are the dose values measured with a pencil ionization chamber, normalized to a 100 mm exposed length and located either at the center or periphery of a tissue-equivalent phantom, and representing thus average internal tissue dose or entrance dose. Either the CTDI or DLP can be expressed in mGy/patient and used in the above formulas.

11.6   Shielding of Discrete Electron Sources

Shielding of electron sources (including beta and positron sources) is basically more straightforward than shielding of photon sources. Electron sources can be completely shielded by choice of a material thickness that corresponds to the electron range. The complication comes, however, with the production of bremsstrahlung radiation. We know to choose low Z materials to shield beta sources, to minimize the production of bremsstrahlung radiation. So we choose a low-Z material such as acrylic or polyethylene. Calculating the range of the electron in this material is not difficult, if the material density is given. From Chapter 4, we recall that the range in mg/cm^2 of electrons of energy E is given as

$$R = 412E^{1.265-0.0954\ln E} \quad 0.01 < E < 2.5 \text{ MeV}.$$

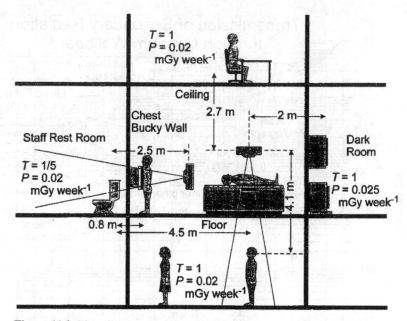

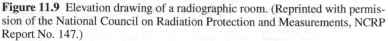

Figure 11.9 Elevation drawing of a radiographic room. (Reprinted with permission of the National Council on Radiation Protection and Measurements, NCRP Report No. 147.)

So the amount of material needed to stop all electrons of energy E can be directly calculated from this formula, and a source shield, either to surround the source during storage, or to form the basis for design of a syringe shield, personnel shield, and the like. (Figures 11.10 to 11.12). Now, we need to know the average atomic number of this absorber, and we can calculate the fraction of the total beta energy released that is converted to bremsstrahlung radiation as

$$f = \frac{6x10^{-4}ZE}{1 + 6x10^{-4}ZE}.$$

Figure 11.10 Beta shields for laboratory work. (From http://www.topac. com/shield.html.)

Figure 11.11 Waste storage container designed to shield beta particles. (From http://www.rpicorp.com/products/.)

Bremsstrahlung radiation is emitted with a spectrum of energies from low energies up to the energy of the electron, or the maximum energy of a beta in a beta spectrum. So it is difficult to characterize and shield by conventional methods. Two approaches would be to shield based on the maximum photon energy to be seen in the spectrum, and neglect the use of a buildup factor, as the shield will be overdesigned, because most of the photons will have a lower energy than the maximum value.

Figure 11.12 Acrylic shielded syringes. (From http://www.rpicorp.com/products/.)

Example

Consider a small 200 GBq source of ^{90}Sr/^{90}Y. You wish to design a Lucite shield for this source and then configure a lead carrier to limit the exposure to bremsstrahlung radiation. Lucite has a density of 1.19 g/cm3 and an atomic number of 6.24. ^{90}Sr has a half-life of ~29 years, and ^{90}Y has a half-life of 64 hours. Thus these nuclides may be considered to be in secular equilibrium. For every decay of ^{90}Sr, there will be a decay of ^{90}Y. The Sr-90 beta has a maximum energy of 0.54 MeV and a mean beta energy of 0.19 MeV, whereas ^{90}Y has a maximum energy beta of 2.27 MeV and an average beta energy of 0.93 MeV. So the beta shield must be designed considering the ^{90}Y beta particles.

We estimate the range as $R = 412(2.27)^{1.265-0.0954\ \ln(2.27)} = 1090\,\text{mg/cm}^2$. The thickness of Lucite needed is thus:

$$\frac{1.09\ \text{g}}{\text{cm}^2}\frac{\text{cm}^3}{1.19\ \text{g}} = 0.917\ \text{cm}.$$

So about a cm of Lucite will make an effective shield for this source. Now, the fraction of source energy converted to bremsstrahlung radiation is:

$$f = \frac{6x10^{-4} \times 6.24 \times 2.27}{1 + 6x10^{-4} \times 6.24 \times 2.27} = 0.00843.$$

The energy flux at 1 m would be:

$$\frac{2x10^{11}\text{dis}}{s}\frac{1.12\ \text{MeV}}{\text{dis}}\frac{0.00843}{4\pi(1\text{m})^2}\frac{\text{m}^2}{10^4\text{cm}^2} = 1.5x10^4\frac{\text{MeV}}{\text{cm}^2\ \text{s}}.$$

Note that here we used the average beta energy, as we want the average energy flux from all of the transitions. For the fraction of energy converted to bremsstrahlung, we needed the maximum energy (that is just what the formula is based on), and in the shielding of the source, we use a trick of using the attenuation coefficient for the maximum energy, with no buildup factor. The unshielded dose rate at 1 m would be:

$$\dot{D} = \frac{1.5x10^4\frac{\text{MeV}}{\text{cm}^2\text{s}}1.6x10^{-13}\frac{J}{\text{MeV}}0.000037\ \text{cm}^{-1}3600\frac{s}{h}}{0.001293\frac{\text{g}}{\text{cm}^3}\frac{\text{kg}}{1000\ \text{g}}}$$

$$= 2.48x10^{-4}\frac{J}{\text{kg-h}} = 2.48x10^{-4}\frac{\text{Gy}}{\text{h}}$$

To shield this to a level of 0.000025 Gy/hr, and finding a value of $\mu = 0.51\ \text{cm}^{-1}$ for lead at 2.27 MeV, we calculate:

$$2.5x10^{-5} = 2.48x10^{-4}e^{-0.51\ \text{cm}^{-1}\times t}$$

$$t = 4.5\ \text{cm lead}$$

So, if the source is spherical, we would want a Lucite container of about 1 cm thickness, surrounded then by a lead shield of 4.5 cm. The weight of the lead would be:

$$\frac{4\pi}{3}(5.5^3 - 1^3)\text{cm}^3 11.3\frac{\text{g}}{\text{cm}^3} = 7800\ \text{g} = 7.8\ \text{kg}.$$

11.7 Shielding of Neutron Sources

Shielding of neutron sources follows a three-step logical chain:

- As formed, neutrons are "fast" neutrons, as defined in Chapter 4. These neutrons are not easily captured; except at a few specific "resonance" energies, their absorption cross-sections are low, but they have high scattering cross-sections in many materials. In low-Z materials, as you know, a larger fraction of the neutron energy may be lost in individual interactions, particularly with hydrogen. So the first step is to place a hydrogenous or other low-Z material near the source, to induce scattering reactions, which slow down, or moderate, the neutrons to thermal energies.
- Thermal neutrons have much higher absorption cross-sections, particularly in a few materials such as boron or cadmium. So, after the fast neutron moderator, we place some material with a high cross-section for thermal neutron absorption.
- After capture, there are often gamma rays emitted, so we then place a certain amount of material that is good for photon shielding, as described above.

Fast Neutron Moderation

Hydrogenous materials are very effective here, as noted above. Materials such as water, paraffin (general formula C_nH_{2n+2}), polyethylene (mixed polymers of ethylene $(CH_2)_n$), or concrete are useful. For small discrete sources, we may use polyethylene or paraffin. For larger sources, concrete is more useful, as paraffin is flammable, and use in extended sources is not too practical. Water shields are good, but have obvious maintenance problems such as leakage and evaporation. For a monoenergetic source, the attenuation of neutrons is mathematically characterized in the same way as for photons, except that instead of linear attenuation coefficients, neutron interaction probabilities are given in terms of cross-sections, which have units of cm^2/atom. When we multiply this by the number of atoms/g and the density (g/cm^3), we again have an exponential term that has units of cm^{-1}, like a photon attenuation coefficient, and our equation works in exactly the same way; that is, we calculate the thickness needed as the natural logarithm of the reduction in intensity required divided by the (here derived) linear coefficient:

$$I = I_0 e^{-(\sigma N \rho)t}$$

$$t = \frac{-\ln\left(\frac{I}{I_0}\right)}{\sigma N \rho}$$

where:

σ = the neutron-scattering cross-section (cm^2/atom, also barns/atom, 1 b = 10^{-24} cm^2)

N = atom density of the material (atoms/g)

ρ = material density (g/cm3)

The quantity σ is referred to as the *microscopic* neutron cross-section for the material. Often you will see the quantity ($\sigma N \rho$) given as the *macroscopic* cross-section, with the symbol Σ, again with units of cm^{-1} (Figure 11.13). Just as for photons, we can use the macroscopic cross-section to define the

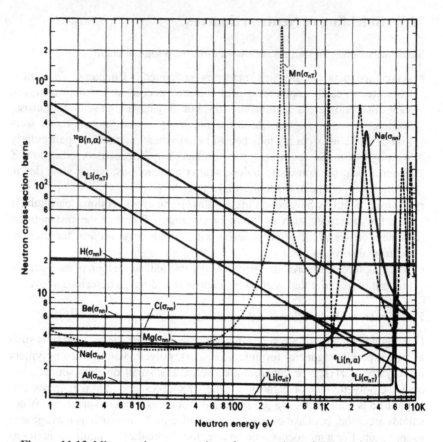

Figure 11.13 Microscopic cross-sections for a number of light materials in the "slowing down" region (From Kaye and Laby Online, *Tables of Physical and Chemical Constants*; from http://www.kayelaby.npl.co.uk/atomic_and_nuclear_physics/4_7/4_7_2.html, courtesy of NPL©Crown, copyright 2006.)

material half value layer, the thickness of material needed to remove half the neutrons from the incident beam, and the attenuation length, the counterpart to the relaxation length for photons, which is $1/\Sigma$, and is the thickness of material needed to reduce the intensity of the incident beam by a factor of e.

The equation above is applicable only to a pure substance, which is almost never the case in practice. For mixtures, even simple mixtures like H_2O, the total cross-section will be given as

$$\sum_s^{H_2O} = \sum_s^H + \sum_s^O = N_H \sigma_s^H + N_O \sigma_s^O.$$

In addition, the cross-sections generally vary with energy, and the results must be integrated over the entire neutron spectrum:

$$\overline{\sigma} = \frac{\int_{E_1}^{E_2} \sigma(E)\, \Phi(E)\, dE}{\int_{E_1}^{E_2} \Phi(E)\, dE}$$

Here $\sigma(E)$ is the microscopic cross-section at mean energy E, $\Phi(E)$ is the neutron flux at mean energy E, and the integration is carried across the energy spectrum from E_1 to E_2.

Example

Design a polyethylene shield for a small spherical source of ^{210}Po-Be neutrons of intensity 10^6 n/sec such that the contact dose is less than 250 μSv/hr (25 mrem/hr).

The average energy of a ^{210}Po-Be neutron spectrum is about 4.2 MeV, and the equivalent dose rate for neutrons of this energy is about 1.56 μSv/hr per neutron/cm^2-s (from Turner[5]). The density of polyethylene is 0.94 g/cm^3. From Figure 11.13, the microscopic cross-sections for C and H are about 4.7 and 2 barns, respectively. Polyethylene is 14% H and 86% C by weight.

$$\frac{0.14 g\,H}{g\,\text{poly}}\,\frac{0.94 g\,\text{poly}}{\text{cm}^3}\,\frac{\text{mol}\,H}{1 g\,H}\,\frac{6.025 \times 10^{23}\,\text{atoms}\,H}{\text{mol}\,H} = 7.93 x 10^{22}\frac{\text{atoms}\,H}{g\,\text{poly}}$$

$$\frac{0.86 g\,C}{g\,\text{poly}}\,\frac{0.94\,g\,\text{poly}}{\text{cm}^3}\,\frac{\text{mol}\,C}{12 g\,C}\,\frac{6.025 \times 10^{23}\,\text{atoms}\,C}{\text{mol}\,C} = 4.06 x 10^{22}\frac{\text{atoms}\,C}{g\,\text{poly}}$$

$$\Sigma = \frac{7.93 x 10^{22}\,\text{atoms}\,H}{g\,\text{poly}}\,\frac{2 x 10^{-24}\,\text{cm}^2}{\text{atom}\,H}\,\frac{0.94\,g\,\text{poly}}{\text{cm}^3}\,|$$

$$\frac{4.06 x 10^{22}\,\text{atoms}\,C}{g\,\text{poly}}\,\frac{4.7 x 10^{-24}\,\text{cm}^2}{\text{atom}\,C}\,\frac{0.94\,g\,\text{poly}}{\text{cm}^3} = 0.328\,\text{cm}^{-1}$$

To calculate the initial dose rate, we have to assume a source size, as the dose depends on the distance from the source. Turner[5] notes that the dose buildup factor (needed for neutrons as well as for photons) for neutrons with low-Z shields of at least 20 cm is around 5. Our solution equation is:

$$250\,\mu\text{Sv/h} = \frac{5 \times 10^6 \frac{n}{s} \times 1.56 \frac{\mu\text{Svcm}^2\text{s}}{hn} \times e^{-0.328 \times T}}{4\pi T^2}.$$

This is not straightforward to solve analytically. By just entering values for T, we find that a thickness of 10 cm produces an equivalent dose rate of 233 μSv/hr. The dose buildup factor applies for a thickness of 20 cm or more, so a 10 cm spherical container should be overdesigned for this source and dose criterion.

Thermal Neutron Capture

Thermal neutrons, typically characterized as having 0.025 MeV of energy, are readily captured by a number of materials. Hydrogen does absorb thermal neutrons, and emits a photon of 2.22 MeV, in the reaction ^{1}H$(n,\gamma)^2$H. The cross-section, however, is only around 0.33 b. It actually does not appear on the graph in Figure 11.14 (also from Kaye and Laby Online), but those for many other neutron absorbers do. Note near the top of the graph the very high cross-sections for cadmium and boron. These are very important materials for thermal neutron shielding, as well as for control of fission chain reactions, which depend on the diffusion and capture of thermal neutrons by ^{235}U, whose $(n,f$; neutron capture, fission) cross-section is shown in the figure. This is covered in more detail in a subsequent chapter, but basically, we want ^{235}U to capture neutrons, fission, and release more neutrons, which will in turn initiate more ^{235}U fissions. If this happens at too rapid a rate, however, the chain reaction can get out of control, so we use these materials as neutron poisons, to either control the reaction to the rate that we want, or in some

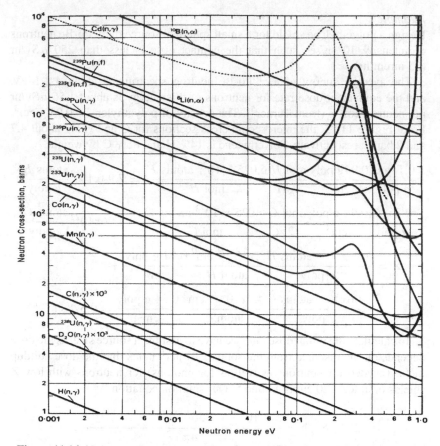

Figure 11.14 Neutron capture cross sections for a number of materials in the "thermal" region. (From Kaye and Laby Online, *Tables of Physical and Chemical Constants*; from http://www.kayelaby.npl.co.uk/atomic_and_nuclear_physics/4_7/4_7_2.html, courtesy of NPL©Crown, copyright 2006.)

circumstances, to rapidly shut down the reaction so that it cannot proceed at all. Cadmium has a thermal neutron cross-section of 2450 b. When it absorbs a neutron, however, it gives off a 9.05 MeV gamma ray, which is difficult to shield. Boron has a thermal neutron cross-section of 755 b, and gives off a 0.48 MeV photon in the reaction $^{10}B(n,\alpha)^7Li$. A popular use of boron as a neutron absorber is as boric acid (H_3BO_3) dissolved in solution. Its solubility is 63 g/L, so one can dissolve about 6×10^{23} atoms of boron in a L of solution:

$$\frac{63\frac{g}{L} 6.025 x 10^{23} \frac{\text{atoms}}{\text{mol}} \frac{1 \text{ atom } B}{\text{molecule}}}{61.8\frac{g}{\text{mol}}} = 6.17 x 10^{23} \frac{\text{atoms B}}{L}.$$

Figure 11.14 shows thermal neutron cross-sections for a number of materials.

Example

Consider a water shield instead of a polyethylene shield for the above ^{210}Po-Be source.

Following the analysis above, we can show that water has 6.7×10^{22} atoms of H (1.9 b cross-section for 4 MeV neutrons) and 3.5×10^{22} atoms of O (1.7 b cross-section for 4 MeV neutrons) and thus has a macroscopic cross-section of 0.184 cm^{-1}. Again, following our analysis above, but for a target equivalent dose rate of 2 μSv/hr, we would estimate that we need around 35 cm of water to shield this source. With a shield this thick, about six times the fast diffusion length $(1/0.184 = 5.4$ cm), the neutrons are thermalized and are thus diffusing in a direction away from the source. This source strength (in units of flux), for thermal neutrons, is given as

$$\frac{10^6 \frac{n}{s} e^{-35/2.9}}{4\pi 35 \text{ cm } 0.16 \text{ cm}} = 0.084 \frac{n}{\text{cm}^2 \text{s}}.$$

Here, 2.9 cm is the thermal diffusion length[6] and 0.16 cm is the diffusion coefficient for thermal neutrons in water.[7] NCRP 38 tells us that an equivalent dose rate of 25 μSv/hr is to be expected from a source of 270 n/cm^2-s; this flux corresponds to a very small equivalent dose rate. Throughout the shield, the ^{1}H(n,γ)^{2}H reaction will be producing 2.26 MeV photons. The strength will be approximately:

$$\frac{10^6 \frac{\gamma}{s}}{\frac{4}{3}\pi (35 \text{ cm})^3} = 5.6 \frac{\gamma}{\text{s cm}^3}.$$

This acts just like a volume source of a gamma emitter, and the exposure rate can be calculated as

$$\dot{X} = \frac{C \Gamma}{2} \frac{4\pi}{\mu} (1 - e^{-\mu r})$$

$$\dot{X} = \frac{5.6 \frac{\gamma}{\text{s cm}^3} 10^{-6} \frac{\text{MBq s}}{\gamma} 2.7 \frac{\text{mGy cm}^2}{\text{MBq h}}}{2} \frac{4\pi}{0.046 \text{ cm}^{-1}} (1 - e^{-0.046 \times 35})$$

$$\dot{X} = 0.0016 \frac{\text{mGy}}{\text{h}}$$

So our total equivalent dose rate is 0.0020 mSv/hr from the fast neutrons, very little ($\sim 8 \times 10^{-6}$ mSv/hr) from the thermal neutrons, and 0.0016 mSv/hr from the emitted gamma rays.

11.8 Performing Radiation Surveys

10CFR20, specifically Sections 20.1501 and 20.1502 requires that

- Licensees perform radiation surveys to ensure compliance with all of the regulations in the code.
- Licensees perform other "reasonable" surveys to evaluate the magnitude and extent of radiation levels, concentrations of radioactive materials, and evaluate potential radiation hazards.
- Licensees must ensure that survey equipment is properly and regularly calibrated.
- Individuals wearing personnel monitors should be suitable for measuring the important radiations involved and accredited by the National Voluntary Laboratory Accreditation Program (NVLAP) of the National Institute of Standards and Technology (NIST).

Personal monitoring is required for:

- Adults likely to receive a dose greater than 10% of the annual limits.
- Minors likely to receive deep dose equivalents in excess of 0.1 rem (1 mSv), a lens dose equivalent in excess of 0.15 rem (1.5 mSv), or a shallow dose equivalent to the skin or to the extremities in excess of 0.5 rem (5 mSv).
- Declared pregnant women likely to receive during the entire pregnancy, from radiation sources external to the body, a deep dose equivalent in excess of 0.1 rem (1 mSv).
- Any individual entering a high or very high radiation area.

Monitoring for internalization of radioactive material and dose calculations are required for

- Adults likely to receive intakes in excess of 10% of the applicable ALI(s) in a given year.
- Minors likely to receive a committed effective dose equivalent in excess of 0.1 rem (1 mSv) in a year.
- Declared pregnant women likely to receive, during the entire pregnancy, a committed effective dose equivalent in excess of 0.1 rem (1 mSv).

Surveys of radiation work areas generally include

- Measurements of ambient radiation levels, using appropriate survey meters
- Measurements of surface contamination, using filter papers or other means to sample the surface, with counting at another time using a benchtop radiation detector appropriate to the radiation of potential interest

Measurements are usually made relative to a floor plan of the facility (Figure 11.15). Or measurements may be made in reference to objective positions near a well-known location or a particular device. (Figure 11.16 is a schematic diagram of a high dose rate brachytherapy afterloader). Reference points in locations where workers or the general public may be present near an important radiation source are also typically included. Table 11.3 also relates to this same brachytherapy afterloader.[8]

Air sampling is also performed, if significant levels of airborne radioactivity may be present; this topic is covered later in this chapter in the section regarding protection of workers from internal contamination. The issue of contamination monitoring and contamination control is also revisited later in this chapter.

11.9   Principles of Optimization

The practice of radiation protection involves three fundamental principles:

- *Justification*: No practice should be undertaken unless sufficient benefit to the exposed individuals will offset the radiation detriment.
- *Optimization*: The magnitude of individual doses, the number of people exposed, and the likelihood of incurring exposures should be kept as low as reasonably achievable (ALARA), economic and social factors being taken into account.

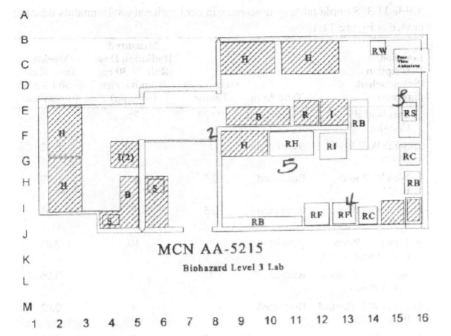

Figure 11.15 Sample of laboratory survey plan.

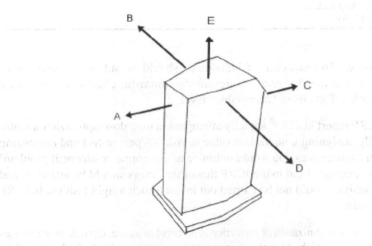

Survey Point	μrem/h @ 1m
A	50
B	55
C	49
D	75
E	55

Figure 11.16 Sample survey plan for specific device.

Table 11.3 Sample table of dose rates in occupational environments near the device in Figure 11.16

Location/ Description (See attached diagram)	Type Area	Occupancy Factor	Measured Radiation Dose Rate @ 30 cm from barrier (μrem/h)	Weekly Dose Rate @ 50 Ci-h/wk (mrem/wk)
A-East Wall B900G Corridor	Unrestricted	0.125	75	0.05
B-South Wall Room B-925	Restricted	1	160	0.82
C-West Wall Room B-926	Restricted	0.5	30	0.08
D-West Wall Room B-923	Restricted	0.5	5	0.01
E-Treatment Room Door Room B-924A	Restricted	1	10	0.05
F-North Wall Control Room B-924	Restricted	1	12	0.06
G-North Wall Control Room B-924	Restricted	1	4	0.02
H-Door Control Room B-924	Unrestricted	0.125	8	0.01
I-Colling	Unrestricted	1	5	0.03
Floor-Not Applicable, Ground Level				

- *Limitation*: The exposure of individuals should be subject to dose limits. These limits are designed to prevent deterministic effects and to reduce stochastic effects to an "acceptable" level.

An ICRP report in 1983[9] actually attempted to treat dose optimization mathematically, assigning a numerical value to risks ($/person-Sv) and optimizing radiation exposures as one would optimize an economic cost/benefit problem! It became later apparent to the ICRP that other factors should be included, and that the analysis could not be carried out in quite such a rigid fashion. In ICRP 55, they said,

"The concept of optimization of protection is practical in nature. Optimization provides a basic framework of thinking that is proper to carry out some kind of balancing of the resources put into protection, and the level of protection obtained, against a background of other factors and constraints, so as to obtain the best that can be achieved in the circumstances[10]".

A new ICRP document on the subject is currently in draft form. In most circumstances, risk/benefit decisions are made rather easily by thoughtful people, considering the risks of routine or emergency situations, and are difficult to categorize uniformly and treat in any rigorous scheme. Some situations can require a bit more thought: a common example is whether respiratory protection should be worn in situations in which airborne radioactivity is present, but the work is difficult, and requires good vision and mobility.

Wearing a respirator may mean that a person needs to work 20% longer in the area, where high gamma dose rates may be present as well, and suffer other untoward effects such as hyperthermia, difficulty in breathing, difficulty in communicating with others, and so on. Thus not wearing a respirator may actually be in the best interests of this worker and other involved workers. If the airborne concentrations are particularly high, however, wearing protection may be necessary. Assigning a $/person-Sv value and optimizing the problem on a piece of graph paper is not the way that most any health physicist would go about solving the problem. Discussions with the workers and a few simple calculations should suffice to make a decision.

11.10   Protection of Workers from Internal Contamination

In the majority of workplace environments, internal contamination should be viewed as an avoidable and rare event. In some environments, for example, uranium and thorium mining and milling operations, routine, low-level intakes of radioactive material is to be expected and controlled through constant vigilance and monitoring. In other situations, the health physicist should think of internalization of radioactive material as an event that can be prevented for the most part by good workplace design, use of proper methods when working with radioactive materials, and regular worker training and reminders of good practice. When active control of radioactive material can be expected to prevent most intakes, the following strategy is utilized.

1. *Control of the source*: The most desirable strategy for preventing intakes of radioactive material is to control the source material so that intakes of radionuclides are unlikely. A number of strategies are applicable:
   - *Storage and control*: Have all unsealed sources carefully stored and controlled such that the chance for intakes of the material, typically by inhalation or ingestion, is minimized. When sources are purchased, the quantity and form must be carefully documented. The activity must be stored in an appropriate type container, possibly a shielded container if necessary and kept in a location designated for the storage of radioactive material. Sources requiring refrigeration or freezing must be stored in cool storage devices that are exclusively used for the storage of hazardous materials (i.e., no food or drinks can ever be stored in these devices).
   - *Limited access*: Access to the material is to be limited to a few individuals who are trained and authorized to access and use radioactive materials. Access is limited by the use of physical locking devices, electronic devices whose password is known only to these few individuals, or by other similar means. If material is carefully stored and locked away and never accessed, there will be zero chance of any intakes of the material. This, of course, is not desirable, as the material was purchased for some intended use. However, limiting the access to the minimum amount possible and limiting the number of users who can access the material will reduce the chance of accidental intakes to the minimum possible. Two intentional radioactive poisonings occurred in 1996, just a few months apart, involving P-32 used for various biochemical experiments.[11] The identity of the persons who caused the poisonings and the motives were not definitively established. No harm came from the incidents, but two

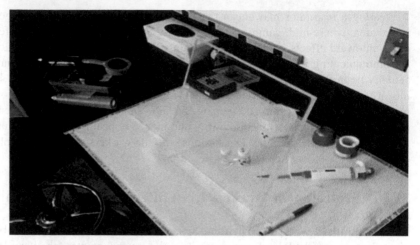

Figure 11.17 Area lined with absorbent mat paper. (From http://www.uos.harvard. edu/ehs/radsafety/gui_wor.shtml.)

individuals did have intakes of very nearly 1 ALI each of P-32 due to these incidents.

- *Continuous tracking*: Whenever radioactive material is removed from storage by these authorized individuals, the date and amount removed must be documented. As the material is used or destroyed, this is to be well documented as well, so that at any time a precise inventory of the remaining material may be calculated by knowledge of the amount received and the various amounts lost via use or radioactive decay.

2. *Control of the environment*: Next in order of desirability is control of activity in the environment in which workers contact the material.

- *Use of gloves and labcoats*: Use of protective gloves and labcoats should be absolutely mandatory in any handling of loose radioactive materials. Monitoring of hands and feet when leaving a "hot" lab environment should be regularly practiced.
- *Contamination control mat paper*: Mat paper may be purchased that is specially designed to limit the spread of liquids. The paper is absorbent on one side, and has a plastic backing that prevents movement of absorbed liquids. This kind of paper may be used to line work areas, ventilated work hoods, trays used to transfer materials within and between labs, and other uses (Figure 11.17).
- *Work within ventilated hoods*: Enclosed areas with controlled air flow are useful for working with any hazardous materials, particularly with radioactive materials that may be volatile. The hoods will have active air flow away from the worker and towards some kind of exhaust (Figure 11.18). They will typically have an access window whose height can be controlled, in order to control the "face velocity," that is, the air flow rate at the face of the work area. The flow rate in units of m/s is equal to the air flow rate (m^3/s) divided by the m^2 of open area = m/s. The flow rate may be variable, as the face window is raised or lowered, or kept constant by using a bypass. The rate should be kept high enough to prevent contamination from leaving the hood but not so high as to create turbulence in the hood working area (potential for spills). Usually a flow rate of about 0.6–1.4 m/s is good.

Figure 11.18 Fume hood. (From http://www.ofm.gov.on.ca/.)

In the ductwork leading from the hood to the exhaust point, it is important to keep the duct at negative pressure, by pulling the air from the hood with an air mover (rather than pushing it away from the hood) so that any leakage that may occur is into the duct, not out of the duct, with radioactive material possibly contaminating the duct, building materials, and so on (Figure 11.19).

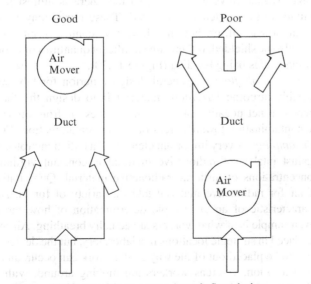

Figure 11.19 Good versus poor duct air flow design.

Figure 11.20 Glove box. (Courtesy of Terra Universal, Inc.)

The airflow from the hood is usually filtered (high-efficiency particulate filter), and may be monitored, using in-line radiation detectors,
removal of filters and counting, in-line sampling of airstream, capture
of particles, gases, tritium vapors, and so on.

- *Work within glove boxes*: A more aggressive approach to personnel protection, appropriate when working with materials of high specific activity, high radiation dose conversion factors (dose to organs or effective
  whole-body per unit activity intake) is to use a "glove box," a sealed
  enclosure into which one may insert one's arms, through closed arm-
  length sealed glove enclosures, so that there is almost no chance for
  activity to be inhaled or contacted. These boxes may be sealed, with
  no air flow at all, or they may have low, controlled air flows. The box
  may also be shielded, if there are significant quantities of photon-emitting
  radionuclides being handled (Figure 11.20).
- *Control of air flow*: A general design criterion for labs working with
  possibly airborne hazardous material is to design the facility so that
  there is a net negative pressure that moves air from areas of lower or
  no contamination towards areas of higher contamination (Figure 11.21).
- *Air sampling*: A very important element of a radiation protection program
  against intakes of radioactive material is constant vigilance over the
  concentrations of airborne radioactive material. Quantitative sampling
  of air for radioactive material takes a variety of forms. An important
  characteristic of an air sample determination of how representative a
  given sample is of what workers are actually breathing. Air monitors may
  be placed in strategic locations in a laboratory, but the devices necessarily
  need to be placed out of the way so that work can occur, and are in some
  fixed location, whereas workers are moving around, with the zone in
  which they breathe being approximately 1.5–2 meters above the ground.
  Air sampling devices need to be strategically placed to provide as

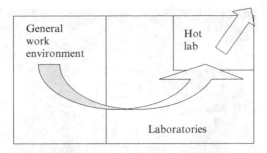

Figure 11.21 General laboratory air flow scheme.

representative a sample as possible. Then, air samplers need to be carefully calibrated so that the registered volume of air given by the device gives a good estimate of the true air volume sampled. The amount of activity collected by the device will then be divided by the reported volume of air to give an average air concentration during the sampling period.

- *Sample types*: Sample types are classically divided into the categories of "grab" or "integrated" samples. As the name indicates, a "grab" sample is a short-term sample that seeks to opportunistically obtain a sample typical of conditions at a point in time, often with a high flow rate sampler ("hi-vol" sampler) (Figures 11.22 and 23). An integrated sample is one collected over a longer period of time, generally using a lower flow rate ("low-vol" sampler). Long-term sampling may also be continuous in nature. A disadvantage to a long-term integrated sample is that a result is obtained that may average out many fluctuations in the ambient concentrations with time. A disadvantage to a grab sample is that it characterizes conditions only in a short period of time. A sampler that is coupled to an in-line detector may permit longer-term sampling, but with continuous detection and readout of the concentration values. This may also permit activation of alarms if certain concentration thresholds are exceeded. There are also small samplers that can be worn on a worker, with low-volume air samplers

Figure 11.22 "Grab" sampler (http://www. laurussystems.com/products/air_samplers.htm).

Figure 11.23 "Hi Vol" air sampler (http://www.laurussystems.com/products/air_samplers.htm).

and small sampling devices that can be placed near the nose and mouth of the worker. These are called "personal" air samplers.

- *Particulates*: Particulate matter is most effectively sampled with various kinds of filters (Figure 11.24). The filter paper must have pore sizes appropriate to stop the size of particle to be sampled. The paper must also not degrade due to the impact of the particles or their corrosiveness. Glass fiber filters, for example, may replace paper filters for air sampling in corrosive environments. The paper must also not become so overloaded with particulates that the filter becomes clogged, and the air flow rate is compromised.
- *Absorbers and adsorbers*: Gases and vapors are not subject to filtration with typical filter apparati. These substances are more effectively absorbed (into liquids, cold traps, etc.) or adsorbed onto the surfaces of specialized materials that specifically react with and trap the materials. Silver zeolite and charcoal are materials particularly useful in the absorption of airborne iodine vapors (Figure 11.25).
- *Particle sizing devices*: Aerosol particles typically have a lognormal distribution of sizes, and are thus characterized by a geometric (as

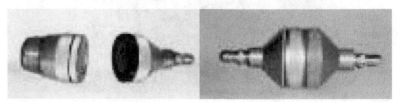

Figure 11.24 Filter holders for particulate sampling (http://www.laurussystems.com/products/air_samplers.htm).

Figure 11.25 Silver Zeolite canisters for adsorption of radioactive iodine (http://www.laurussystems.com/products/air_samplers.htm).

opposed to an arithmetic) mean, or we may use a median to characterize the distribution with a single parameter. The size distribution of a radioactive aerosol has a profound impact on its radiation dosimetry. Smaller particles will be able to penetrate deeper into the bronchial tree and be deposited in the lung spaces, whereas larger particles will tend to be trapped in the upper airways and not reach the lung tissues. Standard dose calculations provided by the ICRP use a default median particle size (1 or 5 μm). Adjustments to the dose estimates may be made if particle size sampling indicates that the aerosol of interest has a different median particle diameter. A number of specialized devices exist for determining the size distribution of aerosol particles, including the use of impaction, condensation, and optical techniques (Figure 11.26).

- *Contamination monitoring/contamination control*: All laboratories where unsealed radioactive materials are used must be periodically monitored for contamination on surfaces, as well as in the ambient air. This is easily done by routine surveys, with portable meters that can measure the radiation of interest, or with the use of filter papers to sample or "smear" surfaces, with the papers counted later in benchtop radiation detectors (see section above on Radiation Surveys). An interesting parameter applied in the measurement of surface contamination levels is the material *resuspension factor*. This quantity is given as

$$f_r = \frac{\text{atmospheric concentration Bq/m}^3}{\text{surface concentration Bq/m}^2} \equiv \text{m}^{-1}.$$

- These factors have been empirically derived for a number of quantities and situations. A guideline (not a regulatory limit) on surface contamination may be thus calculated:

$$\text{permissible surface contamination} = \frac{\text{DAC Bq/m}^3}{f_r \ \text{m}^{-1}} \equiv \frac{\text{Bq}}{\text{m}^2}.$$

11.11 Air Sampling Calculations

When an air sample is taken, the resulting sample is subjected to some form of radioactive counting. One obtains a certain number of counts in the counting time, which can be converted to an activity in Bq and then divided by the

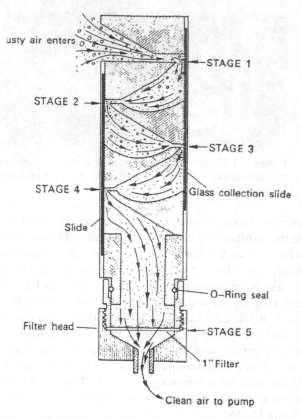

usty air enters

STAGE 1

STAGE 2

STAGE 3

STAGE 4

Glass collection slide

Slide

O–Ring seal

Filter head

STAGE 5

1" Filter

Clean air to pump

Figure 11.26 Cascade impactor for determining particle size distributions, from Cember, with permission, McGraw-Hill, 1996.

measured air volume to obtain the average air concentration over the period of sampling. To obtain the correct answer, one must remember:

- The collecting device has a certain efficiency of collection, which is probably less than 1.0. Particulate filters can have efficiencies near 100%, or 1.0 for particles of a given size or larger, but other devices may have finite efficiencies that need to be considered.
- You then count it on a detection system that has a defined efficiency of detection for the particles in question, as with any radioactive assay.
- Remember also to consider the number of these particles per disintegration of your nuclide, as with any radioactive assay.
- Consider also the system background, as with any radioactive assay.

Short-lived radionuclides will have some additional considerations:

- Some time may have elapsed from when the sample is taken from the sampling apparatus and counted on the detection system during which the activity has decayed.
- Consider also that the sample may have been decaying while it is being collected on the sampling device.

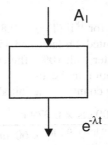

For a device collecting activity at a constant rate, if A_I is the rate of activity intake into the collecting device, which is the airborne concentration times the flow rate (Bq/min = Bq/m^3 × m^3/min), and the rate constant for loss (due to radioactive decay) is λ, the expression that describes the activity as a function of time on the collecting device is:

$$A(t) = \frac{A_I}{\lambda}(1 - e^{-\lambda t})$$

This expression was used in the ICRP II system of worker protection, in which activity intake was assumed to be constant each day, and removal from the body was with an effective removal constant λ_{eff}. The derivation of this expression is not particularly difficult. The activity is entering at the constant rate A_0 and being removed with the rate $\lambda A(t)$:

$$\frac{dA}{dt} = A_0 - \lambda A$$

$$\int_0^A \frac{dA}{A_0 - \lambda A} = \int_0^t dt$$

$$A(t) = \frac{A_0}{\lambda}(1 - e^{-\lambda t})$$

• Consider also that the sample may have been decaying while it is being counted. We obtain just an integrated number of counts during the counting time N. The activity at the beginning of the counting time A_0 is given as

$$N = \int_0^{t_C} A_0 e^{-\lambda t} dt = \frac{A_0}{\lambda}(1 - e^{-\lambda t_C})$$

$$A_0 = \frac{N\lambda}{(1 - e^{-\lambda t_C})}$$

• Consider correction for the presence of short-lived airborne contaminants. In particular, ^{222}Rn and its progeny are often present in many indoor environments. These nuclides collectively have a half-time of about 30 minutes. So if the sample is just left to decay for a few hours, these counts will die off. If not, however, these counts may be included in the overall number that is attributed to the nuclide in question, and may result in an overestimate of the airborne concentration.

Example

An environment is sampled for ^{18}F ($T_{1/2}$ 109.8 min) for 1 h at a flow rate of 5 L/min. The sample is counted 2 h later and 4500 net counts were registered in a 30 min count on a counter with 10% efficiency for the ^{18}F photons. What was the original F concentration in the air?

Considering decay during counting, the initial count rate would be:

$$A_0 = \frac{4500 \text{ cts} \times 0.379 \text{ h}^{-1}}{(1 - e^{-0.379 \text{ h}^{-1} \times 0.5 \text{ h}}) \times 60 \text{ min h}^{-1}} = 165 \text{ cpm}$$

With a 10% counting efficiency, we saw 1650 photons/min, ^{18}F emits 1.92 photons per disintegration, so the activity counted was $(1650 \, \gamma/\text{min})/(1.93 \, \gamma/\text{dis}) = 853$ dis/min = 14.2 Bq.

Correcting the sample for radioactive decay back to the end of the sampling period, we obtain:

$$A_0 = \frac{14.2 \text{ Bq}}{e^{-0.379 \text{ h}^{-1} \times 2.0 \text{ h}}} = 30.3 \text{ Bq}$$

We can then calculate the rate of activity entering the filter from:

$$A(t) = \frac{A_I}{\lambda}(1 - e^{-\lambda t})$$

$$A_I = \frac{30.3 \text{ Bq} \, 0.379 \text{ h}^{-1}}{(1 - e^{-0.379 \text{ h}^{-1} \times 1 \text{ h}})} = 36.4 \frac{\text{Bq}}{\text{h}}$$

The concentration is just:

$$C = \frac{36.4 \text{ Bq}}{\text{h}} \frac{\text{h}}{5000 \text{ ml}} = 0.0072 \frac{\text{Bq}}{\text{ml}}$$

1. *Control at the Worker*: If loose contamination is going to be present, particularly at higher levels, the worker may be protected through use of protective clothing (more than a lab coat and pair of gloves) and active respiratory protection.

Figure 11.27 Use of protective clothing. (Courtesy of Nato Photos, http://www.nato.int/multi/photos/2003/m031007a.htm.)

a. *Protective Clothing*: Anticontamination garments (Anti-Cs) may be full
   body or partial body (Figure 11.27):
   - Lab coats
   - Coveralls
   - Caps
   - Gloves (one or more layers)
   - Respiratory protection
   - Shoes
   - Shoe covers
   - Heavy taping of junctions
b. *Respiratory Protection*: The air that a worker breathes may be protected
   through the use of small simple filters covering the nose and mouth,
   larger, partial face or full face respirators, or completely enclosed sys-
   tems with supplied air (Figures 11.28 and 11.29). Technology includes:
   - Masks
   - Air-purifying respirators
   - Air-supplying respirators
   - Self-contained breathing apparatus (SCBA)

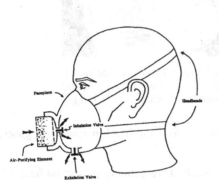

FIGURE 2-7. Typical quarter-mask respirator

FIGURE 2-8. Typical half-mask respirator

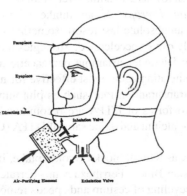

Figure 11.28 Partial face and full face respirators, with filters.

Figure 11.29 Images of full and partial face respirators. (From www.hcn.org and http://www.iaea.org/About/ Policy/GC/GC47/images/ protectiveclothing_240 × 160.jpg.)

- Filters and cartridges designed to trap and block particulates, gases, and vapors.

Respirators are supplied with so-called "respiratory protection factors," which are simply the ratio:

$$\text{Protection factor} = \frac{\text{Ambient airborne concentration}}{\text{Concentration inhaled}}.$$

The National Institute for Occupational Safety and Health provides tests and ratings for many commercially available respirators. PFs vary from values as low as 5 to some over 10,000.

  c. *Decorporation Therapy*: If an intake of radioactive material has occurred, as an absolute last resort, sometimes the removal of activity from the body can be accelerated. Strategies include:
  - *Chelation*: Diethylenetriamine pentaacetic acid (DTPA) is thought to be the chelating agent most effective for mobilizing transportable forms of transuranium ions such as plutonium (Pu) and americium (Am). Two forms of DTPA are recommended for therapeutic use; trisodium-calcium and –zinc salts of DTPA (CaDTPA and ZnDTPA).
  - *Blocking*
    - Iodine, as KI will block thyroid uptake, if given early.
    - "Prussian Blue:" Ferrihexacyano-Ferrate (II) blocks enterohepatic recycling of cesium and speeds removal from the body.

- *Increased fluid intake*: H-3 as ^3H_2O (T_2O or HTO) turnover is related to body turnover of water. Increasing fluid intake will decrease the effective half-time in the body.
- *Lavage*: Nasopharyngeal, gastric, or lung lavage can actively remove accessible activity. Not a fun day at work when this is needed!

11.12 Methods for Gathering Bioassay Data

We estimate internal doses using mathematical models (Chapter 10) which eventually give the absorbed dose to individual organs of the body and effective whole-body doses per unit intake of radioactive material by inhalation or ingestion. If we have a very accurate estimate of how much activity has entered the body (because the person somehow knows how much was taken in, or knows how much was being worked with and we can assay how much is left, if we have very accurate air monitoring results and good knowledge of the time that a person was in an area with a well-defined concentration, etc.), we can immediately calculate the person's committed doses from an intake event. In most cases, however, the greatest unknown to our analysis is the amount of activity involved in the actual intake. We can then look back and try to estimate, from some reconstructions of the incident, how much activity was likely to have been taken in. But if at all possible we want to make a direct evaluation of how much activity is in the worker. Thus we perform an assay of the worker, known as a *bioassay* study.

"Bioassay" is defined as: Determination of the kind, quantity, location, and/or retention of radionuclides in the body by direct (in vivo) measurements or indirect (in vitro) analysis of material removed from the body.

Bioassays are thus conducted using in vivo monitoring or in vitro measurements: we may measure what is still in the body of the worker, or what is being excreted from the worker's body. Both types of measurements can be used with our standard kinetic models to back-calculate an estimate of the original intake, and then use our standard dose coefficients (dose/unit intake) to estimate the organ and effective whole-body doses. This has some significant advantages over other reconstruction methods:

- The main advantage is that we are working with data that we know to be true about the individual. Other modeling efforts, using air sampling data, reconstruction of intake scenarios, and the like usually provide only rough estimates of intake. Bioassay data have a number of uncertainties which we discuss.[12] But if we have a reasonably accurate estimate of the activity in the worker's body or excreta, we have a solid piece of data with which we can work.
- The metabolic models used to derive the dose coefficients in our dosimetric models assume standard kinetic models. With multiple measurements in a worker, we can make a direct evaluation of the biological half-time for clearance in that particular worker, which may differ from the standard models. If we observe data that vary significantly from the standard model, we can make adjustments to the dose coefficients to account for this. If not, we have more evidence confirming the adequacy of the

standard models. Either outcome is useful to the radiation protection community to know, so, if possible, it is always advisable to publish the results observed.

In vivo monitoring involves measuring nuclides with significant penetrating emissions, and/or that are strongly retained in the whole-body or certain organs. Interestingly, this can include bremsstrahlung measurements from internalized beta emitters, if the energy is high enough. In the ^{32}P intake incident at MIT,[11] a very good characterization was made of the whole-body retention of ^{32}P through measurement of its bremsstrahlung emissions. In vitro monitoring involves the measurement of any type of radiation in excreta samples taken from the body.

11.12.1 In Vivo Counting

In vivo counting is referred to by many as "whole-body counting," as a large proportion of counting is done in "whole-body counters," in which the activity in the entire body of a person is measured with external detectors. However, much "partial-body" external counting is done as well, as we show shortly. Detector systems may be:

- Single unshielded detectors
- Shielded multiple detectors

Almost any configuration of system can be envisioned and implemented. Some of the most common types are:

- *Bed-type counters*: The person reclines in a shielded room with one or more detectors viewing the body.
- *Moving bed counters*: The person reclines on a bed that moves past one or more detectors.
- *Chair-type counters*: The person sits in a chair and one or more detectors evaluate the whole-body or parts of the body.
- *Stand-up "quickie" detector systems*: The worker stands in an apparatus that involves an array of detectors and a rapid approximate evaluation is made of body content, usually more for screening purposes than for final determination of body content (Figure 11.30).

Detector types include:

- NaI
- Organic scintillators
- Germanium semiconductor
- Gas-filled detectors

To evaluate the activity within individual organs, we may study (Figures 11.31 to 11.33):

- *The pulmonary region*: This is very common in the study of the possible inhalation of insoluble materials. Detector types include NaI, gas proportional, phoswich, HPGE, and others. This is important for the study of Pu, U, Am, and other nuclides that may have low-energy emissions.

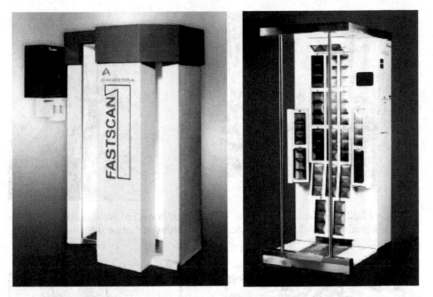

Figure 11.30 'Quickie" whole body counters. (From http://www.canberra.com/about/Capabilities.pdf.)

- *The thyroid*: Most commonly with a NaI probe, for the study of uptakes of the iodines (^{125}I,^{131}I).
- *Skull, other bones*: For the uptake of bone-seeking radionuclides, for example, ^{210}Pb, typically with NaI detectors.

Figure 11.31 Hand and foot monitor. (From http://www.canberra.com/about/Capabilities.pdf.)

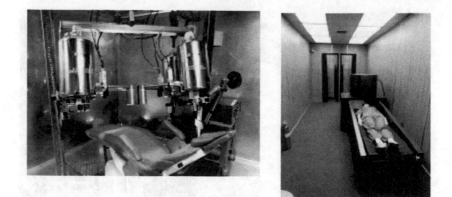

Figure 11.32 Two in vivo counters. (From http://www.pnl.gov/eshs/pub/ivrrf/pub/pnnl574/pnnl-ma-574-3.pdf, courtesy: Pacific Northwest National Laboratory.)

Figure 11.33 Specialized in vivo counters. (From http://www.ortec-online.com/systems/wbc.htm, courtesy of ORTEC, a brand of Advanced Measurement Technology, AMETEK.)

11.12.2 In Vitro Measurements

In vitro measurements involve nuclides that may not have sufficient penetrating emissions to be detected externally, and/or are strongly excreted via urine or feces.

- *Urine*: Most nuclides are eventually completely or partially excreted in the urine. This kind of sample is particularly easy to analyze, particularly using liquid scintillation methods (although the cloudiness of urine can interfere with detection efficiency). These data may be somewhat harder to interpret than a whole-body measurement, but we discuss the mathematics of the analysis shortly. A 24 hour sample is best for interpretation, as urine output and concentration vary during the day.
- *Feces*: Fecal samples are harder to obtain and analyze (especially for α and β emitters). Fecal sampling is unpleasant for all parties involved, including the worker, and fecal excretion of many nuclides is not particularly strong. However, in some cases (notably thorium intakes), fecal sampling may provide the only reliable estimates of intake.
- *Blood*: Blood samples are easy to obtain and analyze, but are very difficult to interpret.
- *Breath*: In some special cases, exhaled radionuclides can be used to infer retained activity in the body. Notable examples are the use of ^{222}Rn to infer ^{226}Ra activity in the skeleton and measurement of exhaled ^{14}CO$_2$ to study the body metabolism of various carbon labeled compounds.
- *Others*: Saliva (and any other body fluid) will have measurable levels of iodine after intakes of radioiodines; these samples are used as indicators, not in any quantitative evaluations. Hair will have measurable levels of heavy metals such as uranium and can be measured (again as an indicator, not a quantitative measure).

11.12.3 Interpretation of Bioassay Data

Once you have obtained one or more reliable measurements from an in vivo or in vitro bioassay system (whole-body count, urinalysis, etc.), you will want to estimate the amount of activity initially taken in by the subject, as this information is directly related to the radiation dose received via dose conversion factors. As noted above, bioassay (either in vivo or in vitro) is the determination of the kind, quantity, location, and/or retention of radionuclides in the body by direct (in vivo) measurements or indirect (in vitro) analysis of material removed from the body.

There are many uses for bioassay analyses, including:

- Detection of exposures that result in internal deposition. These can be baseline, routine, or accident. When workers first arrive at a facility in which intakes of radioactive material may occur, it is very important to obtain a baseline bioassay measurement, both to establish what burdens of activity they may be carrying with them from a previous employer and to establish what naturally occurring radionuclides may be present that could interfere with later measurements.
- Verification of the adequacy of radioactive materials control program or air monitoring program.
- Confirming the character and amount of nuclides taken in.

- Estimation of the radiological consequences of intakes.
- Evaluation of the need for work restrictions or medical attention.
- Demonstration of compliance with regulatory limits.

Several factors influence the need for a bioassay monitoring program. The NRC notes that not all facilities that work with unsealed sources require a bioassay program. The RSO might indeed perform a formal analysis showing that a program is not required, rather than designing an elaborate program to detect and monitor trivial potential exposures. The need for a program will depend on the:

- Radionuclides
- Quantities present
- Concentrations routinely encountered
- Degree of containment of the nuclides
- Training level of personnel
- Prevailing air and surface contamination levels
- Radiotoxicity: energy per decay and biokinetics
- Regulatory requirements

If you are dealing only with small quantities of radionuclides, in levels that are several orders of magnitude lower than annual permissible intake levels (ALIs), you may not need a program at all. Some references for tritium and iodine bioassay programs spell out situations in which a bioassay program is not needed. In addition, if the material is well contained (all in sealed sources, only used in hoods), is handled by well-experienced personnel, and so on, the need for a program might be less than in other circumstances. There are some regulatory limits that may apply, but often it is a matter of professional judgment to decide how much of a program is needed, how many types of measurements, how frequent, how many people are involved, and so on.

Participation

The need for a formal bioassay program is determined by the time-weighted monthly average air concentrations expected to be encountered in a facility. If the weighted monthly average air concentration is <10% of the Derived Air Concentration (DAC) or is the maximum <30% of DAC, a program may not be required. If it is required, the fraction of the workforce that should be monitored will depend on how much above these levels the measurements are. Of course, the RSO may wish to monitor her personnel even if it is not strictly required, for her own peace of mind and vigilance of the use of unsealed sources.

Frequency of Measurements

How often bioassay measurements are taken is determined by a number of factors, including:

- The effective half-life (in the body) of the nuclide to be detected
- Fluctuations in daily concentrations, especially for accidents
- Sensitivity of detection apparatus
- Potential for intake

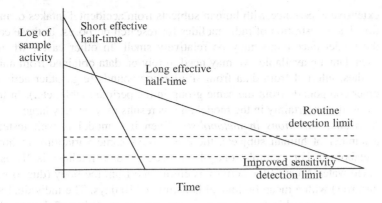

Figure 11.34 The "missed dose" concept.

As demonstrated by Figure 11.34, the value of any bioassay measurement (whole-body count, urine count, etc.) will generally decline over time, as material is eliminated from the body. At some point in time (after intake), the measured value will drop below the detection limits for your counter. For materials of short effective half-time, this point will obviously occur sooner, and sampling must be performed more frequently. For longer-retained nuclides, sampling may be performed at less frequent intervals. At the point where the sample activity falls below the detection limit, the bioassay monitoring system (be it an in vivo or in vitro counting system) can no longer "see" the remaining activity, so you cannot state with confidence that an intake has occurred.

With a longer counting time, larger bioassay sample, or other means, you may be able to improve your detection sensitivity (second dashed line). In either case, one may take the background counting rate of the system and back-calculate from this point to the time of intake, knowing the time since the last measurement and the fraction of activity expected to be in this sample at this time (defined shortly). One can then calculate the amount of activity, and thus committed dose, that may have been missed by not being able to detect the activity. Thus, when monitoring intervals are established, one is establishing the amount of dose that one is willing to overlook in the analyses, based on the detection limits of the system and the kinetics of the radionuclide in the body. This is the "missed dose" concept, and is an essential element of the design of a bioassay program.

Limitations

Various factors influence the uncertainties associated with the interpretation of the results of bioassays including:

- *Data uncertainty*: Any measurement involving measurement of radiation naturally involves the uncertainty of the measured value, as defined in Chapter 8. This is an absolute lower limit on the uncertainty that cannot be removed or reduced.
- *Model uncertainty*: As noted above, standard models have been established for most nuclides in order to calculate dose conversion factors, and bioassay data interpretation functions, as discussed shortly. These models are based on any number of sources. In the best case, we may have

extensive experience with human subjects from accidental intakes or intentional administration of radionuclides for research purposes. In these cases, the model uncertainty may be relatively small. In other cases, when no such data are available, we may need to rely on data obtained from animal studies, inferred from data from similar compounds (e.g., other actinides, other compounds from the same group in the periodic table, etc.). In these cases, the uncertainty in the model and its results may be very large.

- *Individual variations in metabolism*: Even if a model is well tested in a number of human subjects, there is always some variability in human metabolism. For example, one of the better studied nuclides is ^{3}H as tritiated water. Most of the activity is eliminated from the body (due to water turnover) with a mean biological half-time of 10 days. The individual variability in most subjects, however, shows a range from about 7 days to about 14 days under normal conditions. If a subject dramatically increases water turnover, by drinking and eliminating more fluids, the half-time may be considerably shorter. This is certainly advisable, to reduce the possible dose received. As noted above, we may also administer chelating or blocking agents in other cases to intentionally alter the subject's metabolism and retention of the compound, to reduce the eventual dose received. This is desirable for the subject, but its influence on the kinetic models and dose conversion factors used must always be borne in mind. Use of such techniques complicates the life of the health physicist in performing accurate dose calculations, but of course is always indicated if it will be helpful.

- *Uncertainty about initial conditions*: In many cases, in addition to uncertainty in the activity levels and model function, there may be uncertainty about exactly when the intake occurred (i.e., there is uncertainty in the abscissa in addition to uncertainty in the ordinate values!). If there was a clear incident, for example, a spill of liquid iodine, or air sampling data that clearly show a spike in ambient air concentrations, the time of intake may be pinpointed rather well. If, on the other hand, workers may have worked for weeks or months with no notable incidents, and just suddenly present a positive bioassay sample, estimation of the time of intake may be difficult. In addition, in some cases, workers or others may intentionally misrepresent information to the health physicist which complicates interpretation of the data. Workers may be embarrassed to discuss an accidental situation, or, in the case of intentional poisonings, reluctant to divulge details of the incident.

Investigation Levels

If the value of intake is regarded as sufficiently important to require further investigation, the specific actions might include more measurements, dose assessment, recording of exposure, notification of authorities, and work restrictions. This is referred to as an investigation level, and is numerically given as

$$\text{Routine Investigation Level} \quad IL_r = 0.3 ALI \frac{T}{365},$$

where:

$$T = \text{period between samples}$$

$$ALI = \text{annual limit on intake (from ICRP 30)}$$

$$\text{Routine Recording Level} \quad RL_r = 0.1 ALI \frac{T}{365},$$

These values will be represented by activity levels in the whole-body, body organs, or excreta. The *Derived Investigational Level* (DIL) is the value of some measurement which indicates that intake has occurred at a level requiring further action. Numerically it is equal to the investigational level multiplied by the predicted fraction of intake in an organ or sample:

$$\text{Derived Investigation Level} \quad DIL_r = IL_r m,$$

where m = predicted activity level per unit intake, usually evaluated at $T/2$. Here, $T/2$ is the midpoint of your regular sampling interval. Most regulatory and advisory bodies accept this assumption, if the actual time of intake is not known. Obviously, if you know when the intake occurred, you should use that information in your analysis. But if the time of intake is not known, one could assume that the intake occurred one day after the last negative bioassay measurement, but this may be overly conservative. So the $T/2$ assumption is often used; for example, if you sample monthly, the intake is assumed to have occurred 15 days ago.

More Definitions
Intake: The quantity taken into the body, for example, by inhalation or ingestion
Uptake: The quantity absorbed into systemic circulation
Deposition: The quantity deposited in some organ

Example
A worker inhales 100 Bq of a 1 μm AMAD aerosol. The ICRP 30 model predicts 63% deposition in lungs. Only 58% of material deposited in the lungs reaches circulation, and 25% of material in circulation is deposited in the thyroid.

Intake = 100 Bq
Lung deposition = 63 Bq
Uptake = (0.58)(0.63)(100) = 36.5 Bq
Thyroid deposition = (0.25)(36.5) = 9.1 Bq

Estimates of Intake from Bioassay Measurements

Although our metabolic models can be solved in the "forward" direction to give dose coefficients per unit intake of activity, we can also use them to calculate the fraction of activity originally taken in which may be present in any part of the model (an organ, the whole-body, or an excreta sample) at any time after intake. This fraction of intake is often referred to as the *Intake Retention Fraction* (IRF) (also shown as 'm' in some publications). These numbers allow us to look "backwards" from a measurement made on an individual to estimate the original intake, and then apply our dose conversion factors (that give committed dose per unit intake).

For example, for ^3H_2O, the whole-body retention can be shown in a simple formula to be $R(t) = A_0\exp(0.693\,t/10)$ (with t in days). The body loses water through four pathways: sweat, moisture exhaled, urine, and feces. The urinary pathway receives 60% of the total clearance. At any given point in time, we can calculate the instantaneous rate of urinary excretion (which is just the negative time derivative of the retention equation):

$$E_{\text{inst}}(t) = 0.6 \times 0.0693 \quad A_0 \exp(-0.0693t) = 0.0416 \quad A_0 \exp(-0.0693t).$$

What is more useful is to integrate the urinary excretion over a 24 hour period:

$$E_{incr}(t) = 0.6 \times A_0[\exp(-0.0693(t-1)) - \exp(-0.0693t)].$$

Then we can give IRFs for whole-body retention or urinary excretion at any time after intake:

For $A_0 = 1$:

t(d)	$R(t)$	Einst(t)	Eincr(t)
1	0.933	0.0388	0.0402
2	0.871	0.0362	0.0375
3	0.812	0.0338	0.0350
7	0.616	0.0256	0.0265
10	0.500	0.0208	0.0215
20	0.250	0.0104	0.0107

We can relate the retention $R(t)$ to activity measured using in vivo measurements in the whole-body, in this case, or to retention in an organ, if we developed the $R(t)$ function for the organ. For the specific case of ^3H_2O, one cannot detect the activity with external measurements, but in other cases this may be possible. Alternatively, we can relate the excretion, $E_{incr}(t)$, to in vitro measurements made over a 24 hour period and relate them to the appropriate IRF. Why would we choose either the in vivo or in vitro approach? Sometimes, as in the case just noted for ^3H_2O, there is no choice in the matter. In other cases, we may have more reliable data from one counting system or another. The magnitude of the IRF also guides this choice. The higher the IRF, in general, the more reliable our interpretation becomes. If we have reliable data from both approaches, we can use both sets of data.

For individual measurements, the estimate of intake I is just the observed value V_i divided by the IRF value:

$$I = \frac{V_i}{IRF_i}$$

For multiple measurements, we can obtain a single estimate of intake as follows. What we would like to do is to minimize the sum of squared differences between the observed values and the expected value, which is the product of the intake and the IRF:

$$\sum_i (V_i - I \cdot IRF_i)^2$$

$$I = \frac{\sum_i V_i \cdot IRF_i}{\sum_i IRF_i^2}$$

This represents an unweighted estimate of intake (i.e., all the values are given the same importance in the least squares calculation. We can also weight each point by a value proportional to the inverse of the variance of the point. At later times, we have less activity, so the relative variance increases. If we weight this equation, then, by weighting factors of $1/IRF$ (the expected activity at a time)

$$I = \frac{\sum\limits_i V_i}{\sum\limits_i IRF_i}$$

This result looks impossibly simple, but it is correct. The derivation is somewhat involved, so we do not go into it.

Example

For a suspected tritium (HTO) intake, the following data were gathered in a worker (24 hr urine sample).

$$\text{Day 2: } 40\,\text{MBq} \quad IRF = 0.0375$$

$$\text{Day 7: } 30\,\text{MBq} \quad IRF = 0.0265$$

$$\text{Day 10: } 22\,\text{MBq} \quad IRF = 0.0215$$

Individual estimates of intake:

$$\text{Day 2: } 40\,\text{MBq}/0.0375 = 1070\,\text{MBq}$$

$$\text{Day 7: } 30\,\text{MBq}/0.0265 = 1130\,\text{MBq}$$

$$\text{Day 10: } 22\,\text{MBq}/0.0215 = 1020\,\text{MBq}$$

An unweighted, least squares estimate would be:

$$\frac{(40 \times 0.0375 + 30 \times 0.0265 + 22 \times 0.0215)\,\text{MBq}}{0.0375^2 + 0.0265^2 + 0.0215^2} = 1080\,\text{MBq}$$

A weighted, least squares estimate would be:

$$\frac{(40 + 30 + 22)\,\text{MBq}}{0.0375 + 0.0265 + 0.0215} = 1080\,\text{MBq}$$

Then, looking up the dose conversion factor for an intake of 1 Bq of ^{3}H, we very easily calculate the committed dose:

$$D(\text{Sv}) = I\,(\text{Bq}) \times DCF\,(\text{Sv/Bq})$$

$$D = 1.08 \times 10^9\,\text{Bq} \times 1.67 \times 10^{-11}\,\text{Sv/Bq} = 0.018\,\text{Sv}.$$

Large tables of IRFs for many nuclides were provided in a recent special issue of the *Health Physics Journal*[13] (Figure 11.35 and Table 11.4).

11.13 Criticality and Criticality Control

A unique area of radiation protection practice involves protection of workers in situations in which *criticality* may occur. This is principally an external dose problem, involving exposures to neutron and photon radiation, as with general external protection. The technical concepts, however, are specialized, and the situations in which criticality may occur may involve sudden, very high exposures to intense radiation fields, so it deserves special treatment and consideration. Criticality is essential to the functioning of a nuclear reactor. Situations involving uncontrolled criticality (criticality accidents) are among

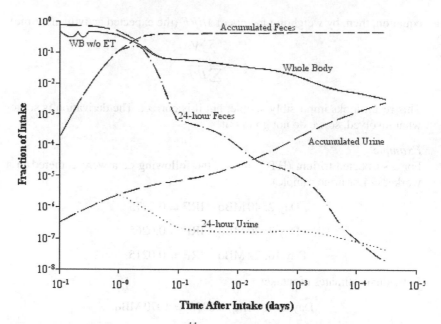

Figure 11.35 Sample plot of IRFs[11]. (Reprinted with permission from the Health Physics Society.)

some of the most serious that a health physicist may ever deal with, as discussed through various examples later in this section.

In order to discuss criticality, a number of definitions and relationships essential to the understanding of the process must be covered. You may recall way back in Chapter 2 that two nuclear reactions can result in the release of energy: the fission of heavy nuclei (e.g., ^{238}U (Z = 92)), which may split into two elements of mass number each around 80–140, or the fusion of light elements (e.g., H) into elements of higher mass number. Fusion reactions are of high interest to power production, but are still at present a matter for interesting research and not widespread practical application. We concentrate thus on fission reactions. Our first definition, *fission*, is the splitting of the nucleus of a heavy atom into two lighter nuclei. It is accompanied by the release of neutrons, X-rays, gamma rays, and kinetic energy of the fission products (Figure 11.36). It is usually triggered by collision with a neutron, but in some cases can be induced by protons, other particles, or gamma rays. Spontaneous fission can also occur in nuclides such as californium-252 (^{252}Cf) and plutonium-240 (^{240}Pu).

- Criticality is the attainment of physical conditions such that a fissile material will sustain a chain reaction.
- Fissionable material is material capable of undergoing fission.
- Fissile material is material that is capable of undergoing fission by interaction with slow neutrons (the definition is more restrictive than "fissionable"). Fissile material is uranium-233, uranium-235, plutonium-239, plutonium-241, or any combination of these radionuclides.

Excepted from this definition are:

1. Natural uranium or depleted uranium that is unirradiated.

Table 11.4 Sample table of IRFs from Ref. 10.

Time	WB w/o ET	WB	Acc. Urine	Acc. Feces	Inc. Urine	Inc. Feces
0.25	4.69×10^{-1}	7.33×10^{-1}	9.95×10^{-3}	1.80×10^{-3}	9.95×10^{-3}	1.80×10^{-3}
0.5	4.50×10^{-1}	6.56×10^{-1}	2.22×10^{-2}	8.58×10^{-3}	2.22×10^{-2}	8.58×10^{-3}
0.75	4.31×10^{-1}	5.91×10^{-1}	3.23×10^{-2}	1.78×10^{-2}	3.23×10^{-2}	1.78×10^{-2}
1	4.13×10^{-1}	5.37×10^{-1}	4.08×10^{-2}	2.76×10^{-2}	4.08×10^{-2}	2.76×10^{-2}
1.25	3.96×10^{-1}	4.93×10^{-1}	4.79×10^{-2}	3.70×10^{-2}	3.80×10^{-2}	3.52×10^{-2}
1.5	3.81×10^{-1}	4.57×10^{-1}	5.41×10^{-2}	4.55×10^{-2}	3.19×10^{-2}	3.69×10^{-2}
1.75	3.68×10^{-1}	4.27×10^{-1}	5.96×10^{-2}	5.29×10^{-2}	2.72×10^{-2}	3.51×10^{-2}
2	3.57×10^{-1}	4.03×10^{-1}	6.44×10^{-2}	5.93×10^{-2}	2.36×10^{-2}	3.16×10^{-2}
2.25	3.48×10^{-1}	3.84×10^{-1}	6.87×10^{-2}	6.46×10^{-2}	2.08×10^{-2}	2.76×10^{-2}
2.5	3.39×10^{-1}	3.67×10^{-1}	7.27×10^{-2}	6.90×10^{-2}	1.86×10^{-2}	2.35×10^{-2}
2.75	3.32×10^{-1}	3.54×10^{-1}	7.63×10^{-2}	7.27×10^{-2}	1.68×10^{-2}	1.98×10^{-2}
3	3.26×10^{-1}	3.43×10^{-1}	7.97×10^{-2}	7.58×10^{-2}	1.53×10^{-2}	1.65×10^{-2}
4	3.07×10^{-1}	3.13×10^{-1}	9.09×10^{-2}	8.35×10^{-2}	1.13×10^{-2}	7.77×10^{-3}
5	2.94×10^{-1}	2.96×10^{-1}	9.99×10^{-2}	8.73×10^{-2}	8.93×10^{-3}	3.82×10^{-3}
6	2.84×10^{-1}	2.85×10^{-1}	1.07×10^{-1}	8.95×10^{-2}	7.36×10^{-3}	2.12×10^{-3}
7	2.77×10^{-1}	2.77×10^{-1}	1.14×10^{-1}	9.08×10^{-2}	6.24×10^{-3}	1.37×10^{-3}
8	2.70×10^{-1}	2.71×10^{-1}	1.19×10^{-1}	9.18×10^{-2}	5.43×10^{-3}	1.01×10^{-3}
9	2.65×10^{-1}	2.65×10^{-1}	1.24×10^{-1}	9.27×10^{-2}	4.82×10^{-3}	8.14×10^{-4}
10	2.60×10^{-1}	2.60×10^{-1}	1.28×10^{-1}	9.33×10^{-2}	4.36×10^{-3}	6.98×10^{-4}
20	2.23×10^{-1}	2.24×10^{-1}	1.60×10^{-1}	9.80×10^{-2}	2.56×10^{-3}	3.75×10^{-4}
30	1.99×10^{-1}	1.99×10^{-1}	1.81×10^{-1}	1.01×10^{-1}	1.83×10^{-3}	2.69×10^{-4}
40	1.81×10^{-1}	1.81×10^{-1}	1.96×10^{-1}	1.03×10^{-1}	1.33×10^{-3}	1.97×10^{-4}
50	1.68×10^{-1}	1.69×10^{-1}	2.08×10^{-1}	1.05×10^{-1}	9.67×10^{-4}	1.45×10^{-4}
60	1.59×10^{-1}	1.59×10^{-1}	2.16×10^{-1}	1.06×10^{-1}	7.12×10^{-4}	1.08×10^{-4}
70	1.52×10^{-1}	1.52×10^{-1}	2.22×10^{-1}	1.07×10^{-1}	5.30×10^{-4}	8.14×10^{-5}

(Continued)

Table 11.5 (Continued.)

80	1.47×10^{-1}	1.47×10^{-1}	2.26×10^{-1}	1.08×10^{-1}	3.99×10^{-4}	6.20×10^{-5}
90	1.43×10^{-1}	1.43×10^{-1}	2.30×10^{-1}	1.09×10^{-1}	3.05×10^{-4}	4.78×10^{-5}
100	1.40×10^{-1}	1.40×10^{-1}	2.32×10^{-1}	1.09×10^{-1}	2.37×10^{-4}	3.75×10^{-5}
200	1.28×10^{-1}	1.28×10^{-1}	2.43×10^{-1}	1.11×10^{-1}	4.87×10^{-5}	7.75×10^{-6}
300	1.24×10^{-1}	1.24×10^{-1}	2.46×10^{-1}	1.11×10^{-1}	2.38×10^{-5}	3.69×10^{-6}
400	1.22×10^{-1}	1.22×10^{-1}	2.48×10^{-1}	1.11×10^{-1}	1.31×10^{-5}	2.01×10^{-6}
500	1.21×10^{-1}	1.21×10^{-1}	2.49×10^{-1}	1.12×10^{-1}	7.31×10^{-6}	1.12×10^{-6}
600	1.20×10^{-1}	1.20×10^{-1}	2.50×10^{-1}	1.12×10^{-1}	4.08×10^{-6}	6.21×10^{-7}
700	1.20×10^{-1}	1.20×10^{-1}	2.50×10^{-1}	1.12×10^{-1}	2.29×10^{-6}	3.46×10^{-7}
800	1.19×10^{-1}	1.19×10^{-1}	2.50×10^{-1}	1.12×10^{-1}	1.28×10^{-6}	1.93×10^{-7}
900	1.19×10^{-1}	1.19×10^{-1}	2.50×10^{-1}	1.12×10^{-1}	7.19×10^{-7}	1.08×10^{-7}
1000	1.19×10^{-1}	1.19×10^{-1}	2.50×10^{-1}	1.12×10^{-1}	4.04×10^{-7}	6.03×10^{-8}
2000	1.19×10^{-1}	1.19×10^{-1}	2.50×10^{-1}	1.12×10^{-1}	1.41×10^{-9}	1.98×10^{-10}
3000	1.19×10^{-1}	1.19×10^{-1}	2.50×10^{-1}	1.12×10^{-1}	6.05×10^{-12}	7.87×10^{-13}
4000	1.19×10^{-1}	1.19×10^{-1}	2.50×10^{-1}	1.12×10^{-1}	3.08×10^{-14}	3.79×10^{-15}
5000	1.19×10^{-1}	1.19×10^{-1}	2.50×10^{-1}	1.12×10^{-1}	1.75×10^{-16}	2.08×10^{-17}
6000	1.19×10^{-1}	1.19×10^{-1}	2.50×10^{-1}	1.12×10^{-1}	1.05×10^{-18}	1.23×10^{-19}
7000	1.19×10^{-1}	1.19×10^{-1}	2.50×10^{-1}	1.12×10^{-1}	6.45×10^{-21}	7.52×10^{-22}
8000	1.19×10^{-1}	1.19×10^{-1}	2.50×10^{-1}	1.12×10^{-1}	4.02×10^{-23}	4.67×10^{-24}
9000	1.19×10^{-1}	1.19×10^{-1}	2.50×10^{-1}	1.12×10^{-1}	2.53×10^{-25}	2.93×10^{-26}
10000	1.19×10^{-1}	1.19×10^{-1}	2.50×10^{-1}	1.12×10^{-1}	1.60×10^{-27}	1.85×10^{-28}
20000	1.19×10^{-1}	1.19×10^{-1}	2.50×10^{-1}	1.12×10^{-1}	1.97×10^{-49}	2.24×10^{-50}
30000	1.19×10^{-1}	1.19×10^{-1}	2.50×10^{-1}	1.12×10^{-1}	3.16×10^{-71}	3.56×10^{-72}

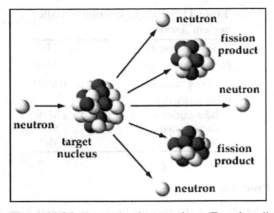

Figure 11.36 Neutron nuclear reactions. (From http://
www.childrenofthemanhattanproject.org/ML/Photo-
Pages/MLP-498.htm.)

2. Natural uranium or depleted uranium that has been irradiated in thermal
   reactors only.

A *fission fragment* is a nucleus resulting from nuclear fission, which carries
kinetic energy from that fission. This definition is used only in contexts where
the particles themselves have kinetic energy and thus could represent a hazard,
irrespective of whether the particles are radioactive. Otherwise, the more usual
term *fission product* is used. A fission product is a radionuclide produced by
nuclear fission. This definition is used in contexts where the radiation emitted
by the radionuclide is the potential hazard.

Energy Released During Fission

The energy released during the fission process is given as the energy equivalent
of the mass difference between the reactants and products:

$$E_f = (M - m_1 - m_2 - m_n)c^2,$$

where:

M = mass of fissioned nucleus
m_1, m_2 = mass of fission fragments
m_n = mass of all neutrons

A key example, in power production and use in nuclear weapons, is the
fission of ^{235}U:

$$^{235}_{92}U + ^{1}_{0}n \rightarrow ^{236}_{92}U$$

$$^{236}_{92}U \rightarrow ^{A_1}_{Z_1}F_1 + ^{A_2}_{Z_2}F_2 + v + ^{1}_{0}n + Q$$

Here F_1 and F_2 are fission products 1 and 2. The value of Q is around 200
MeV and is distributed approximately as shown in Table 11.5.

Most of this energy is dissipated as heat within the critical assembly. The
heat energy is converted to electrical energy through the boiling of circulating
water, the steam from which drives turbines that generate electricity.

Table 11.5 Energy distribution after nuclear fission.

Fission fragments	167 MeV
Neutron kinetic energy	6 MeV
Fission gamma rays	6 MeV
Radioactive decay:	
Beta particles	5 MeV
Gamma rays	5 MeV
Neutrinos	11 MeV

The Fission Products

- The atomic numbers of the various fission *fragments* vary from about 30 (Zn) to about 64 (Gd).
- These species are all radioactive, and decay to other fission products by a variety of reactions.
- The species have a variety of half-lives, from microseconds to thousands of years.
- The species may be solid (e.g., Zn and Gd), gaseous (e.g., Kr, Xe), or volatile (e.g., I). See Figure 11.37.

The collective activity of the activity in a core inventory is given as

$$A = 3.81 \times 10^{-6} T^{-1.2} \text{ Bq/fission, or}$$

$$A = 1.03 \times 10^{-16} T^{-1.2} \text{ Ci/fission,}$$

where T is the time in days after fission.

Criticality

Note from Figure 11.37 that a neutron induces the fission reaction, and one or more neutrons are released from the fission reaction. If these released neutrons cause other fission reactions, a chain reaction is set up. If every fission reaction

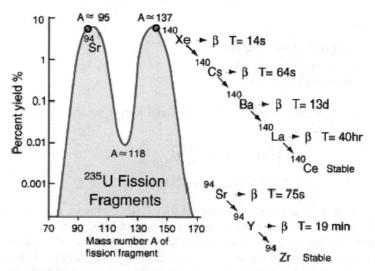

Figure 11.37 Distribution of fission fragments by mass number. (From http://hyperphysics.phy-astr.gsu.edu/.)

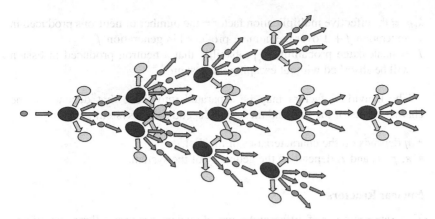

Figure 11.38 A nuclear chain reaction. (From http://resourcefulphysics.org/ freematerial/foundation/text/Fission_and_Fusion.doc.)

induces exactly one more fission reaction, we say that we have achieved criticality; that is, the system is critical.

If more neutrons are absorbed in the system (not in fissile nuclei) than are produced in fission reactions, the critical reaction is not self-sustaining, and the system is said to be *subcritical*. If the rate of fission neutron production exceeds the rate of loss, the chain reaction will be accelerating, and the system is said to be *supercritical* (Figure 11.38).

The Mathematics of Criticality

In order for a system to undergo sustained criticality, the number of neutrons being produced in fission reactions must be equal to that being absorbed by the fissile material. Therefore, the ratio of (neutrons out)/(neutrons in) must be 1.0. We can treat this mathematically and study the various factors that influence the production and loss of neutrons. This is modeled in the famous "four factor formula" used in nuclear engineering. The four factor formula is given as

$$k_\infty = \eta\varepsilon pf$$

where:

k_∞ = infinite multiplication factor

η = number of neutrons emitted per thermal neutron absorbed in the fuel (value perhaps 1.0–3.0)

ε = fast fission factor = the total number of fast (fission) neutrons produced divided by the total number of fast neutrons produced by thermal-neutron induced fission (value between 1.0–1.29)

p = resonance escape probability = fraction of the fast, fission-produced neutrons that are eventually thermalized (value between 0.0–1.0)

f = thermal utilization factor = the fraction of thermalized neutrons that are actually absorbed in the fuel (value between 0.0–1.0)

We can then define the *effective multiplication factor*, k_{eff}:

$$k_{eff} = k_\infty \times L$$

$$k_{eff} = \frac{N_{f+1}}{N_f}$$

where:

k_{eff} = the effective multiplication factor = the number of neutrons produced in generation $f + 1$ over the number produced in generation f

L = nonleakage probability = probability that a neutron produced in fission will be absorbed without escaping the system[14]

The individual factors may be separated into those that depend on the characteristics of the fuel and those that depend on the system geometry:

- η depends on the characteristics of the fuel.
- ε, p, f, and L depend on the geometry of the system.

Nuclear Reactors

There are a number of different designs of nuclear reactors. All reactors of any type share a number of common features. The materials used in these features can vary considerably, however. The features are:

- *Fuel*: Fissile material that generates heat.
- *Moderator*: Material that moderates the fast neutrons formed in the fission reaction to thermal energies.
- *Coolant*: Substance that removes heat from the reactor core, both to maintain a safe operating temperature of the system and transfer the heat eventually to the turbine/generator system for the production of electricity.
- *Control rods*: Fuel is maintained in long thin tubes arranged into a critical assembly (core); within the core other rods containing neutron poisons (discussed above, typically cadmium or boron) are used to control (including rapidly shutting down) the nuclear reaction.
- *Reflector*: Material placed around the core to reflect escaping neutrons back into the core (water, graphite, or beryllium, e.g.).

Currently in the United States, there are 104 operating commercial nuclear reactors NRC. Sixty-nine (69) of the total are Pressurized Water Reactors (PWRs) generating 65,100 net megawatts (electric) and 35 units are Boiling Water Reactors (BWR) generating 32,300 net megawatts (electric). World-wide, there are 441 nuclear power reactors operating in 31 different countries. In total, they account for approximately 17% of worldwide electricity generation and provide around half of the electricity for some industrialized countries.

As the name implies, PWRs use water maintained under high pressures (to prevent boiling) as the core coolant. As steam is needed to turn the turbines, PWRs thus contain a separate system for transferring heat from the pressurized water circuit to a second water circuit that is allowed to boil (steam generator). In the BWR the steam from the coolant is transported directly to the turbines.

Canada employs a number of commercial reactors that use heavy water (D_2O) instead of normal H_2O, and are called Heavy Water Reactors (HWRs), and often referred to as CANDU (CANadian-Deuterium-Uranium) reactors. Heavy water absorbs fewer neutrons than ordinary water, so more thermal neutrons survive in the reactor environment. In these reactors, natural uranium fuel, consisting of (99.3%) ^{238}U and (0.7%) fissionable ^{235}U may be used, avoiding the expensive ^{235}U fuel enrichment step (light water reactors need uranium fuel enriched to several percent ^{235}U, and the separation of ^{235}U from ^{238}U is expensive and time and energy consuming).

Gas-Cooled Reactors (GCRs) may use carbon dioxide or helium as the coolant instead of water, and are used in some systems in the United Kingdom and France. Other reactor designs used outside the United States include Fast Breeder Reactors (FBRs) and Pressurized Heavy Water Reactors (PHWRs). In FBRs, some of the ^{238}U in the fuel is converted to ^{239}Pu isotope, and actually converts more ^{238}U to usable fuels than the reactor consumes.

A number of new reactor designs have been under development recently (Figures 11.39 and 11.40). Due to the expense of the process and political resistance, no new nuclear reactors have been ordered in the United States for decades. It now appears that some new traditional reactors will be built, and some of the new design reactor types may also reach commercial viability:

- The *Pebble Bed Reactor* uses pyrolytic graphite as the neutron moderator, and an inert or semi-inert gas such as helium, nitrogen, or carbon dioxide as the coolant, at very high temperature, to drive the turbines, thus eliminating the steam management system and increasing the overall heat transfer efficiency. Also, unlike water, gases do not dissolve and transport contaminants.
- *Very-High-Temperature Reactor* (VHTR) systems use either a Gas Turbine-Modular Helium Reactor (GT-MHR) or the pebble fuel of the Pebble Bed Modular Reactor (PBMR). The primary circuit is connected to a steam reformer/steam generator to deliver process heat
- The *SuperCritical Water-cooled Reactor* (SCWR) would use either a thermal neutron spectrum. or a closed cycle with a fast-neutron spectrum. Both approaches use a high-temperature high-pressure water-cooled reactor that operates above the thermodynamic critical point of water (22.1 Mpa, 374°C) to achieve a thermal efficiency of 40% or more.
- *Gas-cooled Fast Reactors* (GFRs) work with fast-neutron spectra and high outlet temperatures (~850°C) to enhance high thermal efficiency.
- The *Lead-cooled Fast Reactor* (LFR) employs a fast-neutron spectrum and a lead or lead/bismuth eutectic liquid-metal-cooling system.
- *Sodium-cooled Fast Reactor* (SFR) system uses a fast-neutron spectrum and liquid sodium as the coolant.

Reactor fuel is organized into fuel rods, which are bundled together in a geometry to facilitate the fission process and maintain criticality. Control rods are interspersed with the fuel rods, as noted above, to control the reaction process. As fuel rods are used, the fissile material is gradually depleted. As new fuel is brought in to replace old rods, partially used rods are then cycled closer and closer to the center of the core (Figures 11.41 and 11.42). When $k_{eff} = 1.000$, we have criticality. When the control rods are fully inserted, $k_{eff} = 0.0$. The rods may be substantially withdrawn, to drive $k_{eff} > 1.0$, and the power level rises. When power is at the desired level, the control rods are reinserted, to bring k_{eff} to 1.000.

Some Other Definitions:

Excess reactivity: $\Delta k = k_{eff} - 1$.

Reactor Period: The time during which the neutron population (and thus the power level) in the reactor increases by a factor of e.

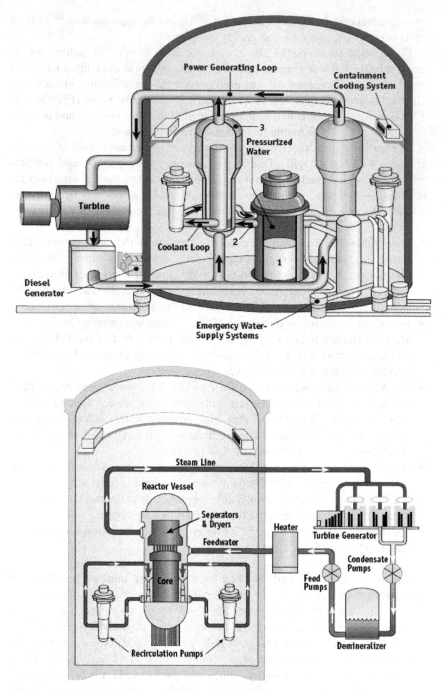

Figure 11.39 Pressurized water reactor (top) and boiling water reactor (bottom). (From http://eia.doe.gov/cneaf/nuclear/page/nuc_reactors/pwr.html, provided by the Energy Information Administration.)

Fission Product Inventory

As a reactor operates and ^{235}U atoms are fissioned, the fission product inventory builds up. The quantities of those with a relatively short half-life will

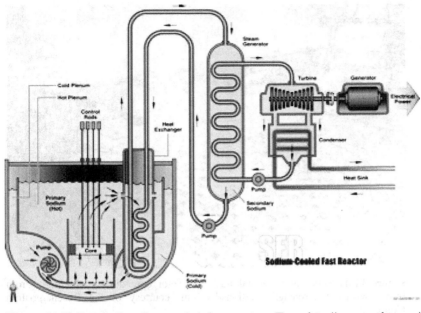

Figure 11.40 Proposed sodium-cooled fast reactor. (From http://www.ne.doe.gov/
genIV/neGenIV4fastSpectrum.html.)

build to an equilibrium level (as production = decay). The quantities of those
with very long half-lives just increase linearly. After the reactor is shut down,
the short-lived products will decay away, but the long-lived species remain
radioactive for very long times, and emit significant quantities of mixed radi-
ations (α, β^-, β^+, γ, etc.) (Figure 11.43). This constitutes "high level waste,"
or "spent fuel".

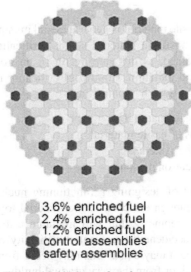

3.6% enriched fuel
2.4% enriched fuel
1.2% enriched fuel
control assemblies
safety assemblies

Figure 11.41 Core of nuclear reactor.
(From http://www.npp.hu/mukodes/
tipusok/aktiv-e.htm.) (see color plate)

Figure 11.42 Photo of an actual nuclear reactor core during refueling. (From http://www.rosatom.ru/english/stations/lwr.htm, courtesy of the Rosenergoatom Concern.)

Activation Products

Activation products are formed when neutrons are absorbed by other materials, causing a transmutation into a radioactive species. Activation products are created in many structural elements of the reactor (particularly in its steel and concrete components) and include tritium (^{3}H), ^{14}C, ^{60}Co, ^{55}Fe, and ^{63}Ni (Figure 11.44). These nuclides create difficulties in the operation and ultimate demolition of the reactor, and affect the extent to which materials can be recycled.

Criticality Control

In facilities where fissile material is generated and manipulated, it is important to very carefully control the material so that criticality is not accidentally achieved and maintained, as this may result in the release of a significant amount of energy and radiation. Control of criticality in materials involving fissile material may be done by controlling:

- The quantity of fissile material
- The concentrations of fissile material
- The geometry of the system
- The presence of neutron poisons

The entire concept of designing a functioning nuclear reactor or nuclear weapon is to bring enough enriched fissile material together in the presence of a moderator in a geometry that will maintain criticality. Should such a configuration occur accidentally, unwanted criticality may occur, and a significant radiation hazard may exist. This may occur from a sudden accidental movement of material, or from the very gradual buildup of material over time (e.g., at the bottom of a storage tank or in the elbow of a pipe through which solutions are flowing).

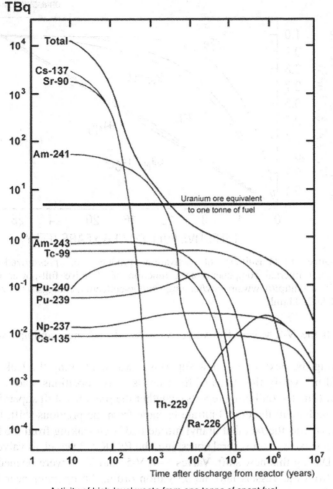

Figure 11.43 Fission product inventory at various times after removal from reactor. (From http://www.uic.com.au/uicphys.htm.)

Criticality Accidents

There are many that could be discussed. We discuss three U.S. incidents. (Accounts and figures from McLaughlin et al.,[15] http://www.csirc.net/docs/reports/la-13638.pdf.)

Y-12 Accident, Oak Ridge, TN, June 16, 1958

In a facility designed to recover enriched uranium from various solid wastes via dissolution in nitric acid, purification, concentration, and conversion to uranium tetrafluoride, a material inventory was ongoing. Three 5 inch diameter vessels used to store uranyl nitrate were undergoing leak testing. During one shift, a supervisor noted that uranyl nitrate was present in a 6 inch diameter glass standpipe that was part of a pH adjustment station (see Figure 11.45). He instructed a worker to drain the standpipe, but later noted that there was still uranyl nitrate solution in the standpipe. The worker confirmed that he had drained the pipe, but the two noticed leakage of the solution into the standpipe

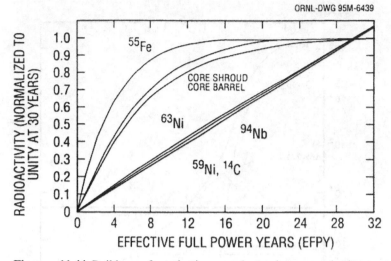

Figure 11.44 Buildup of activation products in pressurized-water reactor internal components as a function of effective full-power years. (From http://www.nrc.gov/reading-rm/doc-collections/nuregs/staff/sr1437/v1/fig020.html.)

through the V-2 valve. The valve was closed and the standpipe was again drained.

During the next shift, a new supervisor was continuing the leak checking but did not verify the status of fluid levels, valve positions, and so on. It is unclear (due to conflicting reports) whether the previous shift supervisor communicated about the uranyl nitrate leakage from the previous shift. Unknown to anyone was the fact that uranyl nitrate had been leaking from the B-1 wing through valve V-1 for several hours, into the FSTK 1-2 vessel, as valve V-3 was open. During this new shift, Valves V-4, V-5, and V-11 were opened to drain water from the vessels into the 55 gallon drum. An operator near the drum saw yellow-brown fumes rising from the drum and saw a blue flash, indicating that an excursion had occurred. The local criticality alarm sounded, and the

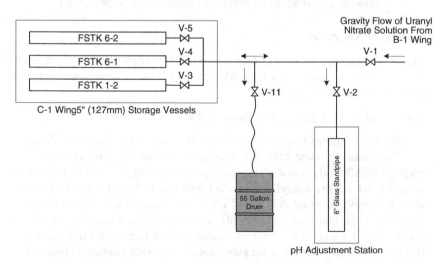

Figure 11.45 Diagram of the materials involved in the Oak Ridge Y-12 Criticality accident.

area was quickly evacuated. Further flow of water into the drum increased the reactivity for about 11 minutes; then decreased it. Eight persons received very high radiation doses (4.6, 4.3, 4.1, 3.4, 3.0, 0.087, 0.087, and 0.029 Sv). None died from their radiation exposures.

Los Alamos Scientific Laboratory, December 30, 1958

In this facility, plutonium was being purified from various residues created in a recovery process. Solids that were unexpectedly rich in plutonium appear to have washed from several vessels into a single large vessel that contained dilute aqueous and organic solutions. About 200 liters of this material were transferred to a 1000 L, 1 m diameter vessel (Figure 11.46). The vessel had a stirrer that mixed the materials inside. When this stirrer was started, the initial movement of material was such that a central region that contained this enriched plutonium was suddenly thickened, causing the system to go super-prompt critical and creating an excursion yield of about 1.5×10^{17} fissions!

An operator observing the process received about 120 Sv to his upper torso and died within 36 hours. Two other persons received doses of about 1.3 and 0.5 Sv and had no ill effects.

SL-1 (Stationary Low-Power Plant No. 1) Reactor Incident, January 3, 1961

In an experimental BWR reactor in Idaho called SL-1, several individuals were working with the reactor when apparently a control rod was suddenly removed manually from the core. This caused the reactor to become critical, with high reactivity. The power surge caused the reactor power to reach about

Figure 11.46 Image of the Los Alamos mixing vessel.

20,000 MW in a fraction of a second, causing melting of the fuel. The molten metal reacted violently with water in the core, causing a steam explosion that caused the entire reactor vessel to rise 3 m in the air and crash back down.

The three individuals working at the site were killed. One was actually impaled on the ceiling on a control rod, and remained there for some days before the building could be safely entered. The others died of their injuries, but were so heavily exposed to radiation that their radiation exposures would have surely killed them. Their bodies were heavily contaminated and their bodies were buried in lead coffins, with their hands buried separately with other highly contaminated radioactive waste from the site.

Nuclear Weapons—Supercritical Devices

Conventional explosives typically force fissile material into a critical geometry, causing a supercritical excursion which is very brief but which liberates huge amounts of energy. Whereas a typical commercial reactor uses ^{235}U enriched to perhaps 3–4%, weapons may use fuel enriched to 90%. When such material is suddenly forced into a supercritical configuration, the energy release is of the highest level known at present to mankind. As is well known, since the discovery of this process, the geopolitical implications have been enormous, and have been the cause of emotional stress among all populations of the world and of political tension between many countries and groups.

The Cold War, between the United States and the U.S.S.R. lasted for several decades, being focused on the nuclear weapons capabilities of the two super-powers. Each had so much to lose, however, in an all-out nuclear exchange, that the likelihood of a real large-scale event was small. Now many smaller countries as well as independent terrorist groups can potentially have access to these weapons of incredible destructive power, and our fear of a catastrophic event continues. These weapons have been used only once in human warfare, in the bombings of Hiroshima and Nagasaki at the end of the Second World War. One can only hope that this level of destruction (caused by devices that are very tiny compared to modern nuclear weapons) is never witnessed again.

The design of weapons varies. A simplified schematic of a multistage thermonuclear weapon is shown in Figure 11.47. The sequence of events in explosion is given briefly below.

Stage 1: Fission Explosion
- Multiple detonators (3) simultaneously initiate detonation of high explosives (4).
- As detonation progresses through high explosives (4), shaping of these charges transforms the explosive shock front to one that is spherically symmetric, traveling inward.
- Explosive shock front compresses and transits the pusher (5) which facilitates transition of the shock wave from low-density high explosive to high-density core material.
- Shock front in turn compresses the reflector (5), tamper (6), and fissile core (7) inward.
- When compression of the fissile core (7) reaches optimum density, a neutron initiator (either in the center of the fissile core or outside the high explosive assembly) releases a burst of neutrons into the core.

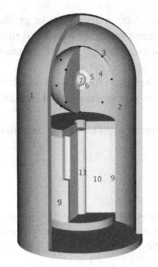

Numbered parts:

1. bomb casing
2. interior filling (plastic material)
3. detonators
4. conventional high explosive
5. pusher (aluminum, others) and reflector (beryllium, tungsten)
6. tamper (uranium-238)
7. fissile core (plutonium or uranium-235)
8. radiation shield (tungsten, others)
9. fusion pusher/tamper (uranium-235 sleeve)
10. fusion fuel (solid lithium-deuteride)
11. sparkplug (uranium-235 or plutonium)

Figure 11.47 Simplified schematic of a multistage thermonuclear weapon. (From http://www.johnstonsarchive.net/nuclear/diagthermon.html.)

- The neutron burst initiates a fission chain reaction in the fissile core (7): a neutron splits a plutonium/uranium-235 atom, releasing perhaps two or three neutrons to do the same to other atoms, and so on; energy release increases geometrically.
- Many neutrons escaping from the fissile core (7) are reflected back to it by the tamper (6) and reflector (5), improving the chain reaction.
- The mass of the tamper (6) delays the fissile core (7) from expanding under the heat of the building energy release.
- Neutrons from the chain reaction in the fissile core (7) cause transmutation of atoms in the uranium-235 tamper (6).
- As the superheated core expands under the energy release, the chain reaction ends.

Stage 2: Fusion Explosion

- Gamma radiation from the fission explosion superheats the filler material (2), turning it into a plasma.
- The vaporized filler material (2) is delayed from expanding outward by the bomb casing (1), increasing its tendency to compress the fusion pusher/tamper (9).
- Compression reaches the fusion fuel (10), which has been partially protected from gamma radiation by the radiation shield (8).
- Compression reaches the fissile sparkplug (11), compressing it to a supercritical mass.
- Neutrons from the explosion of Stage 1 reach the fissile sparkplug (11) through the channel in the radiation shield (8), initiating a fission chain reaction.
- The sparkplug (11) explodes outward.
- The fusion fuel (10) is now supercompressed between the fusion pusher/tamper (9) from without and the sparkplug (11) from within, turning it into a superheated plasma.

- Lithium and deuterium nuclei collide in the fusion fuel (10) to produce tritium, and tritium and deuterium nuclei engage in fusion reactions: nuclei fuse by pairs into helium nuclei, producing a large energy release of gamma rays, neutrons, and heat.
- The large release of neutrons from fusion in the fusion fuel (10) causes transmutation of uranium-235 atoms in the fusion pusher/tamper (9), releasing additional energy.
- All reactions end as the superheated remnants expand under the energy release; the entire weapon is vaporized.
- Total elapsed time: about 0.00002 seconds.

Endnotes

1. D. J. Strom, R. Harty, E. E. Hickey, R. L. Kathren(a), J. B. Martin, and M. S. Peffers, Collective dose as a performance measure for occupational radiation protection programs: Issues and recommendations. Pacific Northwest National Laboratory Report PNNL-11934/UC-610 (1998).
2. A. B. Chilton, J. K. Shultis, and R. E. Faw, *Principles of Radiation Shielding* (Prentice-Hall, New York, 1984).
3. B. T. Price, C. C. Horton, and K. T. Spinney, *Radiation Shielding* (Pergamon, Oxford, 1957).
4. National Council on Radiation Protection and Measurements, NCRP Report 147, *Structural Shielding Design for Medical X-ray Imaging Facilities* (NCRP, Bethesda, MD, 2004).
5. J. Turner, *Atoms, Radiation, and Radiation Protection* (Pergamon Press, New York, 1986).
6. Thermal diffusion length is the characteristic distance between the point at which a neutron becomes thermal and the point of its final capture.
7. H. Cember, *Introduction to Health Physics*, 3rd ed. (McGraw-Hill, New York, 1983).
8. All survey diagrams graciously provided by personal communication from David Burkett, Vanderbilt University Radiation Safety Office.
9. International Commission on Radiological Protection, ICRP 37 Cost-Benefit Analysis in the Optimisation of Radiation Protection (1983).
10. International Commission on Radiological Protection, ICRP 55 Optimisation of Decision-making in Radiological Protection (1989).
11. U. S. Nuclear Regulatory Commission, NUREG 1535, Ingestion of Phosphorus-32 at MIT, Cambridge, MA Identified on August 19 (1995).
12. Dr. Ken Skrable, a well-known expert in internal dose models and calculations, was once heard to quip that if he had a measurement of activity from a personal air sampler, "Well at least I have a good estimate of what activity *was not* taken up by the worker!"
13. C. A. Potter, Intake retention fractions developed from models used in the determination of dose coefficients developed for ICRP Publication 68- particulate inhalation. *Health Physics* **83** (5), 593–789 (2002).
14. So, in the reality of practice, the four factor formula becomes a five factor formula, but who's counting?
15. T. P. McLaughlin, S. P. Monahan, L. Norman, N. L. Pruvost, V. V. Frolov, B. G. Ryazano, and V. I. Sviridov, *A Review of Criticality Accidents*, 2000 revision (Los Alamos National Laboratory, 2000).

12

Environmental Monitoring for Radiation

Our basic definition of health physics included concepts of the protection of humans and the environment from the harmful effects of radiation while permitting its beneficial uses. It is practically a universally accepted axiom that protection of the environment is in the best interests of the human race, and even with this consideration aside, most reasonable persons recognize an ethical responsibility for people to protect the environment that they affect by their very existence and their activities. Protection of human beings remains at the center of our concerns when monitoring the environment for the presence and migration of radionuclides. People live best when living in harmony with nature, of course, so a clean and healthy environment has indirect benefits for all of us as well. We are always concerned about potential carcinogens moving through environmental pathways and possibly reaching human receptors.

Radiation is, by comparison with other contaminants, a weak carcinogen, but it is a well-known carcinogen. The general public has a heightened awareness of the carcinogenic effects of radiation, and so releases of radioactive materials are always of concern. Minor amounts of environmental releases or releases of radionuclides with very short half-lives (such as some of the gaseous fission products released from nuclear reactors) may be routinely tolerated. Releases of high concentrations of more long-lived or otherwise high-risk radionuclides must be given a high priority for monitoring and interdiction.

Besides aesthetic problems with badly contaminated or otherwise damaged segments of the environment, the long-term effects of negligent care of our environment may have direct effects on human health. One of the best-known examples of this was the tragedy at Love Canal in New York. The Hooker Chemical Corporation, a subsidiary of Occidental Petroleum, dumped over 20,000 tons of chemical pollutants into a clay-lined canal, then simply covered it with a layer of dirt, and sold the land to the local government for 1 dollar. Some years later, around 100 homes and a school were built on the site. A large rainfall triggered an explosion of much of the buried material, and significant leaching from the site ensued.

One observer wrote that

"Corroding waste-disposal drums could be seen breaking up through the grounds of backyards. Trees and gardens were turning black and dying. One entire swimming pool had been popped up from its foundation, afloat now on a small sea of chemicals.

Puddles of noxious substances were pointed out to me by the residents. Some of these puddles were in their yards, some were in their basements, others yet were on the school grounds. Everywhere the air had a faint, choking smell. Children returned from play with burns on their hands and faces[1]".

In this small population of working families, there were high rates of miscarriage, birth defects, cancer, and other illnesses. The president of the United States declared a state of emergency, and the residents either fled their homes or were relocated. Eventually the site was cleaned and residency restored, and the incident sparked legislation and funding that led to the evaluation and cleanup of thousands of other sites across the nation. Regardless of the ultimate benefits, this incident was a striking example of how poor environmental stewardship may be not only embarrassing from an ethical standpoint but also highly undesirable from a human health standpoint.

12.1   Types of Environmental Assessment Programs

The goals of any sampling and analysis program are accurate and precise determination of the amounts of specific radionuclides and/or radiation levels in any relevant environmental media. If monitoring is performed near a facility that processes a limited amount of materials, it may be easier to know on which nuclides to focus. If a complex set of radionuclides may be present, or if multiple facilities may be contributing to the presence of radionuclides at a location, sample analysis may be fairly complicated. Specific types of programs that are typically performed include the following.

1. *Site Characterization Studies*: A site characterization study is an initial assessment of a potentially contaminated site. Typically, the site characterization consists of two parts: a records search to evaluate what sources of contamination may have contributed to radionuclide levels in an area, and a field investigation to inspect and sample areas of potential concern based on these findings.
2. *Risk Assessment*: An assessment of radionuclide transport to human receptors and evaluation of radiation and dose to human populations.
3. *Remedial Technology Assessment*: Evaluation of contamination levels at a site to repair or contain environmental damage and eliminate unacceptable health risks.
4. *Cleanup Verification*: Assessment of the adequacy of remedial measures and of the suitability of a site for further use or release for access by the general public.
5. *Compliance Status*: Measurement of contaminant levels to assure compliance with applicable regulations for environmental contamination and/or human exposure.

The key to the success of any of these kinds of studies is that the sampling and analysis approaches answer the questions that are being asked in an analysis program. A site characterization study is designed to gather enough information about a site to guide decisions about a more thorough environmental study. Therefore, sampling will typically be less extensive and more focused on areas of suspected contamination, rather than employing a more systematic and broad sampling design. Types and numbers of samples and sample detection levels may be very different in a remedial assessment, risk assessment, and compliance status study.

12.2   Types of Facilities Monitored

Basically any facility that may release radionuclides into the environment should have some monitoring plan for detecting and quantifying these releases, and quantifying their potential impacts on living species nearby. Some examples, certainly not an exhaustive list, follow.

- The first type of facility that comes to mind for many people is a nuclear power plant. These facilities obviously manage very large amounts of radioactive material (see Chapter 11), and have the potential for significant releases of radioactive material to the atmosphere and to surface and groundwater. These facilities routinely release small quantities of activity through various routes and carefully monitor the nearby environment for these routine releases as well as any unplanned releases that may occur.
- Smaller research and radionuclide production reactors function in the same basic manner as a power reactor and manage the same kinds of radionuclide inventories, just in smaller absolute quantities. Releases from these facilities are generally more rare and of lower magnitude, but some release of noble gases are not uncommon.
- National laboratories (see Chapter 7) are large facilities that involve dozens of buildings often spread out over large areas, and these facilities manage a diverse range of research, production, and other activities which may involve the use of significant quantities of radioactive material. These facilities, like nuclear power sites, generally have fairly large and well-trained health physics staff that carry out a range of routine environmental monitoring in and around the facilities.
- Radionuclide and radiopharmaceutical production facilities manage a finite number of radionuclides, but often in very significant quantities and concentrations, and in loose forms that are readily transported.
- Radioactive waste disposal facilities take many different forms of radioactive material and place them in contained and monitored sites. They are quite strict regarding the forms, packaging, and other characteristics of the waste that they accept. Releases from the site are carefully and regularly monitored, as the material is expected to be in place for many decades (or even millennia, in the case of high-level waste facilities, discussed below), and the potential is high for the release of a wide variety of different radionuclides into surface or groundwater that may be directly consumed by the general public.

12.3   Types of Samples and Sampling Strategies

The goal of any sampling program is to provide an accurate characterization of the types and concentrations of radionuclides in a particular set of locations, either at a point in time or over a period of time. Naturally the first step in the process is assurance of accurate and precise analytical methods and equipment.

This is not a minor component, but is a given starting point in any program. Radiation counting equipment must be carefully calibrated for all nuclides that may be encountered. If many different radionuclides may be present, either high-quality spectroscopy detectors (gamma or alpha spectroscopy) must be employed, or elaborate radiochemical separation techniques will be needed for all potential contaminants of concern, with reliable counting methods and equipment employed for the separated nuclides. For any equipment, after establishment of the initial calibration, a regular and rigorous quality assurance program must be maintained, with:

- Frequent (usually daily) checking of energy and activity calibration
- Plotting of behavior on control charts
- Establishment of action levels for suspicious or noncompliant behavior
- Random and blind introduction of blank and spike samples (samples known to be free of activity or with known levels of contaminants being tested in the program)
- Training and retraining of personnel performing procedures
- Extensive documentation of all methods, procedures, and training
- Regular program evaluation by internal or external auditors

Almost any form of environmental media may be sampled. The sampling strategy generally will focus on the pathways most important to human exposure. The most basic types of samples are air, soil, and water samples, which are generally quite easy to obtain.

Figure 12.1 is an example showing a qualitative analysis of pathways from a national laboratory facility to a general member of the public assumed to be exposed off-site by releases from this facility. People breathe air and consume water, vegetation, and foods that all may be contaminated with radionuclides. People may also be exposed to penetrating radiations (gamma rays and possibly electrons/betas) due to passing clouds of radioactive materials or large ground areas or groundwater shorelines that have become contaminated. The following types of samples are thus routinely monitored at various locations between the source and the target population.

1. Direct gamma exposure readings
   - Instantaneous or cumulative over fixed times
   - At the facility site boundary
   - At strategic locations at various distances from the site
2. Airborne concentrations of radionuclides
   - In ducts and stacks
   - Near the facility site boundary
   - At strategic locations at various distances from the site
3. Concentrations of radionuclides in water
   - Surface waters (lakes, rivers, ponds)
   - Groundwater
   - Seawater
4. Concentrations of radionuclides in soil. Soil is usually studied as an indicator of radionuclide transport, but soil is also directly consumed at times, either as a contaminant on vegetable foods or directly by infants or impoverished cultures.
5. Concentrations of radionuclides in biological species (biota)

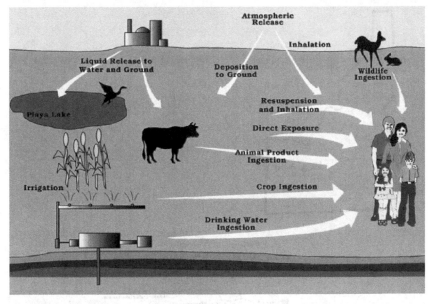

Figure 12.1 Potential pathways for human exposure. (From http://www.pantex.com/aser/1996/ch6-96.pdf).

- Animal species that are directly consumed (e.g., fish).
- Animal species that may act as pathways to human consumption by means other than being consumed (e.g., cows eating contaminated grass and producing milk that may be contaminated).
- An important component of this pathway is an understanding of foodstuff consumption habits and amounts. This is often a very difficult variable to characterize. Much of the U.S. population purchases foods from commercial distributors, but others have a large portion of their diet comprised of foods that they directly consume from their own land.

The absolute value of the measurements made in the various components of the environment are used to directly indicate the dose potential for humans, but are also useful in studying the transport of radionuclides through the environment.

12.3.1 Direct Gamma Exposure Readings

Surveillance networks of ThermoLuminescent Dosimeters (TLDs) are often used to monitor gamma doses over fixed periods of time. TLDs, as noted in Chapter 8, integrate radiation exposure over possibly long periods of time, providing a signal that is quite linear with both photon energy and dose over wide ranges. TLDs are often worn by radiation workers to actively monitor their radiation doses over fixed periods. TLDs are also placed in protective containers around hospitals and other facilities to measure integrated exposures over time, and are used in the same way in environmental monitoring stations to systematically monitor cumulative (lower level) doses over fixed periods of time. Figure 12.2 shows an example of a series of monitoring stations placed around a national laboratory facility.

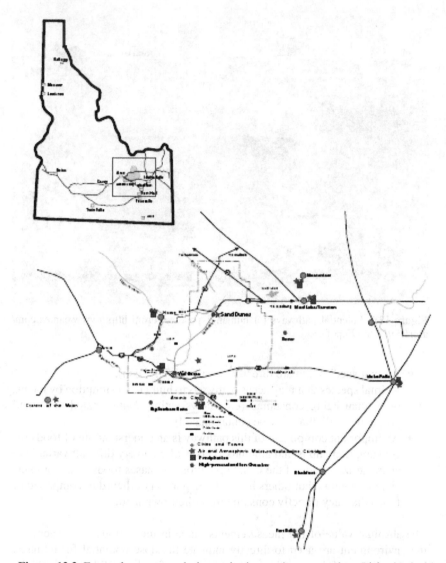

Figure 12.2 External gamma and air monitoring stations around an Idaho National Engineering and Environmental Laboratory (INEEL) facility. (From http://www.deq. idaho.gov/inl_oversight/index.cfm.)

Sodium iodide survey meters and pressurized ionization chambers (see Chapter 8) can be used to directly measure ambient exposure rates from photon radiation in the environment. As with the selection of soil or other physical samples (discussed below), the choice of sampling points when surveying a large outdoor area may be done in a number of different ways.

Sensitive survey meters may also be mounted in airplanes or helicopters and used to perform rapid monitoring of very large areas, cities, and so on (Figure 12.3). As with the use of walk-over surveys, currently detectors may include the use of Global Positioning System (GPS) technology to automatically send measurements to a computer system for storage and representation, without the need for technologists to write down individual data as they are gathered.

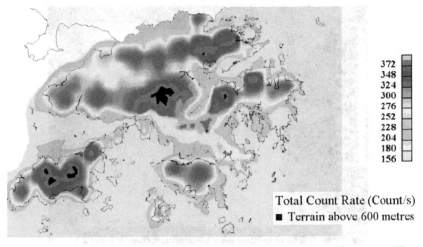

372
348
324
300
276
252
228
204
180
156

Total Count Rate (Count/s)
■ Terrain above 600 metres

Figure 12.3 Results of an aerial survey recorded by the Hong Kong Observatory (From http://www.hko.gov.hk/radiation/ermp/survey/. Courtesy of the Hong Kong Observatory of HKSAR.) (see color plate)

12.3.2 Airborne Concentrations of Radionuclides

General Ambient Sampling

As is true within a facility, "grab" air samples may be obtained outside a facility at strategic locations at any time in general or after a suspected release of radioactivity. The same basic technologies, discussed for the workplace in Chapter 11, will be employed, as for any other general air sample.

Stack Sampling

In addition, one may monitor the air moving through ducts and stacks to quantify the concentrations of radionuclides likely to be emitted from the facility. Sampling of air moving through ducts or stacks at high velocities and possibly at high temperatures (due to association with some industrial processes) is not a trivial task. Obtaining a representative average air sample from air moving at high velocities through a duct is difficult for several reasons:

- First, there is usually a strong velocity profile across the cross-sectional area of a circular or square duct. The velocity is highest in the center of the duct, but drops off with some profile to lower values near the walls of the duct. One must choose a sampling point within the duct and be able to know what the average velocity is at that point, in order to know what the air flow was during the time of sampling, in order to infer the concentration measured.
- Introduction of a sampling device into the air stream perturbs the air stream. The device introduced has some flow resistance that the air experiences when entering and passing though the various filtering or other sampling devices inside. It is essential that this device be engineered in a way that the velocity within the device is the same as that at the point chosen for sampling. This is termed *isokinetic sampling*, which means:

Any technique for collecting airborne particulate matter in which the collector is so designed that the airstream entering it has a velocity equal to that of the air passing around and outside the collector. The advantage of isokinetic sampling consists in its freedom from the uncertainties due to selective collection of only the larger,

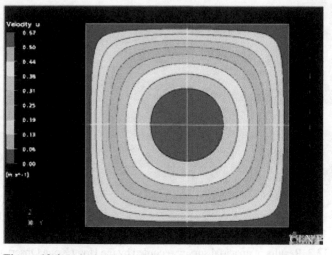

Figure 12.4 Velocity profile of laminar flow through a square duct. (From http://www.ansys.com/.) (see color plate)

less easily deflected particulates. In principle, an isokinetic sampling device has a collection efficiency of unity for all sizes of particulates in the sampled air.

- The device must be able to obtain a representative sample within its sampling apparatus while dealing with the high velocities, possibly high temperatures, corrosive materials, or other effects introduced by the sampled gas (Figures 12.4 and 12.5).

12.4 Long-Term Off-Site Monitoring

For long-term environmental monitoring, there are sampling stations that are specifically designed for long-term sampling of airborne radionuclides. These stations, as shown in Figures 12.6 and 12.7, usually have several different forms of sampling apparatus, for measuring particulate forms of contamination, gases, perhaps tritium, and other species. They may also contain, or be complemented by other stations that contain, devices for measuring and charting wind speeds and directions at a point of interest. As shown shortly, an important element of predicting downwind air concentrations that people

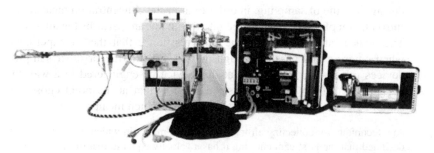

Figure 12.5 Isokinetic sampler and sampling train. (From http://www.apexinst.com/intross.html.)

Figure 12.6 Mobile air monitoring station. (From www. americanecotech.com.)

Figure 12.7 Alpha detector and continuous air monitor. (From http://www.bhi-erc.com/.)

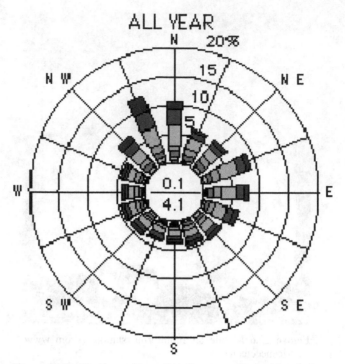

Figure 12.8 Wind rose. (From http://www.pref.kagoshima.jp/.)

may be exposed to is knowledge of the pattern of wind speeds and directions, which may be displayed in a graphical plot known as a "wind rose" (see Figure 12.8).

12.4.1 Concentrations of Radionuclides in Water

Radionuclide contamination of water has important implications, as people may directly ingest such water (surface waters that are treated and distributed, well water that individuals drink directly), or the water may act as a transporting medium, causing the radionuclides to be dispersed far from their point of release and possibly distributed in other segments of the environment that may ultimately lead to human consumption (e.g., deposition in soils or sediments, from which plants may take up the nuclides and concentrate them in edible portions of the plant or fruit, use in irrigation of crops that are consumed, or transfer to animal species that may be consumed or transfer the activity to other species that are consumed). Several types of water samples may be of interest:

- *Surface water*: Lakes, rivers, ponds, and the like. May be contaminated by direct deposition of passing plumes or runoff from contaminated surface materials. May be directly consumed by humans or animals.
- *Groundwater*: May be contaminated by leaching of rainwater or other liquids through soils. May be directly consumed by humans or used for irrigating crops.
- *Well water*: In rural areas, some households are directly served by their own wells which draw from a groundwater reservoir.

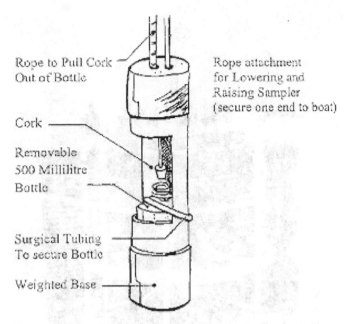

Rope to Pull Cork
Out of Bottle

Rope attachment
for Lowering and
Raising Sampler
(secure one end to boat)

Cork

Removable
500 Millilitre
Bottle

Surgical Tubing
To secure Bottle

Weighted Base

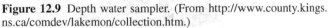

Figure 12.9 Depth water sampler. (From http://www.county.kings. ns.ca/comdev/lakemon/collection.htm.)

- *Tap water*: In urban or rural areas, water taken from surface or groundwater reservoirs and treated is received by consumers.

Obtaining a surface water sample from a lake is not particularly difficult; one must choose sampling points within the lake area and depths of interest, often with most any simple sampling device (as simple as a bucket or milk container). If sampling near the lake bottom, either at shallow depths or in the middle of the lake while sampling considerably below the surface, care must be taken not to entrain too much sediment in the sample, as radionuclides in the sediment may contribute to the eventual counts obtained and give a result that is not representative of the water concentration. Sediment may be sampled separately, as noted in the next section.

Sampling of river water carries with it some of the same difficulties as duct sampling, discussed above. The velocity profile across the open area through which the water passes is nonuniform, and sampling must take into account the flow rate at the point sampled. Sampling of ocean water at shallow depths, directly by human efforts, is more similar to sampling of lake water: a representative sample should be taken, with the location and depth noted, while limiting the entrainment of sand or sediment from the ocean bottom. Deeper samples may be obtained using mechanical sampling devices (Figures 12.9 through 12.11).

Sampling of groundwater is more difficult. As with obtaining soil samples at significant depths below the ground surface (discussed in the next section), obtaining a groundwater sample requires drilling into the ground surface to a depth at which the water table is found and extraction of a water sample from the groundwater aquifer.

Figure 12.10 "Rosette" system for ocean water sampling. (From http://oceanexplorer.noaa.gov/ explorations/02hudson/background/watercolumn/ watercolumn.html.)

12.4.2 Concentrations of Radionuclides in Soil or Sediment

Soil or sediment samples give important indications of contamination of the site, as soils and sediments (in the ocean, river banks, and lake beds) will often trap many kinds of contaminants through various mechanisms (e.g., ion exchange, adsorption). Contaminated soil will be an important source of external radiation exposure to present or future inhabitants of a region. Soil may contribute radionuclides to internal pathways as well, most commonly via plant uptake, with the plants directly consumed by humans or by animals that are subsequently consumed by humans. Soil may also be directly ingested, either inadvertently with harvested crops, or intentionally in unusual cases (children, malnourished populations). Contaminated soils near the surface can also become resuspended, resulting in inhalation of radionuclides. Water passing through contaminated soil can reach surface water or groundwater reservoirs and be consumed. As with above-ground exposure rate monitoring, discussed above, soil samples may be obtained using random, systematic, or other approaches, at a site that must be characterized. For samples near the surface, simple sampling methods may be used, such as the use of a trowel to gather the soil and place it into a container for holding. For sampling beneath the surface, more aggressive methods may be needed. In particular, for depth characterization of soil radioactivity, careful sampling using borehole cutting and sampling tools is necessary (Figures 12.12 and 12.13).

Figure 12.11 Buoy equipped with a detector for measuring caesium-137, and also carries instruments for measuring physicochemical parameters such as current velocity, salinity, and temperature. (Courtesy Radiological Protection Institute of Ireland.)

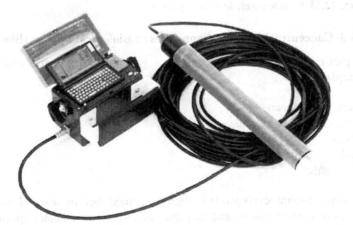

Figure 12.12 Gamma-ray spectrometer for surface and borehole surveying. May use NaI or BGO scintillators. (From http://www.giscogco.com/pages/radgyg2k.html.)

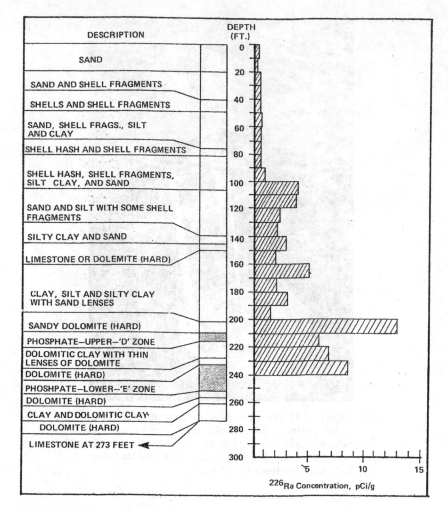

Figure 12.13 Sample borehole sampling result.

12.4.3 Concentrations of Radionuclides in Biological Species (Biota)

Samples may be taken of plants and/or organisms at different trophic levels. Examples:

- Benthic organisms
- Plants, edible or inedible
- Birds
- Fish
- Mammals

The lower trophic levels are less often consumed, but are often closer to the sources of contamination, and thus may be important indicator species used to track contaminant movement through food chains. It is often important to sample separate portions of higher species—for example, soft tissue, bone, individual organs—in order to track the movement of particular nuclides and evaluate their potential impact on human health. Contaminants entering food chains may be concentrated through increasing trophic levels (accumulation) or may have higher concentrations in lower trophic levels (discrimination). A number of terms are defined to express this phenomenon:

- *Bioaccumulation* refers to the concentrating of material by an organism from its direct environment or diet. A BioAccumulation Factor (BAF) is defined as the concentration of a contaminant in the tissue in the species to the concentration in a specified element of its environment or diet (e.g., Bq/kg of tissue per Bq/kg of fresh water or sediment). This may also be referred to as *bioconcentration*, with an associated BioConcentration Factor (BCF).
- *Biomagnification* is considered to result from the direct uptake of a substance by an organism via food and the accumulation of a contaminant at increasingly higher levels in higher trophic levels, that is, the "food chain effect."

Examples

- Cesium (^{137}Cs primarily) from weapons testing fallout was shown to be trapped and retained by the lichen and moss of the Canadian tundra, which are major sources of food for caribou and reindeer. Caribou flesh was found to contain up to 100 times the quantities found in meat in mid-latitudes. Concentrations (Bq/kg) were successively higher at each step in the food chain, as cesium is a potassium analogue, and concentrates in edible tissues of the body. ^{137}Cs was widely dispersed throughout Eastern Europe after the Chernobyl disaster, and caused banning of the entry of many crops and meats into other countries, thus significantly affecting the livelihoods of many in the agriculture, farming, and food service industries for many years.
- The marine food chain concentrates iron isotopes (e.g., ^{55}Fe) due to the low concentration of stable iron in sea water.[2]
- Strontium isotopes, most notably ^{90}Sr, and radium isotopes, notably ^{226}Ra, may be taken up by plant structures and thus consumed directly, but are discriminated against in food chains, because it is a calcium analogue and concentrates in the skeleton, and also is discriminated against (in favor of calcium) in cattle diets during milk formation.[3]

12.5 General Sampling Strategies and Techniques

Any samples taken should provide data that are representative of the true radionuclide levels or concentrations in the areas sampled. If you have a very good analysis of a sample that poorly represents the system under study, the results will not provide a good evaluation of the conditions. Obtaining a representative sample is therefore the cornerstone to a good environmental assessment. Several types of samples are taken in these analyses.

- *"Grab" sample*: A single sample representative of a "point" location at a specified "instantaneous" time.
- *Composite sample*: A single sample composed of several grab samples combined together. The individual samples may be take from the same location at different times (temporal composite) or from different locations at the same time (spatial composite).
- *Background samples*: Before the significance of samples taken from the site can be appreciated, one must understand the naturally occurring levels of the

radionuclides of interest to the study and possible interfering radionuclides. Thus samples must be taken at locations outside the potentially contaminated area, but that are as close as possible to the area in both location and character. The goal is to obtain a quantitative assessment of the conditions that may have existed before the contamination occurred, for reference and comparison to samples from the contaminated site. Background samples are:

o Outside the area of contamination: soil samples
o Upstream of the source of contamination: water and sediment samples (care must be taken in tidal environments due to the oscillatory nature of flow in these regions)
o Upgradient of the source of contamination: groundwater samples
o Upwind of the source of contamination: air samples

Important interferences with obtaining valid background samples include:

o Naturally occurring radionuclides identical to those of interest from the site (e.g., ^{137}Cs or ^{90}Sr due to previous weapons testing fallout) or which mimic emissions of the nuclides of interest to the study (e.g., both ^{226}Ra and ^{238}U have a gamma ray at 186 keV and both occur naturally in most soils; evaluation of a depleted uranium facility must take into account both potential interferences in background materials)[4].
o Other potential sources of radionuclide contaminants. For example, other facilities may be located nearby and may be introducing contaminants into the air or river water which then flows into the area of concern to the survey.

• *Core sample*: For deeply buried or possibly leaching contaminants, a core may be dug into the earth, with soil samples removed from various depths and/or radiation measurements made with a gamma probe at different depths. Concentrations of material in soil adjacent to the core may sometimes be measured by lowering a neutron source into the core and measuring the activation products via gamma spectroscopy.

In choosing actual samples, often some balance must be struck between a number of factors, including:

• Accessibility of terrain
• Number of samples desired for statistical reasons
• Number of samples possible due to budgetary limitations
• Amount of sample material that is available at the site
• Amount of sample material that can be transported to a laboratory
• Amount of sample needed to meet desired detection limits, and so on

A variety of sampling strategies has been developed, based on the overall goals of the program and other constraints (Figures 12.14 and 12.15)[5,6].

• *Judgmental*: This approach uses existing knowledge, technical judgment, and/or visual observations to select sample locations.
• *Random*: This approach uses a random process to identify sample locations within the overall area/volume of concern.
• *Stratified random*: This approach uses existing information to divide the sampling area into smaller areas (strata) that are relatively homogeneous. Random samples are collected within each stratum; for example,

different subsurface soil formations could provide the basis for (depth) stratification.

- *Systematic grid*: This approach places a square or triangular grid over the area of concern and specifies sample collection locations to be the nodes or centers of the individual grid components.
- *Systematic random*: This approach uses a random process to select sample locations within a predetermined grid or pattern.
- *Search*: This approach utilizes either a systematic grid or systematic random approach to "search" for areas of relatively high concentrations or concentrations that exceed standards ("hot spots").

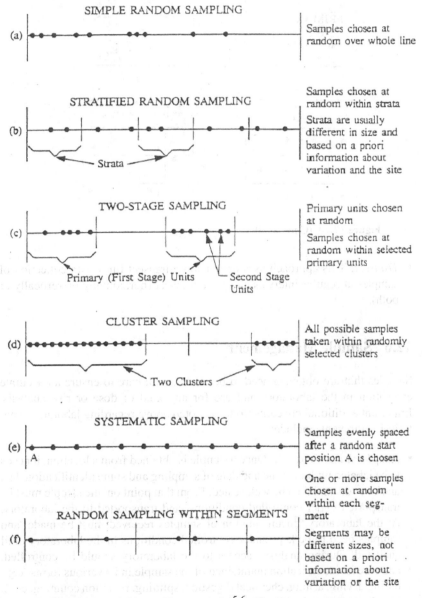

Figure 12.14 Various field sampling strategies[5,6]. (*Continued on next page.*)

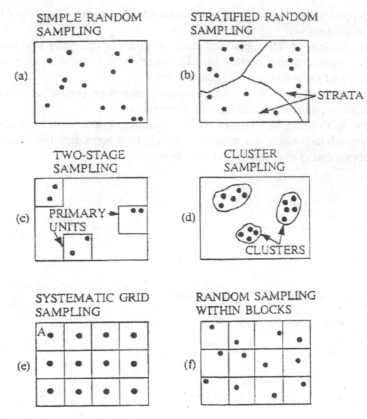

Figure 12.14 (Continued).

- *Transect*: This approach uses one or more transect lines with collection of samples at regular intervals along the transect (horizontally or vertically or both).

12.6 Sample Management

Samples that are obtained need to be treated with care to ensure an accurate evaluation in the laboratory and use for any kind of dose or risk analysis. Important additional concerns (to those noted above regarding laboratory analytical equipment) include:

- *Sample chain of custody*: Once a sample is obtained from a location, it needs to be labeled with the time and date of sampling and some identification data and/or code, and then be well sealed. From that point on, the sample must be transported under controlled conditions, and transported to the laboratory. At the laboratory, documentation of samples received must be made, and samples need to be kept under controlled conditions in the laboratory until analyzed. During analysis, access to the laboratory should be controlled, and careful identification maintained of the sample in its various forms (e.g., during drying, ashing, chemical digestion, splitting, radiation counting, etc.). The important principles are prevention of tampering with samples and maintenance of sample identity, with reporting of appropriate results for each sample.

- *Sample Preservation*: Some samples may need to be kept cooled or frozen, to maintain the sample integrity or prevent loss of volatile elements. Samples may also need to be kept at a particular pH to prevent plating out of some

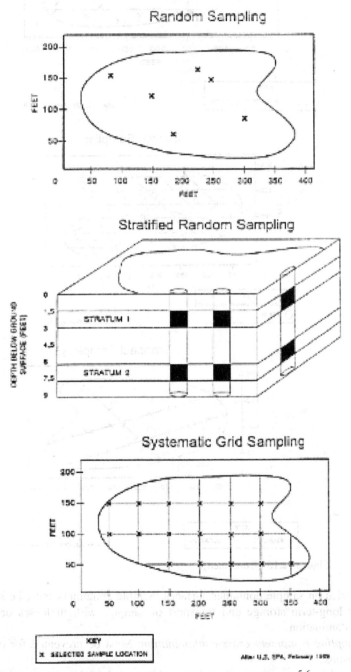

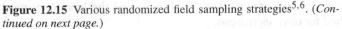

Figure 12.15 Various randomized field sampling strategies[5,6]. (*Continued on next page.*)

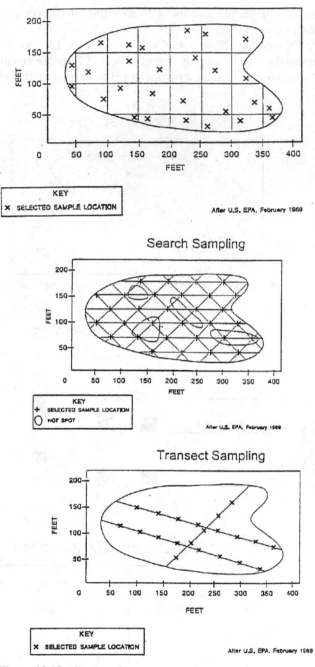

Figure 12.15 (Continued).

species on sample container surfaces. Sample containers must be suitable for long-term storage and transport of samples without losses or cross-contamination.

- *Sampling equipment cross-contamination*: Must be prevented for obvious reasons.
- *Sample cross-contamination*: In the field and the laboratory must be prevented for obvious reasons.

12.7 Instrumentation

A variety of instrumentation is employed in environmental monitoring, in both field and laboratory. As the levels being measured are typically very low, equipment must have high sensitivities. For laboratory equipment, most counting will be done at or near detection limits, so much attention is needed to this area, and long counting times are typical. A problem is sometimes encountered when highly contaminated sites are being evaluated, however, as dose and activity levels can be so high that the routinely used instruments cannot handle them.

- Field Instruments
  - *Survey meters*: For above-ground surveys of radiation exposure levels, NaI scintillation meters will be used. These are sensitive down to background levels of exposure (typically 5–10 μR/hr and measure up to several thousand μR/hr, i.e., several mR/hr). If higher rates are observed in special cases, ionization chamber or GM survey meters may be employed.
  - *Borehole logging meters*: NaI scintillation detectors will normally be used to just log exposure rates as a function of depth. Neutron activation/prompt gamma emission spectroscopy may be done with NaI or bismuth germanate (BGO) scintillators, and has application as well in exploration for coal, ores, oil, and gas, to determine elemental contents of phosphates, coals, and other materials, and in explosives detection.
  - *Thermoluminescent dosimeters* (TLDs): Placed in fixed locations and periodically read to monitor ambient dose levels.
  - *Continuous monitors*: Active or passive systems that are always in place, accumulating samples (typically air) to monitor for various contaminants.
    - *Passive devices* (e.g., sticky paper): Catching deposited particulate matter or monitoring gamma exposure rates.
    - *Active devices*: Continuously pulling air through various devices at a continuous low rate, to count the monitored air directly or trap and store radionuclides for later collection and analysis (e.g., "cold traps" for ^{3}H, charcoal to adsorb iodine). May include gas proportional counters (α/β), alpha spectrometers, and other detector types. Unless you are monitoring for environmental radon (and you might be), radon will be an interference in most samples that needs to be corrected for, either by waiting a few hours for radon progeny on the sample to decay away or by mathematical corrections to the measured data.
- *Laboratory instruments*: Really may include any kind of radiation detector system that can provide low-level counting capabilities. Some common systems include:
  - *Gas flow proportional counters*: For counting air filters, or water, sediment, or other samples for gross α/β activity. Water samples will be evaporated to dryness, with the residue counted. Self-absorption corrections will be needed. Standard systems may not have a low enough background, and more expensive, low-background systems may be needed. Some provide automated sample changing capabilities for dozens or hundreds of samples.

○ *Gamma spectroscopy systems*: For most any kind of sample, various configurations of sample containers are available. Some provide very low energy detection capabilities.

○ *Alpha spectroscopy systems*: Usually some extensive radiochemical separation is required before analyzing, as alpha energies of many species of interest (actinides) may be fairly similar.

12.8 Evaluation of the Data

In addition to simple characterization of the radioactivity in individual samples, other obvious endpoints in the analysis of the data collected include some sort of calculation of potential radiation doses received by people from environmental transport of the nuclides. The basic approach is:

- Identify the most important pathways by which radiation exposure of the public may occur.
- Combine measured results with model predictions of radionuclide dispersion and migration.
- Collect samples and exposure data to determine the most important radionuclide concentrations or radiation levels in the identified pathways.
- Use predetermined dose conversion factors (which give the dose to individuals per unit activity inhaled or ingested) to calculate dose estimates for members of the public.

The modeling of radionuclide dispersion and environmental migration is a complex subject. Very sophisticated models exist for dispersion of pollutants in air or water, considering both physical transport and particle diffusion from areas of higher to lower concentration. Depending on the "stability" of the air column near a release point, which has to do with the relationship between the local air temperature profile and the *adiabatic lapse rate* (natural rate at which the air temperature decreases with altitude), the behavior of plumes of pollutants will vary (Figure 12.16). The downwind concentration profile may be treated mathematically (Figure 12.17).

The equation that calculates the downwind steady-state air concentrations of contaminants from a point source is given by

$$C(x, y, z) = \frac{Q}{2\pi u \sigma_y \sigma_z} \cdot \exp\left(\frac{-y^2}{2\sigma_y^2}\right) \cdot \left\{\exp\left[\frac{-(z - H)^2}{2\sigma_z^2}\right]\right.$$

$$\left. + \exp\left[\frac{-(z + H)^2}{2\sigma_z^2}\right]\right\}$$

where:
$C(x, y, z)$ = contaminant concentration at the specified coordinate $[ML^{-3}]$
x = downwind distance [L]
y = crosswind distance [L]
z = vertical distance above ground [L]
Q = contaminant emission rate $[MT^{-1}]$
σ_y = lateral dispersion coefficient function [L]
σ_z = vertical dispersion coefficient function [L]

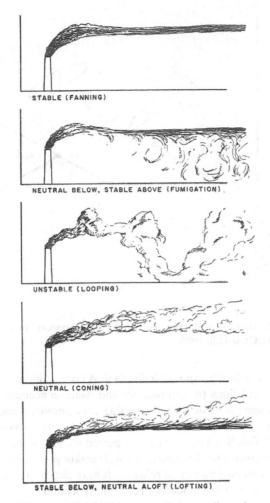

STABLE (FANNING)

NEUTRAL BELOW, STABLE ABOVE (FUMIGATION)

UNSTABLE (LOOPING)

NEUTRAL (CONING)

STABLE BELOW, NEUTRAL ALOFT (LOFTING)

Figure 12.16 Airborne plume dispersion patterns. (From http://cities.poly.edu/environment/gradedu/ce752/ce752_airmodeling.html.)

u = wind velocity in downwind direction $[LT^{-1}]$
H = effective stack height [L]

The values of σ_x and σ_y are obtained from graphs that give the values as a function of distance from the source for different stability categories (Figure 12.18). Implementation is most effectively done with computer codes, as the basic calculations are quite complex, and ultimately have to be folded over wind rose data for a site to give long-term average values. Similar expressions have been derived for liquid sources (moving streams, oscillatory flow in estuaries, etc.).

The models are quite elegant, but are really only reasonably accurate in wide flat areas with relatively constant air flow. In real situations, the models cannot effectively handle disturbances due to irregular terrain, buildings, and the like, and give only approximate results (Figure 12.18).

Two important approaches in the use of dose conversion factors include calculation of:

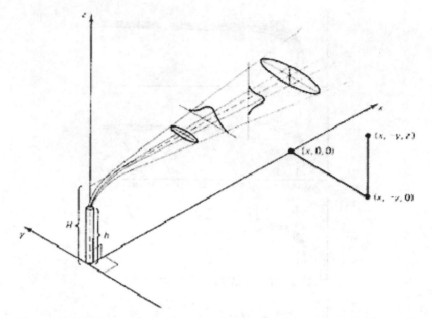

Figure 12.17 Mathematical treatment of plume migration, from Cember, with permission, McGraw-Hill, 1996.

- *Dose to the maximally exposed individual*: A combination of environmental transport calculations (to estimate air and water concentrations) and conservative assumptions about radionuclide concentrations and consumption habits is used to estimate the radiation dose to the "maximally exposed individual." This is not necessarily the person who lives closest to a facility; it may be a person who, because of lifestyle and/or proximity, has the highest exposures from environmental releases. It may also be a subject who is the most radiosensitive or vulnerable, particularly a child.
- *Collective dose to a population*: A combination of environmental transport calculations (to estimate air and water concentrations) and reasonable assumptions about radionuclide concentrations and consumption habits is used to estimate the radiation dose to all members of a population affected by environmental releases. Collective dose (person-Sv) and collective risk may be estimated and used for descriptive purposes and decision making.

Treatment of Data Below the Detection System MDA

As noted in Chapter 8, treatment of "less than" values is important to do correctly. The most correct method is:

- Show all measured values
- Show the values' associated uncertainty
- Show the system MDA for comparison

Assigning all "less than" values to be equal to the MDA or to be zero activity will bias the results too high or too low, respectively. Showing the true values (including "negative" values) the average sample concentrations and the distribution of concentrations may be understood.

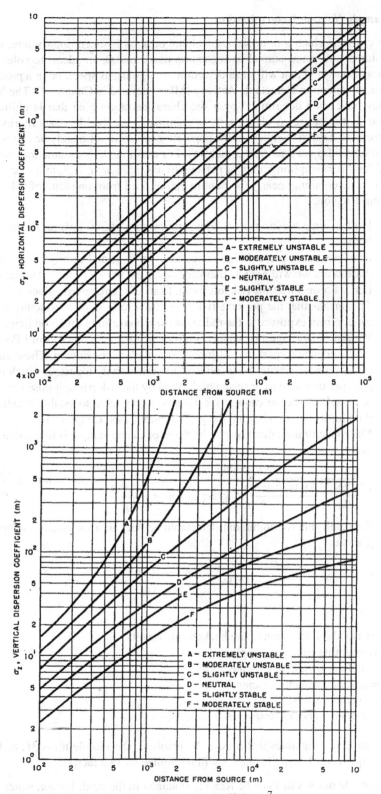

Figure 12.18 Gaussian plume dispersion coefficients.[7]

Sample Volume Determination

In cases in which a finite sample volume or mass is taken (air volume, water volume, soil mass, animal or vegetation tissue mass), an adequate volume or mass is targeted that will meet detection limit goals, as specified in a predetermined Minimum Detectable Activity (MDA; defined in Chapter 8). The MDA may have been determined by a site characterization goal that is ultimately tied to the desire to characterize a minimally acceptable risk criterion for assessing dose to members of the population who will ultimately be exposed to radionuclides via a particular pathway. For example, let's assume the analysis of food consumption habits of persons living in an area, and, in a worst case scenario, obtaining nearly their entire diet of food from one kind of fish taken from a region.

1. Let's assume a conservative consumption rate of 300 kg/year of this food item.
2. Now, assume a person's life expectancy is 75 y, and during every one of these years we will assume that she consumes 300 kg of this food item.
3. Let's choose a lifetime risk threshold, from EPA suggested values, of 10^{-6}. This means that the person's risk is one in a million of contracting a fatal cancer from exposure to the radiation from consuming this food item.
4. We can obtain ingestion cancer morbidity risk coefficients from EPA Federal Guidance Report No. 13 for a given nuclide considered. These coefficients give the risk to the individual per unit activity ingested, which is the dose per unit activity ingested multiplied by the risk per unit dose received, ultimately based on cancer risk models, as are used to establish radiation dose limits for workers and the general public (see Chapter 7).
5. The concentration that needs to be detected in any sample is thus calculated as

$$\frac{\text{acceptable risk}}{\dfrac{\text{risk}}{Bq} \times \dfrac{\text{kg ingestion}}{y} \times \text{years of life} \times \dfrac{1000 \text{ g}}{\text{kg}}} = \text{concentration } \frac{Bq}{\text{g}}$$

6. We recall that the minimum detectable activity in a radioactive sample is given as

$$\text{MDA} = \frac{4.66 \, \sigma_b + 2.71}{TY\varepsilon Mk}$$

where:

σ_b = standard deviation of the background counts
T = counting time per sample
Y = radiation yield per disintegration
ε = absolute detector efficiency
M = sample size (g)
k = unit conversions (cts/sec to Bq, etc.)

We have two variables that need to be resolved, the sample mass M, and the counting time T. Sample masses will be limited by two factors:

• Sample masses that can be feasibly obtained in the field. For air, water, and soil and sediment samples, these values are generally unlimited. One can sample air for as long as is needed, and one can generally find an abundance of water, soil, and sediment to sample (in some cases there may be limitations, but generally this is limited only by the size of available sampling containers). When sampling biota, however, sample masses may be limited by the number and size of the available species. Some avian species, for example, are very small. If one needs to obtain samples of only the kidneys of a very small avian species, these samples may be very small and cannot be improved, except by "pooling" of samples from several different individuals.

- Standard analytical sample containers may come in particular fixed sizes/volumes. One can calibrate a NaI or Ge system for any sample geometry, theoretically. In practice, however, for large samples, particular industry-standard container sizes are available, and these match the sizes of the volumes of calibration standards provided by important standards organizations (e.g., the National Institute for Standards Technology, NIST). Containers and standards for Ge detectors, for example, come in 0.5 L and 1.0 L sizes. One may fill a sample container only partially, but then the efficiencies as a function of energy derived for the full container will not be exactly correct. Thus, a full container is needed to obtain an accurate assessment of radionuclide content. See Figure 12.19. Thus, even if 3 kg of a sediment sample is available, the sample size may be limited to the size of the sample container that is available for use in transporting the samples.

The needed counting time T may be found by solving this equation, setting the MDA = concentration limit established in step 5. For example, on an HPGe detector, we may obtain about 1 count per minute background in the ^{137}Cs region (around the 662 keV photopeak), and have an efficiency of about 3%. If we express σ_b on the basis of the background count rate (R), and assume a

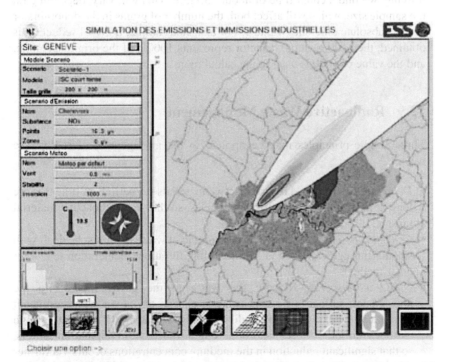

Figure 12.19 Pollutant transport simulation model. (From http://ecolu-info.unige.ch/recherche/eureka/AIDAIR/SPPA_english.html.) (see color plate)

100 g sample:

$$\text{MDA} = \frac{4.66\sqrt{\dfrac{R}{T}}}{60 \times \varepsilon \times Y \times 100\ \text{g} \times 2.22\dfrac{\text{dis}}{\text{min} - \text{pCi}}}$$

$$+ \frac{2.71}{60 \times \varepsilon \times Y \times T \times 100\ \text{g} \times 2.22\dfrac{\text{dis}}{\text{min} - \text{pCi}}}$$

Letting $Q = \text{MDA} \times 60 \times \varepsilon \times 100 \times 2.22$, we end up with:

$$Q\,T = 4.66\sqrt{R\,T} + 2.71$$

Squaring both sides, we obtain a quadratic equation that can be solved for T. Example from the spreadsheet: ^{137}Cs, assuming a photopeak efficiency of 3%, and using $Y = 0.85$:

$$Q = 1.19 \times 10^{-3}\ \text{pCi/g} \times 60 \times 0.03 \times 0.85 \times 100 \times 2.22 = 0.404.$$

In the quadratic equation,

$$a = Q^2$$

$$b = -(5.42Q + 21.72) \quad (5.42\ \text{is}\ 2 \times 2.71;\ 21.72\ \text{is}\ 4.66^2)$$

$$c = 7.34 \quad (2.71^2).$$

Solving, we find a count time of about 150 min. This will vary, depending on the sample size, which will affect both the number of grams in the denominator and the absolute efficiency value ε. For an ashed sample, if a 10:1 reduction is obtained, then 100 g on the detector represents 1000 g of the original sample, and the value of 1000 is used in the calculation.

12.9 Radioactive Waste Management

Three simple principles may be applied in the management of radioactive wastes.

- *Delay-and-decay*: Hold the radioactive materials for periods of time that are long compared to their physical half-lives, so that less radioactive material must be disposed of.
- *Concentrate-and-contain*: Make the volume of waste that must be dealt with smaller through compaction, compression, filtration, incineration to ash, or other methods.
- *Dilute-and-disperse*: Release the waste to the environment, into the atmosphere, surface waters, seawaters, and so on, along with significant quantities of clean medium for dilution, also releasing the material at high altitudes (airborne wastes) and/or at long distances from population centers, so that significant reduction in the medium concentrations occurs via further dilution and dispersion (see section above on pollutant transfer).

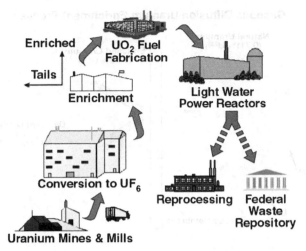

Figure 12.20 Schematic drawing of the nuclear fuel cycle. (From http://www.nrc.gov/materials/fuel-cycle-fac/stages-fuel-cycle.html; courtesy U.S. Nuclear Regulatory Commission.)

12.9.1 The Nuclear Fuel Cycle

Before discussing wastes in detail, a brief overview of the nuclear fuel cycle is given. The nuclear fuel cycle accounts for a large percentage of waste in all categories. Certainly other important forms of waste exist, particularly in the medical and research uses of radionuclides. The forms and activity levels of such wastes overlap with those of the nuclear fuel cycle. The functioning of the nuclear fuel cycle (Figure 12.20) is relatively complex, and the kinds and levels of radioactive waste produced at each step should be understood before considering waste reduction and storage strategies.

1. *Uranium mining and processing*: Ores rich in uranium occur at a number of locations around the world. Almost all soils contain some measurable levels of naturally occurring uranium, which is about 99.3% ^{238}U, 0.7% ^{235}U, and 0.005% ^{234}U; uranium ores contain more substantial concentrations. Uranium ore may be mined by surface or underground mining techniques. Mined uranium ore will be crushed and ground to a fine slurry, then leached with sulfuric acid to separate the uranium from the surrounding matrix. The separated uranium is recovered from the leach solution and precipitated as a concentrated form of uranium oxide (U_3O_8). This material is also known as *yellowcake*. A technique known as "In Situ Leaching" (ISL) may be employed, in which the uranium is extracted from the ore body by injecting the leaching solution underground and bringing the uranium up to the surface in solution for later uranium recovery. A few hundred tons of U_3O_8 are needed to provide fuel for a year for a reasonably large (1000 MW electric) nuclear reactor.

2. *Conversion*: Before being usable as a power plant fuel or in an explosive device, the levels of ^{235}U need to be increased significantly (to 3–4% for power plant fuel, or ~90% for a weapon). Thus, *enrichment* of the ^{235}U must occur. Enrichment methods, described next, require the uranium to be in a gaseous form. Thus, in this step of the process, yellowcake is converted

Gaseous Diffusion Uranium Enrichment Process

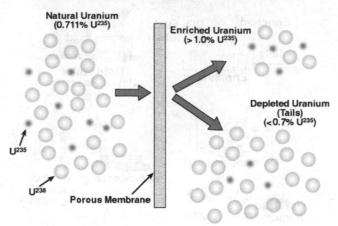

Figure 12.21 Gaseous diffusion of uranium. (From http://www. nrc.gov/materials/fuel-cycle-fac/ur-enrichment.html; courtesy U.S. Nuclear Regulatory Commission.)

to UF_6 (uranium hexafluoride, or "hex"), which can be turned into a gas. There are a couple of processes by which this conversion is accomplished. One, for example, may react nitric acid with the yellowcake to produce uranyl nitrate, which is decomposed to UO_3, then reduced to UO_2, then reacted with HF (hydrofluoric acid) to produce UF_4 and then UF_6.

3. *Enrichment*: Natural uranium ore contains about 0.7% ^{235}U, as noted above. For an operating reactor, we would like fuel with a level of 3–5%. This means that we need to enrich the level of ^{235}U in the material somehow. We cannot perform a chemical separation of ^{235}U from the ^{238}U, as they will act similarly in chemical reactions. At present, two methods are used for commercial enrichment:

   (a) *Gaseous diffusion*: Gaseous UF_6 is passed through special barriers or porous membranes that have very small holes (Figure 12.21). Enrichment of ^{235}U over ^{238}U occurs when the lighter $^{235}UF_6$ gas molecules diffuse faster through the barriers than the heavier $^{238}UF_6$ gas molecules. Actually, the gas must diffuse through hundreds of barriers, one after the other, before the ^{235}U enrichment reaches the level needed for use in power reactors. After the diffusion process is complete, the enriched UF_6 gas is condensed back to a liquid and allowed to cool and solidify. It can then be made into reactor fuel assemblies. Some laser-based approaches were used for enrichment for a time, but then have fallen out of use.

   (b) *Centrifugation*: Similar to the diffusion process, the gaseous UF_6 may be centrifuged at high rates (50–70,000 rpm), with the slight mass difference between ^{235}U and ^{238}U causing the heavier ^{238}U to migrate towards the edges. Each rotor is about 1–2 m in length and perhaps 15–20 cm diameter (Figure 12.22). The enriched and depleted gases are then extracted from the rotors.

4. *Fuel Fabrication*: The enriched UF_6 is moved to a fuel fabrication plant and converted to uranium dioxide (UO_2), in powder form. It is pressed

Figure 12.22 A bank of gas centrifuges. (From http://www.uic.com.au/nfc.htm.)

into small pellets, which are inserted into long thin tubes, made of stainless steel or perhaps a zirconium alloy (sometimes called "zircalloy") to form fuel rods. The rods are then sealed; groups of rods are organized into fuel assemblies for use in the nuclear reactor core.

5. *Reactor*: Several hundred fuel assemblies will be in the core of a normally operating power reactor. A reactor with a 1000 megawatt electric (MWe) capacity will have perhaps 75 tons of low-enriched uranium in its various fuel assemblies. As discussed in Chapter 11, the ^{235}U in the fuel fissions, producing heat when placed in this critical assembly in the presence of a moderator such as water or graphite. Interestingly, some of the ^{238}U in the reactor core is converted to ^{239}Pu. Around half of the ^{239}Pu will fission, providing about one-third of the reactor's output. As was briefly covered in Chapter 11, the reactor spent fuel is moved about, with about a third being removed every year or so, to be replaced with fresh fuel. The two-thirds that remain, the more "burned-up" fuel, is moved towards the center, and eventually is removed and replaced as well.

6. *Spent fuel storage*: Spent fuel assemblies taken from the reactor core are stored in storage areas near the reactor site, in water-filled or air-cooled tanks, to allow both their heat dissipation and decay of the fission products. In water-filled spent fuel pools, the water provides cooling as well as some shielding. Spent fuel theoretically can be stored safely for long periods of time (although many worry about targeting of these sites by terrorist groups). This storage, however, is only an interim step; ultimately the spent fuel will be either reprocessed or sent to a long-term disposal site.

7. *Reprocessing*: Spent fuel still contains around 95% of the uranium that was in the original fuel rods, although the ^{235}U content has been reduced to less than 1%. Small percentages of the spent fuel are comprised of waste product plutonium (Pu) that was created during use of the fuel. Spent fuel can be reprocessed to recover a good portion of the still usable uranium and plutonium. The fuel rods are cut up and dissolved in acid to separate the uranium from the waste products, cladding, and the like, and the uranium is chemically isolated. Any recovered uranium (which will contain both ^{238}U and ^{235}U) can be sent back to a conversion plant for conversion to uranium hexafluoride and enrichment. Reactor-grade plutonium can be blended with

enriched uranium to produce a substance called Mixed OXide (MOX) fuel, in a special plant. Remaining high-level radioactive wastes (perhaps 750 kg/y from a 1000 MWe reactor) can be disposed of (see next section). Reprocessing of spent fuel was halted in the United States in the 1970s, over concerns that terrorists could intercept shipments of spent fuel and gain access to highly enriched uranium that might be used to make nuclear weapons. Reprocessing is performed in Europe and Russia.

8. *Long-term disposal*: We discuss waste disposal technologies shortly. Ultimately, however, spent fuel, whether having undergone reprocessing or not, needs to be disposed of with great care to the confinement of the long-lived fission and activation products, in a very stable underground environment that does not contain groundwater that may be consumed by the human population. Some geological formations will be deep underground (Figure 12.23) and should show such long-term (i.e., thousands of years!) stability, for example, granite formations, volcanic emissions, or salt or shale environments. In the United States, as noted in Chapter 7, the Nuclear Waste Policy Act of 1982 mandated that a solution to the storage of high-level wastes be found, and the Yucca Mountain project in Nevada was indicated as a suitable solution. Yucca Mountain is a ridge comprised of layers of volcanic rock, called "tuff", which is made of ash that was deposited by a series of volcanic eruptions between 11 and 14 million

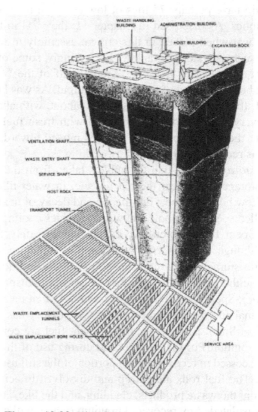

Figure 12.23 High-level waste deep underground storage facility. (From http://www.nea.fr/html/brief/ brief-03.html.)

years ago. Political struggles, however, have prevented final approval of this project, and high-level wastes at this time remain in temporary storage near operating reactors.

12.9.2 General Waste Types

As noted in the chapter on radiation regulations, the legal definitions of waste in the United States are a bit strange, permitting fairly high-intensity sources to be classified as low-level wastes. In other countries, and in a daily practical sense, we may define wastes in a more logical way for dealing with the different types of wastes as low-, intermediate-, or high-level.

Low-Level Waste (LLW)

Low-level waste, made up of contaminated gloves, clothing, rags, tools, syringes, and other random objects slightly contaminated with radioactive material, is generated in substantial volumes every day from hospitals, laboratories, and various industrial activities, including elements of the nuclear fuel cycle. In general, the radionuclides involved are short-lived, so a lot of this waste can be eliminated from treatment as radioactive waste by simply allowing it to sit around in a controlled location until nearly all the radioactivity in the waste decays to background levels. Some longer-lived nuclides are not amenable to this solution, so quantities of this kind of waste must be routinely disposed of or removed from the waste (e.g., contaminated tools may be subject to cleaning and stripping solutions that remove the material from the surfaces and allow the tools to be returned to normal use, with the waste streams from the cleaning operations then becoming the focus of attention for disposal). This kind of waste will ultimately be buried in some shallow landfill sites (again, see Chapter 7 for a discussion of low-level waste sites currently operating in the United States). Such waste may also be compacted or incinerated to reduce its volume prior to disposal. Around the world LLW makes up about 90% of the total volume of radioactive waste, but contains only 1% of the total radioactivity.

Low-level waste disposal occurs at commercially operated facilities that must be licensed by the NRC or the appropriate agreement state. The facilities need to be well characterized prior to operating, to demonstrate their safety for siting and operation. Such characterization is usually done with long projection times into the future. The sites use shallow land burial techniques to store the waste. Unlike a sanitary landfill, however, the material is not just dumped on open space and eventually covered. The waste is first stored in sealed drums or other sealed containers. The material is shipped by truck, rail, or ships to the site, then placed in concrete vaults in the shallow land trenches, which are typically lined (Figures 12.24 and 12.25).

At uranium mines, considerable quantities of dust and radon (^{222}Rn) are produced. The dust is controlled as much as possible to minimize workers' inhalation of the radioactive materials. Much of the dust created can be collected and fed back into the process, however, significant quantities of radon gas may be released to the atmosphere. The use of large volumes of air can help reduce these concentrations ("dilute and disperse"). There are a number of residual waste types from the uranium processing that contain other radioactive materials from the ore, such as radium. These wastes may be placed into "tailings" areas or structures. They may be put injected back into the mine

Figure 12.24 Containers for shipment and disposal of low level waste. (From http://www.cs.virginia.edu/~jones/tmp352/projects98/group14/disposal.html.)

or perhaps covered with materials such as rock and clay, with new vegetative growth established over the area (this is called site reclamation).

Intermediate-Level Waste (ILW)

This contains higher amounts of radioactivity and will require more shielding than low-level wastes. This may include ion exchange resins from reactors (used in water purification), chemical sludges, and other reactor components. This kind of waste may be solidified in concrete or bitumen to facilitate safe disposal. Generally short-lived waste (mainly from reactors) is buried as LLW, but long-lived waste (from reprocessing nuclear fuel) will be disposed of deep underground as HLW.

High-Level Waste

This may include spent fuel and reprocessing wastes. HLW comprises only 3% of all radioactive waste by volume, but has about 95% of the total radioactivity.

Figure 12.25 Disposal of waste packages into a disposal pit, and large scale view of a low level radioactive waste disposal facility. (From http://www.rwmc.or.jp/jigyou/je2.html.)

Figure 12.26 Borosilicate glass product. (From http://www.uic.com.au/wast.htm.)

As noted above, this material generates heat and must be cooled. The material is highly radioactive, containing many beta and gamma emitters, and thus requires significant shielding during all phases of handling and transport. HLW may be incorporated into borosilicate (Pyrex) glass (called *vitrification*). The vitrified waste is then placed in sealed stainless steel canisters for eventual deep underground burial (Figures 12.26 and 12.27).

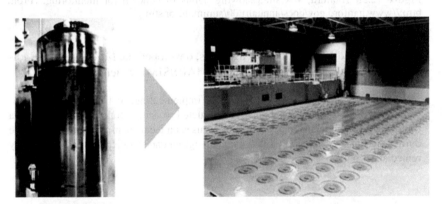

Figure 12.27 Container holding vitrified waste and storage facility with vitrified waste casks. (From http://www.rwmc.or.jp/jigyou/je2.html.)

12.10 Site Evaluation

Complete treatment of an environmental monitoring plan for a site, whether an operating facility, waste storage facility, or other site, begins with an analysis of the significant pathways to humans, as noted at the beginning of the chapter; identifies all important sample types; develops a plan and timetable for sampling, analysis, and interpretation of the data; and then implements this plan and executes it, perhaps with periodic modification. Figure 12.28 demonstrates a study of the presence of contaminants in groundwater beneath an area near the Department of Energy Hanford site, where the presence of contaminants is a concern because the adjacent riverbed is used by salmon for spawning.

Figure 12.28 Hanford, WA site, showing points of concern for monitoring. (From http://www.hanford.gov/docs/annualrp00/summonitor.stm.)

The EPA and NRC, with other agencies, developed the Multi-Agency Radiation Site Survey Investigation Manual (MARSSIM), which

...provides detailed guidance for planning, implementing, and evaluating environmental and facility radiological surveys conducted to demonstrate compliance with a dose- or risk-based regulation. MARSSIM focuses on the demonstration of compliance during the final status survey following scoping, characterization, and any necessary remedial actions.[8]

Here is the MARSSIM table of contents.

No.	Title
1.	Introduction
2.	Overview of the Radiation Survey and Site Investigation Process
3.	Historical Site Assessment
4.	Preliminary Survey Considerations
5.	Survey Planning and Design
6.	Field Measurement Methods and Instrumentation
7.	Sampling and Preparation for laboratory Measurements
8.	Interpretation of Survey Results
9.	Quality Assurance and Quality Control
	References
	Glossary
	Index
	Bibliographic Data

Appendices

A. Example of MARSSIM Applied to a Final Status Survey
B. Simplified Procedure for Certain Users of Sealed Sources, Short Half-life Materials, and Small Quantities
C. Site Regulations and Requirements Associated with Radiation Surveys and Site Investigations
D. The Planning Phase of the Data Life Cycle
E. The Assessment Phase of the Data Life Cycle
F. The Relationship Between the Radiation Survey and Site Investigation Process, the CERCLA Remedial or Removal Process, and the RCRA Correction Action Process
G. Historical Site Assessment Information Sources
H. Description of Field Survey and Laboratory Analysis Equipment
I. Statistical Tables and Procedures
J. Derivation of Alpha Scanning Equations Presented in Section 6.7.2.2
K. Comparison Tables Between Quality Assurance Documents
L. Regional Radiation Program Managers
M. Sampling Methods: A List of Sources
N. Data Validation Using Data Receptors

Standardized laboratory methods for sample analysis are contained in the Multi-Agency Radiological Laboratory Analytical Protocols Manual (MARLAP).[9] Here is the MARLAP table of contents.

Volume I

Chapter	Title
	Title Page, Foreward, Contents, Acknowledgements, Abstract, Acronyms, Units
1.	Introduction to MARLAP
2.	Project Planning Process
3.	Key Analytical Planning Issues and Developing Analytical Protocol Specification
4.	Project Plan Documents
5.	Obtaining Laboratory Services
6.	Selection and Application of an Analytical Method
7.	Evaluating Methods and Laboratories
8.	Radiochemical Data Verification and Validation
9.	Data Quality Assessment

Appendix	Title
A.	Directed Planning Approaches
B.	The Data Quality Objectives Process
C.	Measurement Quality Objectives for Method Uncertainty and Detection and Quantification Capability
D.	Content of Project Plan Documents
E.	Contracting Laboratory Services
	Glossary

Volume II

Volume III

As with the computer codes mentioned in Chapter 9, useful computer codes for modeling radionuclide transport and performing population dosimetry will change often. At present, two examples of useful codes that are available for this purpose include RESRAD (http://web.ead.anl.gov/resrad/home2/) and GENII (http://www-rsicc.ornl.gov/codes /ccc/ccc6/ccc-601.html).

Endnotes

1. http://www.epa.gov/history/topics/lovecanal/01.htm.
2. C. D. Jennings and S. W. Fowler. Uptake of 55Fe from contaminated sediments by the polychaete Nereis diversicolor. *Marine Biology* **56**(4):277–280, 1980.
3. Booth, R.S. Burke, O.W., Kaye, S.V. Dynamics of the Forage-Cow-Milk pathway for Transfer of Radioactive Iodine, Strontium, and Cesium to Man. *Oak Ridge National Lab* CONF-710818–2; from Meeting on nuclear methods in environmental research;Columbia, Mo. (23 Aug 1971).
4. F. W. Whicker and J. E. Pinder. Food Chains and Biogeochemical Pathways: Contributions of Fallout and other Radiotracers. *Health Phys.* **82**(5):680–689, 2002.

5. U. S. Environmental Protection Agency, Superfund Program Representative Sampling Guidance Volume 1: Soil, EPA 540/R-95/141, December 1995

6. U. S. Environmental Protection Agency, Methods for Evaluating the Attainment of Cleanup Standards Volume 1 Soils and Solid Media, EPA 230/02-89/042, February, 1989.

7. R. H. Slade, ed. Meteorology and Atomic Energy. US Atomic Energy Commission, Oak Ridge, Tennessee, 1968.

8. http://www.epa.gov/radiation/marssim/index.html.

9. http://www.epa.gov/radiation/marlap/manual.htm.

13

Nonionizing Radiation

Treatment of the safety issues related to uses of nonionizing radiation is not part of the routine practice of health physics by the vast majority of professionals. In those instances in which these applications are encountered, however, the health physicist may be responsible for understanding and managing the issues involved. Thus, it is essential to have a basic understanding of many aspects of the use of these forms of radiation, namely:

- Theory of operation and basic principles
- Mechanisms of radiation interaction (mainly with human tissue, but also with other materials, for shielding, detector design, and other uses)
- Exposure guidelines
- Safety procedures

As is obvious from the name, the forms of radiation considered here do not cause interactions via the ionization of atoms. Thus, other forms of interaction are involved, and are discussed individually.

13.1 Ultraviolet Radiation

Theory of Operation and Basic Principles

Recall from Chapter 2 the various elements of the total electromagnetic spectrum (Figure 2.3). We tend to think of the spectrum in relation to the (extremely narrow) portion that is visible to our eyes: immediately below, at lower energies (thus lower frequencies, thus longer wavelengths), is infrared radiation, and just above, at higher energies (thus higher frequencies, thus shorter wavelengths), is ultraviolet (UV) radiation. The name means "beyond violet;" violet is the color of the most energetic (shortest wavelength) forms of visible light. Some animals (birds, reptiles, and some insects) can see into the near ultraviolet range. Some birds have segments of their feathers that are not seen in normal visible light but are seen under ultraviolet light sources.

UV radiation can be subdivided into the *near UV* (wavelengths between 380–200 nm), *far* (also *vacuum*) *UV* (wavelengths between 200–10 nm; abbreviated as FUV or VUV), and *extreme UV* (wavelengths between 1–31 nm; abbreviated EUV or sometimes XUV). Another designation of UV radiation, related to health effects, subdivides UV into *UVA* (wavelengths between 380–315 nm, also called "long wave" or "blacklight"), *UVB* (wavelengths between 315–280 nm, also called medium wave); and *UVC* (wavelengths between 100–280 nm, also called short wave or "germicidal" UV). The sun emits UVA, UVB, and UVC; due to absorption in the ozone layer of the earth's atmosphere nearly all of the ultraviolet radiation that reaches the Earth's surface is just UVA. UV light is used in a number of applications, including:

- *Black lights*: Black lights are commonly encountered as consumer products that produce an interesting visual effect when illuminating certain substances, and have more serious applications as well. These lights emit principally long wave UV radiation and give off very little visible light. Fluorescent black lights often employ deep blue colored glass called Wood's glass, which is made of a cobalt impregnated nickel oxide, that blocks almost all visible light above 400 nm. The UV radiation itself is not visible to the human eye, but when certain materials are irradiated with UV radiation, fluorescent and phosphorescent radiations given off can cause the objects to have a more intense than usual color appearance. Black light testing is also used to evaluate the authenticity of antiques and banknotes. UV is used as well to analyze minerals and gems for authenticity due to natural fluorescence under UV light. It is also used in nondestructive testing. Fluids with fluorescent properties may be illuminated with a black light when applied to metal objects, and cracks and other artifacts may be detected. UV is used in forensic investigations to reveal the presence of trace quantities of blood, urine, semen, and saliva, as substances within these fluids have natural fluorescence.
- *Fluorescent lamps*: These lamps produce UV radiation using low-pressure mercury (Hg) gas, mainly in the UVC range.
- *Treatment of skin conditions*: Skin conditions such as eczema and psoriasis may be treated by the application of a skin-sensitizing agent (e.g., "psoralen"), then exposing the skin to UVA (Psoralen + UVA = PUVA treatment).
- *Pest control*: Small flying insects are attracted to UV radiation, and thus may be killed using an electric shock or trapped. The former are affectionately called "bug zappers" in southern regions of the United States; inexpensive but relaxed and entertaining weekend evenings in the summer months are shared by many in these regions, observing the routine functioning of such devices while imbibing refreshing fermented malt beverages from suitable metallic containers.
- *Spectrophotometry*: Ultraviolet-Visible Spectroscopy or Ultraviolet-Visible Spectrophotometry (UV/VIS) measures the absorbance of substances within laboratory samples as a function of wavelength, and the concentration of the substances can be derived using the Beer–Lambert Law (no relation to UV pest control):

$$A = -\log\left(\frac{I}{I_0}\right) = \varepsilon c L.$$

Here, A is the measured absorbance, I_0 is the intensity of the incident light at a given wavelength, I is the transmitted intensity, L the path length through the

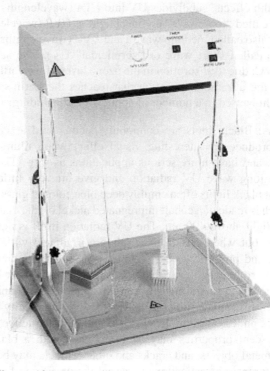

Figure 13.1 UV sterilization cabinet. (From http://www. ehs.unr.edu/.)

sample, and *c* the concentration of the absorbing species. ε is a constant called the *extinction coefficient*, and is known for many common substances.

- *Photolithography*: UV light is used to harden materials (called *photoresists*) used in the fabrication of semiconductor devices.
- *Sterilization*: Low-pressure mercury-vapor lamps emit a large portion of their UV radiation around 254 nm, which is an important region of the *germicidal effectiveness curve* (effectiveness for UV absorption by DNA). These lamps thus may assist in killing microorganisms in accessible locations in laboratories. Germicidal cabinets (Figure 13.1) may be used to sterilize laboratory equipment. UV also kills viruses and bacteria, so is used in wastewater treatment applications and drinking water treatment. It is also used to attempt to sterilize some foods and to pasteurize fruit juices.
- *Transilluminators*: Electrophoretic gels containing nucleic acids or proteins may be stained with fluorescent dyes. Specific bands of activity in nucleic acids or proteins may then be detected by excitation using an UV source in a device called a transilluminator (Figure 13.2).
- *Crosslinkers*: DNA or RNA may be bound to specific membranes via exposure to UV radiation (this was done previously with vacuum oven baking methods which took much longer Figure 13.3).
- *Curing of adhesives and coatings*: Some adhesives and coatings (e.g., glass and plastic bonding, optical fiber coatings, the floor coats, and dental fillings) harden under UV irradiation.

Figure 13.2 Transilluminator. (From http://www.ehs.unr.edu/.)

Mechanisms of Radiation Interaction and Biological Damage

Prolonged exposure to solar UV radiation is known to be associated with acute and chronic health effects involving the skin, eye, and immune system.[1]

• In the skin, UV radiation excites DNA molecules, causing the formation of covalent bonds between adjacent thymine bases, thus producing thymidine dimers. These dimers do not bond correctly with base pairs on the alternate DNA strand, causing distortion of the DNA helix, possibly leading to delayed replication, gaps in the DNA, and incorporation of incorrect base pairs. This in turn may lead to mutations, which can lead to cancer. All forms of UV (UVA, UVB, and UVC) can damage collagen fibers in the skin and accelerate skin aging. UVA is the most penetrating form, and can cause

Figure 13.3 Crosslinker. (From http://www.ehs.unr.edu/.)

aging, DNA damage, and cancer. It does not cause skin reddening, however. Exposure to UVB has been linked to the formation of skin cancer. Our skin darkens in response to UV exposure, producing melanin, but more protection is generally needed and is provided in many commercially available sun-blocking products.

- Exposure to UVB at high intensities, such as from a welder's arc (but also suntanning beds) can damage the eyes, possibly causing photokeratitis, also called keratoconjunctivitis (an irritating or painful eye condition) and may ultimately lead to cataracts, pterygium (a benign growth of the conjunctiva), and other problems. The condition usually becomes apparent about six hours after an exposure, and will usually heal itself. In very severe cases, however, permanent corneal clouding may occur and corneal transplants may be needed to restore vision. Protective eyewear is effective in blocking the occurrence of such effects.

- Effects on the immune system are actually positive, as exposure to UV stimulates production of vitamin D in the skin. Many thousand premature deaths occur each year in the United States from cancer related to insufficient UVB exposures! Vitamin D deficiency may also lead to a condition called osteomalacia ("rickets," soft bone disease), which can result in bone pain, fractures, and an inability to bear normal weight in the bones.

Detectors

Silicon carbide photodiodes are often used to detect and quantify UV radiation, which can be coupled to amplifiers to produce measurable output (Figure 13.4).

Ultraviolet Detectors

Figure 13.4 Silicon carbide photodiodes. (From http:// optoelectronics.perkinelmer.com/content/RelatedLinks/ UltravioletDetectors.pdf.)

Gallium nitride or aluminum nitride semiconductors can also be used to detect UV radiation.

Exposure Guidelines

Workers' exposure to ultraviolet radiation produced by any equipment or industrial processes are subject to "threshold limit values" specified by the American Conference of Governmental Industrial Hygienists. The exposure limits are wavelength-dependent:

- For UVA, the ACGIH recommends 1.0 J/cm^2 for periods lasting less than 1000 seconds, and 1.0 mW/cm^2 for periods lasting greater than 1000 seconds.
- For UVB, TLV values are 3.4 mJ/cm^2 at 280 nm and 500 mJ/cm^2 at 313 nm.
- For UVC, TLV values are 250 mJ/cm^2 at 180 nm and 3.1 mJ/cm^2 at 275 nm.

Safety Procedures

Procedural controls, limiting the number of personnel using the equipment, limiting times and distances from the equipment, and so on, are reasonable and fairly easy to implement regularly. Engineering controls, such as sealed enclosures with interlocks are helpful in preventing exposures entirely if they are used properly. UV may be shielded by materials such as polycarbonate, metal, wood, and cardboard. Skin can be easily covered with normal or specially designed clothing. As noted briefly above, protective eyewear, including polycarbonate safety glasses or face shields can very effectively protect the eyes from exposure.[2]

13.2 Lasers

Theory of Operation and Basic Principles

The word "LASER" is an acronym, meaning "Light Amplification through Stimulated Emission of Radiation". Laser devices produce very highly coherent beams of radiation in the visible, UV, or infrared (IR) regions of the electromagnetic spectrum. Coherence here means that the waves of the beam are very tightly in phase due to the means in which they were produced. Often, visible or other forms of electromagnetic radiation contain waves in various phases, both in space and in time. In a laser device, a lasing medium (a material that can be excited to a metastable state by the introduction of energy in an appropriate form) is contained within an optical cavity that is completely mirrored at one end and partially mirrored at the other end.

When energy is "pumped" into the cavity (e.g., an intense light source that contains light of the appropriate wavelength to excite the atoms in the cavity), electrons in the atoms of the lasing medium are excited to higher energy levels, and then de-excite to their original states, releasing light of a particular wavelength. In the normal lasing medium, the vast majority of atoms are in the ground state and few are in an excited state. The well-known Boltzmann principle states that the number of atoms in a given excited state (N_i), relative to those in a ground energy state (N_0), in a system at temperature T, is

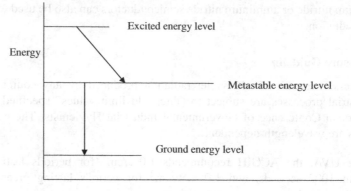

Figure 13.5 Energy levels in lasing medium.

given by

$$\frac{N_i}{N_0} = e^{\frac{-E_i}{kT}}.$$

Here E_i is the difference in energy between the excited and ground state and k is Boltzmann's constant ($1.38 \times 10^{-23} J/K$).

By pumping energy into the cavity, one can induce a condition in which the majority of the atoms are in the excited state; this is called a *population inversion*. Atoms that are excited by radiation waves already in the lasing medium will emit radiation that is in phase with the existing radiation of the appropriate wavelength and thus will add constructively to the radiation. These photons may induce other atoms to emit radiation, also in phase, thus continuing to amplifying the signal (Figure 13.5). The process of one atom causing another to emit more radiation is termed *stimulation*, which is greatly amplified by this multiplication process, thus the origin of the acronym. Actually, to function effectively as a lasing medium, the material must have at least one excited state in which electrons become trapped for a short period (microseconds to milliseconds) before dropping back to the ground state. If the trapping time is not sufficiently long, the population inversion cannot be maintained and lasing action will not occur.

Within the optical cavity, the radiation is reflected back and forth between the mirrors on either end, so that even more stimulation and amplification occurs. On the partially mirrored end, however, some of the highly intense, highly coherent radiation is emitted; this is the useful beam of radiation (Figure 13.6).

Types of Lasers

The lasing medium may be a gas, solid, or liquid. The first operational laser was made of a ruby crystal rod. Lasers are usually designated according to the type of lasing material employed.

- *Solid-state lasers* employ a solid matrix as the lasing material, for example, the ruby or neodymium-YAG (yttrium aluminum garnet) lasers. The neodymium-YAG laser emits infrared light at 1.064 micrometers.

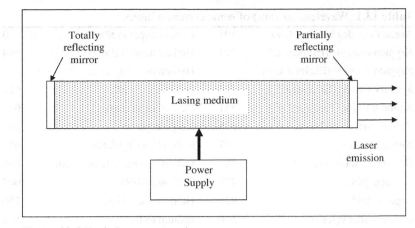

Figure 13.6 Basic laser construction.

- *Gas lasers* (e.g., helium and helium-neon, HeNe) have output in the visible red range. CO_2 lasers emit radiation in the far-infrared region, and may be used for cutting materials.
- *Excimer lasers* (the name comes from the terms "excited" and "dimer") have reactive gases such as chlorine or fluorine mixed with inert gases such as argon, krypton, or xenon. Dimers are produced during the excitation process; these devices produce radiation in the ultraviolet range.
- *Dye lasers* use complex organic dyes (e.g., rhodamine 6G in a liquid solution). These lasers may be tunable over a range of output wavelengths.
- *Semiconductor lasers*, also called diode lasers, are generally small and have low power outputs. They are used in some laser printers and compact disk players.
- Special research lasers have been built on the *Free Electron Laser* (FEL) principle, in which the wavelengths emitted are completely "tunable" (selectable) from a broad range of frequencies (in standard lasers, once the medium is chosen, the wavelength of the emitted radiation is fixed). In a FEL, the lasing medium is actually a beam of electrons moving at relativistic energies. The beam is made to pass through a group of magnets that present alternating poles along the beam path. The array of magnets is called a "wiggler" because the alternating magnetic field causes the electrons in the beam to follow a sinusoidal path. As the electrons are accelerated due to the deflection in path, photons are released. Varying the parameters in the magnetic field permits the release of photons of any energy within the available range.

The laser output may be continuous, pulsed, or "Q switched". Continuous wave (cw) lasers typically use gas media; the first operational cw laser was a He–Ne laser. Pulsed lasers deliver laser output in brief intense pulses of about 0.1–10 ms duration. Ruby lasers are operated as pulsed output lasers. A Q-switched laser is a laser that produces periodic pulses when a "shutter" device is opened; the operation of the "shutter" is controlled so that large population inversions are attained, and thus, short, very high power bursts of radiation are produced. Table 13.1 shows the wavelengths of some common lasers.

Table 13.1 Wavelengths (nm) of some common lasers.[3]

Argon fluoride (Excimer-UV)	193	Copper vapor (yellow)	570
Krypton chloride (Excimer-UV)	222	Helium neon (yellow)	594
Krypton fluoride (Excimer-UV)	248	Helium neon (orange)	610
Xenon chloride (Excimer-UV)	308	Gold vapor (red)	627
Xenon fluoride (Excimer-UV)	351	Helium neon (red)	633
Helium cadmium (UV)	325	Krypton (red)	647
Nitrogen (UV)	337	Ruby ($CrAlO_3$) (red)	694
Helium cadmium (violet)	441	Gallium arsenide (diode-NIR)	840
Krypton (blue)	476	Nd:YAG (NIR)	1064
Argon (blue)	488	Helium neon (NIR)	1150
Copper vapor (green)	510	Erbium (NIR)	1504
Argon (green)	514	Helium neon (NIR)	3390
Krypton (green)	528	Hydrogen fluoride (NIR)	2700
Helium neon (green)	543	Carbon dioxide (FIR)	9600
Krypton (yellow)	568		

Mechanisms of Radiation Interaction and Biological Damage

Laser radiation may be very intense and may cause direct damage to the skin and eyes. We are all accustomed to science fiction movies in which individuals carry around handheld laser guns and land- and air-based vehicles contain laser weapons that are used to attack people and destroy objects. High-intensity laser light can indeed burn holes in skin or other materials. Handheld weapons are still the stuff of science fiction at present.

Lasers of lower power are a part of everyday life now, as lasers are used in entertainment (light shows at planetariums, music concerts, and other outdoor affairs), as scanners in grocery store checkout counters and other locations in which laser-read bar codes are used, as pointers during oral presentations, and other routine applications. Lasers are used to bleach teeth and control gum disease, to remove hair, and in ocular surgery (to remove tumors, cataracts or blood vessels, and to correct vision problems). Laser radiation needs to be very seriously respected and carefully used, because its effects on human tissue, particularly the eye, can be quite serious, and can occur with very brief exposures. Even the low-power lasers that are easily purchased by any member of the public should not be used carelessly in a way that might strike the eye of another person, as people's sensitivity to such exposures varies, and mild temporary blindness, flashes, afterimages, and other effects have been known to occur. Airplane pilots have reported being temporarily blinded by lasers being shone into the cockpits of commercial airliners.

The lens of the human eye focuses incoming light on the retina, where the recorded information is translated into sensory information and passed to the brain (Figure 13.7). The lens and cornea will absorb most radiation with wavelengths less than around 400 nm. Radiation in the 315–340 nm range will be transmitted and absorbed by the retina, and radiation of 400–1400 nm

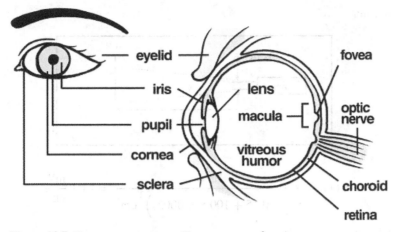

Figure 13.7 Human eye structure. (From www.mvrf.org.)

(0.4–1.4 μm) is almost totally transmitted through the lens and cornea and may be absorbed by the retina.

Permanent retinal damage due to direct heating can easily occur from exposure to laser light, due to its high intensity and the ability of the lens to focus that energy in a small spot on the retina, thus increasing its intensity per unit area. Continuous exposure to visible light lasers at around 6 W/cm^2, or pulsed exposures of 0.1–1 J/cm^2 have been shown to produce retinal damage.[4] Most light (and other radiation) sources diverge substantially as they move away from their source. The example we have worked with mostly in this text has been an isotropic point source of radiation, for which the student knows very well by now that the radiation decreases with an inverse square relationship to distance from the source. As laser light is highly focused and coherent, however, very little beam divergence occurs, over large distances. For circular beams, for example, the power density is calculated as the energy delivered per unit time over the area of the circular beam. Divergence of a laser may be specified in *milliradians*, not radians (abbreviated rad, not to be confused with rad of ionizing radiation). Consider a simple He–Ne laser with a 1 mW continuous wave output. In a device with a 1.5 mm aperture, the radiation intensity at the device output is

$$E_0 = \frac{1\text{mW}}{\pi \left(\frac{0.15}{2}\right)^2 \text{cm}^2} = 56.6 \frac{\text{mW}}{\text{cm}^2}.$$

Reduction of the intensity occurs by both attenuation and due to beam divergence. Attenuation is usually negligible unless air is smoky or foggy, thus the reduction in intensity at different distances can be calculated just via study of the beam size. If we assume a divergence (ϕ) of 0.0002 radians, the intensity at distance D is given as

$$E(D) = \frac{1\text{mW}}{\pi \left(\frac{1}{2}(0.15 + D \times 0.0002)\right)^2 \text{cm}^2}$$

At 1 m:

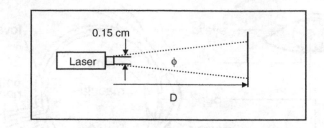

$$E(1) = \frac{1\text{mW}}{\pi\left(\frac{1}{2}\left(0.15 + 100 \times 0.0002\right)\right)^2 \text{cm}^2} = 44.1 \frac{\text{mW}}{\text{cm}^2}.$$

At 10 m:

$$E(10) = \frac{1\text{mW}}{\pi\left(\frac{1}{2}\left(0.15 + 1000 \times 0.0002\right)\right)^2 \text{cm}^2} = 10.4 \frac{\text{mW}}{\text{cm}^2}.$$

At the 10 m distance, let us evaluate what would happen if this beam were to fall on an open eye of an observer. The focusing action of the eye lens creates an image on the retina that is much smaller than the opening of the pupil. If the light has wavelength λ, the eye has pupillary diameter d_p with a lens of focal length l_f, the diameter of the image on the retina, d_r, is given by

$$d_r = \frac{2.44\,\lambda\,l_f}{d_p}.$$

A reasonable value for l_f is 2.2 cm and d_p is 7 mm. For a He–Ne laser at 610 nm,

$$d_r = \frac{2.44 \times 6.1x10^{-5}\,\text{cm} \times 2.2\,\text{cm}}{0.7\,\text{cm}} = 4.7x10^{-4}\,\text{cm}$$

and the power level experienced by the retina is:

$$10.4 \frac{\text{mW}}{\text{cm}^2}\left(\frac{0.7\,\text{cm}}{4.7x10^{-4}\,\text{cm}}\right)^2 = 2.3x10^7 \frac{\text{mW}}{\text{cm}^2}\,!\,!$$

For reference, the intensity of sunlight at the Earth's surface is about 100 mW/cm^2, and the intensity on the retina is thus about 10^4 mW/cm^2 if the sun is viewed directly.

Exposure Guidelines
Lasers are grouped into four broad categories (I to IV) based on their power level (which relates directly to potential hazard level) (Figures 13.8 through 13.10).

- *Class I lasers* do not emit radiation at hazard levels known to cause biological damage (cw 0.4 μW in the visible wavelength range), or may have the laser located in a place where no exposure can occur (e.g., CD and DVD players). Class I laser products are generally exempt from enforced radiation hazard controls. *Class I.A laser* have a special designation that is based on a 1000-second exposure. This category applies only to lasers that are "not

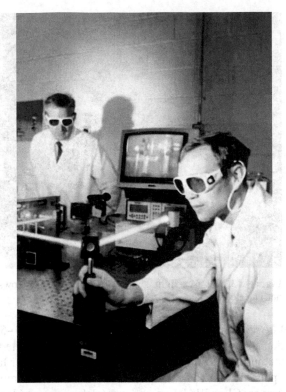

Figure 13.8 Laboratory laser source. (From www. wikipedia.com.)

intended for viewing" such as supermarket laser scanners. The upper power limit of a Class I.A laser is 4.0 mW.

- *Class II lasers* emit low-power visible radiation at levels above Class I levels but below 1 mW. Limited controls are exerted over these devices, and it is assumed that natural human aversion reactions to bright light will protect the eye from damage, as exposure times will be short. This may include low-powered laser pointers, data and telecommunications equipment, and some laboratory lasers. These products are not capable of causing permanent eye injuries, but may still cause glare and temporary flash blindness.

Figure 13.9 Laser pointer, for oral presentations. (From www.quirkle.com.)

Figure 13.10 Laser light show. (From www.lasershows. net.) (see color plate)

- *Class IIIA lasers* are intermediate power lasers (cw perhaps 1–5 mW). These are hazardous only if intrabeam viewing occurs. Some limited controls are usually recommended. *Class IIIB lasers* are moderate power lasers (cw: 5–500 mW, pulsed: ~10 J/cm^2). Specific controls are recommended. Most laser pointers are Class IIIa lasers, and many laboratory and low power laser display systems are Class IIIb.
- *Class IV lasers* are high-power lasers (cw: 500 mW, pulsed: 10 J/cm^2), are hazardous to view under any condition (directly or reflected), and are a potential fire and skin hazard. Significant controls are required.

Thresholds for Radiation Damage

Damage to the Eye
Damage can occur to the cornea, lens, or retina (see Table 13.2). For visible light lasers, the principal hazard is to the retina. The main factor involved in determining thresholds for damage is the rate at which heat energy can be removed.

Damage to the Skin
Damage may be due to burns and necrosis (see Table 13.3). Lasers denature proteins, vaporizing water and carbonizing tissue. UV lasers are also thought to be able to cause cancer, as we know that UV light itself is carcinogenic.

Laser exposure limits are given as *Maximum Permissible Exposure* (MPE) values. The numerical values were specified in an American National

Table 13.2 Retinal damage thresholds.

Type	Wavelength	Pulse	Level
Continuous wave	White light		6 W/cm^2
Pulsed	694 nm	200 ms	0.85 J/cm^2
Q-switched pulse	694 nm	30 ns	0.07 J/cm^2

Table 13.3 Skin damage thresholds.

Type	Wavelength	Exposure Time	Area (cm^2)	MRD* (J/cm^2)
Ruby, normal pulse	694 nm	0.2 ms	0.003	14–20
Argon	500 nm	6 s	0.095	13–17
CO$_2$	1060 nm	4–6 s	1	4–6
Ruby, Q-switched	694 nm	10–12 ns	0.33–1	0.5–1.5

*Minimal Reactive Dose.

Standards Institute (ANSI) document entitled American National Standard for Safe Use of Lasers.[5] The values depend on the laser wavelength and the duration of the exposure, and are shown in Tables 13.4 through 13.6.

Safety Procedures

MPE calculations are thus based on the type of laser involved and the exposure duration and type. For example, for a 694 nm ruby laser with a single pulse of 10 μs, the MPE is given simply from Table 13.4 as 5×10^{-7} J/cm^2. If the pulse is instead 15 ms, the MPE is given as $1.8\ t^{3/4} \times 10^{-3}$ J/cm^2 = 1.8 $(0.015)^{3/4} \times 10^{-3}$ J/cm^2 = 7.7×10^{-5} J/cm^2. When lasers are repetitively pulsed, other MPE calculations come into play, as thermal injury from radiation delivered in narrow repeated pulses is more damaging than radiation delivered continuously. If laser light is reflected, the reflected light will be transmitted at a different angle and at reduced intensity (as only a fraction of the incident light is reflected). Further calculational procedures are involved, but are not covered in detail here. The purpose of this chapter is basically to introduce the concepts and quantities involved. The reader is referred to more comprehensive texts on this subject for more in-depth application of this specialized field.[6]

Most of laser safety with the higher-power lasers that can cause biological damage involves keeping laser sources locked so that only authorized (and supposedly educated and experienced) users can manipulate them, and by posting appropriate warning signs (Figure 13.11). One can also use protective goggles when using lasers (see Figure 13.8). These goggles will be rated with protection factors, given as optical density, which is the logarithm of the ratio

Table 13.4 MPE limits for direct ocular exposures.[5]

λ, nm	Exposure Time (s)	MPE
400–700	10^{-9} to 1.8×10^{-5}	5×10^{-7} J/cm^2
400–700	1.8×10^{-5} to 10	$1.8\ t^{3/4} \times 10^{-3}$ J/cm^2
400–550	10 to 10^4	10×10^{-3} J/cm^2
550–700	10 to T_1	$1.8\ t^{3/4} \times 10^{-3}$ J/cm^2
550–700	T_1 to 10^4	$10 \cdot C_B \times 10^{-3}$ J/cm^2
400–700	10^4 to 3×10^4	$C_B \times 10^{-6}$ W/cm^2

$C_B = 1$ for $\lambda = 400$–550 nm, $C_B = 10^{[0.015(\lambda - 550)]}$ for $\lambda = 550$–700 nm, $T_1 = 10 \times 10^{[0.02(\lambda - 550)]}$ seconds for $\lambda = 550$–700 nm.

Table 13.5 MPE limits for particular lasers.[5]

Laser type	Wavelength (μm)	MPE Level (W/cm^2)			
		0.25 s	10 s	600 s	30,000 s
CO$_2$ (CW)	10.6	—	100.0×10^{-3}	—	100.0×10^{-3}
Nd: YAG (CW)	1.33	—	5.1×10^{-3}	—	1.6×10^{-3}
Nd: YAG (CW)	1.064	—	5.1×10^{-3}	—	1.6×10^{-3}
Nd: YAG (Q-switched)	1.064	—	17.0×10^{-6}	—	2.3×10^{-6}
GaAs (Diode/CW)	0.840	—	1.9×10^{-3}	—	610.0×10^{-6}
HeNe (CW)	0.633	2.5×10^{-3}	—	293.0×10^{-6}	17.6×10^{-6}
Krypton (CW)	0.647	2.5×10^{-3}	—	364.0×10^{-6}	28.5×10^{-6}
	0.568	31.0×10^{-6}	—	2.5×10^{-3}	18.6×10^{-6}
	0.530	16.7×10^{-6}	—	2.5×10^{-3}	1.0×10^{-6}
Argon (CW)	0.514	2.5×10^{-3}	—	16.7×10^{-6}	1.0×10^{-6}
XeFl (Excimer/CW)	0.351	—	—	—	33.3×10^{-6}
XeCl (Excimer/CW)	0.308	—	—	—	1.3×10^{-6}

Table 13.6 MPE limits for skin exposures.[5]

λ, nm	Exposure Time (s)	MPE
315–400	10^{-9} to 10	$0.56\,t^{1/4}$ J/cm^2
315–400	10 to 1000	1 J/cm^2
315–400	10^3 to 3×10^4	$0.001\,t$ J/cm^2
400–1400	10^{-9} to 10^{-7}	$0.002\,C_A$ J/cm^2
400–1400	10^{-7} to 10	$1.1\,C_A t^{1/4}$ J/cm^2
400–1400	10 to 3×10^4	$0.02\,C_A$ W/cm^2
1400–10^6	10^{-9} to 10^{-7}	0.01 J/cm^2
1400–10^6	10^{-7} to 10	$0.56\,t^{1/4}$ J/cm^2
1400–10^6	>10	$0.1\,t$ J/cm^2

$C_A = 1$ for $\lambda =< 700$ nm, $C_A = 10^{[0.002(\lambda-700)]}$ for $\lambda = 700$–1049 nm, $C_A = 5$ for $\lambda = 1050$–1400 nm,

of the exposure experienced to the MPE level:

$$OD = \log \frac{E \ (\text{or } H)}{\text{MPE}}.$$

E and H are the radiance (W/cm^2) and radiant exposure (J/cm^2). Starting with a particular laser and viewing distance, given the beam divergence, one can calculate the expected radiant exposure and calculate the OD rating needed for working with such a laser. It is important to remember that the OD rating applies only to a narrow range of radiation wavelengths and may provide little or no protection at other wavelengths. It is also important not to use goggles that make it difficult to see the work being done, so that there is no temptation to peek around the goggles or simply not use them. There are also many other physical hazards of working with lasers, including electrical, chemical, and fire hazards, which all must be borne in mind to work safely with a laser system.

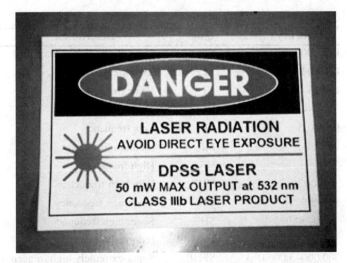

Figure 13.11 Sample laser warning sign. (From www.laserharp.manuel-schulz.com.)

13.3   Radiofrequency Radiation, and Microwave Sources

Theory of Operation and Basic Principles

Radiofrequency (RF) radiation is defined as electromagnetic radiation between 3 kHz and 300 GHz, with microwave radiation comprising the region between 300 MHz and 300 GHz.[7] Many different devices use RF radiation, with some of the most important applications being in communication and industrial processing (mostly heating processes). All members of the public are affected by the presence of these technologies, in radio and television broadcasting, satellite communications, mobile telephone applications, and home microwave ovens. In addition, shortwave, police, fire, and taxi communication radios; medical diathermy devices; radar; and radionavigational devices will be familiar to broad segments of the population. Specialized industrial applications, such as RF heaters, dryers, wood laminators, and plasma etchers will be encountered by a limited number of workers on a regular basis. Common radiofrequency band designations, as given in NCRP 119[7], are shown in Table 13.7.

Mechanisms of Radiation Interaction

Microwaves are reflected by metals, but transmitted by glass and plastics and absorbed by polar materials (such as water). The principal means of interaction with human tissue (which is mostly water) that is well understood is molecular agitation, leading to localized heating. This is the main principle behind the routine operation of microwave heating devices (Figures 13.12 through 13.14). Microwaves passing through matter lose energy to the medium by inducing ionic currents in the medium and causing vibration of polar molecules in the medium. The microwaves lose energy continuously, in an exponential fashion as they penetrate matter, having an effective penetration depth of a few cm in water or tissue.

In a conventional oven, the chamber must be heated to the desired temperature, and then the food inside is slowly heated from the outside, via

Table 13.7 RF frequency bands.

Frequency (MHz)	Band	Description
0–0.00003	SELF	Sub-extremely low frequency
0.00003–0.0003	ELF	Extremely low frequency
0.0003–0.003	VF	Voice frequency
0.003–0.03	VLF	Very low frequency
0.03–0.3	LF	Low frequency
0.3–3.0	MF	Medium frequency
3–30	HF	High frequency
30–300	VHF	Very high frequency
300–3,000	UHF	Ultra high frequency
3,000–30,000	SHF	Super high frequency
30,000–300,000	EHF	Extremely high frequency
300,000–3,000,000	SEHF	Supra-extremely high frequency

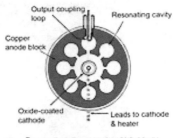

Figure 13.12 Magnetron images and cross-sectional operational view. (From www. wikipedia.com.)

convection and radiation. Microwaves are generated by some device, typically a magnetron. In a magnetron, electrons flow from a cathode to an anode. The electron flow causes a magnetic field to be set up. On their way to the anode, they encounter the magnetic fields from two permanent magnets whose magnetic fields are oriented parallel to the cathode. The electrons are thus deflected into circular orbits as they move towards the anode. The whirling pattern of electrons is similar to spokes moving around a wheel and the rotating space charge sets up an alternating current flow in the "resonant cavity" of the magnetron. This creates a high-frequency electromagnetic field in the cavity, and some of this field is extracted with an antenna into a waveguide (a metal tube, often of rectangular cross-section) and into the cooking chamber (of a microwave oven) or an antenna (for communications use).

Microwaves transmitted over large distances are generally sent from and received by parabolic reflectors (see Figure 13.13). Significant beam divergence does occur with such devices, and beamwidths are given as the width of the beam at the half-power points (measured in degrees). Measurements are made within two defined fields:

Figure 13.13 Microwave communication tower. (From www.wikipedia.com.)

- *Near Field*: At distances less than $\lambda/2\pi$, here the electric and magnetic fields are not perpendicular to each other, as part of the field energy (the reactive energy) is stored, recovered, and re-emitted during successive oscillations.
- *Far Field*: At distances greater than $\lambda/2\pi$, the electric and magnetic fields are perpendicular to each other, and meaningful measurements of power density are possible.

For measurement and hazard assessment purposes, we define the beginning of the far-field region as

$$R_{ff} = \frac{D^2}{2.83\,\lambda}$$

where D is the diameter of the antenna aperture and λ is the wavelength of the radiation. After we reach the far-field distance, the beam begins to diverge substantially, taking on a conical shape.

Exposure Guidelines

The transfer of energy from electric and magnetic fields to charged particles in a medium is given as a *Specific Absorption Rate* (SAR). The SAR is the time rate of change of energy deposition at a point, per unit mass of the medium at that point. To establish the intensity of a radiation field at a point, one may measure the electric field, the magnetic field, or the SAR. Electric and magnetic

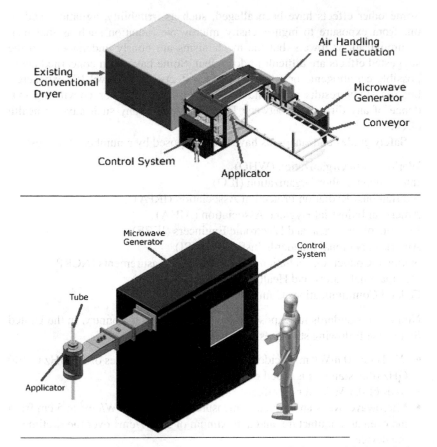

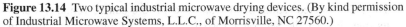

Figure 13.14 Two typical industrial microwave drying devices. (By kind permission of Industrial Microwave Systems, L.L.C., of Morrisville, NC 27560.)

field measurements may be made with an antenna coupled to a detector that reads the signal. SAR measurements are more difficult to make in the field, and are usually performed only in laboratory environments. One may place electric field probes at a few locations inside experimental animals, or measure local heating changes. Field measurements of RF radiation are made with crystal diodes (which produce an output voltage proportional to the power measured) or thermistors (devices whose resistance changes with temperature changes) read by portable survey meters.

In humans, documented damage from microwaves has been demonstrated in the eyes and testicles (both are poorly perfused tissues and do not efficiently dissipate energy absorbed at a rate greater than about 10–15 mW/cm^2).

- *Eyes*: The lens of the eye is avascular and encapsulated. Microwave radiation can induce cataracts. These cataracts will occur on the posterior surface of the lens capsule, and so are distinguishable from senile cataracts, which form on the anterior surface of the lens.
- *Testicles*: Normally they are at a temperature about 2°C lower than core body temperature (37°C). Elevation to even 37°C will reduce spermatogenesis (reversibly).

Some other effects have been alleged, such as irritability, agitation, and so on, from exposure to high-intensity microwave radiation such as that near communications towers, but the mechanisms are poorly understood and the suggested effects are difficult to document. Some have been concerned about possible carcinogenic or other genotoxic effects of low-level RF exposures, but the study results so far have shown no consistent or convincing evidence of any direct link between RF exposure and any such adverse health effects.[8]

Safety guides and standards have been proposed by a number of bodies:

World Health Organization (WHO)
International Labor Organization (ILO)
International Radiation Protection Association (IRPA)
American Industrial Hygiene Association (AIHA)
Institute of Electrical and Electronic Engineers (IEEE)
American National Standards Institute (ANSI)
National Council on Radiation Protection and Measurements (NCRP)
Occupational Safety and Health Organization (OSHA)
Federal Communications Commission (FCC)

Numerical standards for exposure vary from country to country. In the United States, the following standards have been developed.

- Workers: 10 mW/cm^2 incident power density, frequencies of 10 MHz to 100 GHz (this standard is based on thermal effects only).
- SAR of 0.4 W/kg for all effects.
- Microwave ovens are limited to measured levels of 1 mW/cm^2 at 5 cm from the oven at manufacture, and a maximum of 5 mW/cm^2 over the lifetime of the device.
- The FCC's stated requirements for RF exposure can be found in Part 1 of its rules in 47 CFR Section 1.1307(b). Exposure limits are specified in 47 CFR Section 1.1310 specified for frequency, field strength, power density, and averaging time for measurements.

Safety Procedures

The basic principles of time, distance, and shielding apply as well.

- Shielding against magnetic fields may be accomplished using ferromagnetic materials (iron or alloys of high permeability).
- Shielding of microwaves may be done with wire mesh shields.
- Attenuation factors and formulas are discussed in specialized texts, but are not treated here.

13.4 EMF

Humans have always been exposed to very low-level time-varying, extremely low-frequency electromagnetic fields (EMF) from natural sources, including the earth's own fields, sources from the sun, and even from other human beings (very small electrical currents exist in the human body due to the electric signals relayed by nerves, biochemical reactions involved in various body processes, and the electrical activity of the heart). Residential and industrial uses of electricity have increased mankind's exposure to EMF dramatically over the last 120 years. There is no known biological mechanism by which EMF can raise an individual's risk of cancer, so even an exposure parameter to measure it is not established. Quoting directly from a recent review of epidemiologic studies on health effects possibly related to EMF exposures:[9]

(a) The quality of epidemiologic studies on this topic has improved over time and several of the recent studies on childhood leukemia and on cancer associated with occupational exposure are close to the limit of what can realistically be achieved in terms of size of study and methodological rigor.

(b) Exposure assessment is a particular difficulty of EMF epidemiology, in several respects:

   (i) The exposure is imperceptible, ubiquitous, has multiple sources, and can vary greatly over time and short distances.

   (ii) The exposure period of relevance is before the date at which measurements can realistically be obtained and of unknown duration and induction period.

   (iii) The appropriate exposure metric is not known and there are no biological data from which to impute it.

(c) In the absence of experimental evidence and given the methodological uncertainties in the epidemiologic literature, there is no chronic disease for which an etiological relation to EMF can be regarded as established.

(d) There has been a large body of high quality data for childhood cancer, and also for adult leukemia and brain tumor in relation to occupational exposure. Among all the outcomes evaluated in epidemiologic studies of EMF, childhood leukemia in relation to postnatal exposures above 0.4 μT is the one for which there is most evidence of an association. The relative risk has been estimated at 2.0 (95% confidence limit: 1.27–3.13) in a large pooled analysis. This is unlikely to be due to chance but, may be, in part, due to bias. This is difficult to interpret in the absence of a known mechanism or reproducible experimental support. In the large pooled analysis only 0.8% of all children were exposed above 0.4 μT. Further studies need to be designed to test specific hypotheses such as aspects of selection bias or exposure. On the basis of epidemiologic findings, evidence shows an association of amyotrophic lateral sclerosis with occupational EMF exposure although confounding is a potential explanation. Breast cancer, cardiovascular disease, and suicide and depression remain unresolved.

13.5 Magnetic Resonance Imaging (MRI)

Theory of Operation and Basic Principles

Magnetic resonance imaging (MRI) is a medical imaging technique that produces very high quality anatomic images of humans and animals. MRI is based on a spectroscopic technique that is exploited to obtain fundamental chemical and physical information at the molecular level in cells. MRI systems use strong magnets that are rated using a unit of measure known as the tesla (T; one may also use the gauss, where 1 tesla = 10,000 gauss). MRI magnets in common imaging systems are in the 0.5–4.0 T range. The earth, for comparison, has about a 0.5 gauss magnetic field. The fields are typically generated using *superconducting* magnets, which have coils of wire through which a current of electricity is passed creating the magnetic field and which are kept at low, generally liquid nitrogen, temperatures.

Any atom with a magnetic moment can be used for imaging; the vast majority of work with MRI focuses on the hydrogen atom. Hydrogen is ideal for MR imaging because its nucleus has a single proton and a reasonably large magnetic moment. Thus when hydrogen atoms are in a strong magnetic field, they will tend to line up with the direction of the magnetic field. The MRI scanner applies a pulsed RF field oriented towards the area of the body to be imaged. Protons in that area absorb the pulse energy and are oriented (precessed) in the direction of the field. The RF pulses are applied through specific coils designed to be ideal for particular body parts such as the abdomen, head, knees, shoulders, wrists, and so on. Then gradient magnets are turned on and off very rapidly in a specific manner to alter the main magnetic field in the region to be imaged.

When the RF pulse is turned off, the protons return to their previous natural alignment, at a particular rate, and release their excess stored energy. In doing so, they give off an energy signal that the coil reads and sends to a computer system for processing. Relaxation times are specified as the time for the magnetized atom to (exponentially) return to 63% of its ground state (T1). There are other relaxation times related to higher-order effects that are also characterized (T2, T2*). *Functional* MRI (fMRI) is a technique that detects subtle increases in blood flow in regions of the brain, via study of T2* changes related to changes in deoxyhemoglobin levels. Magnetic Resonance Spectroscopy (MRS) exploits the principle that nuclei (such as ^{13}C, ^{31}P) in different chemical structures may have different characteristic resonance patterns, depending on the structure of adjacent molecules and their interactions. MRS can detect the chemical composition of individual tissues and thus may allow visualization of molecular processes and not only structural and anatomic information, as with MRI (Figures 13.15 and 13.16).

Mechanisms of Radiation Interaction and Biological Effects

Strong static magnetic fields cause translation and orientation of biological molecules, exert electrodynamic forces on moving electrolytes (e.g., ions in blood vessels), and affect the electron spin states of various chemical reaction intermediates.[10] No short-term adverse biological effects have been reported in human subjects exposed to fields up to 8T, but longer-term studies are desirable for populations of medical patients and hospital workers exposed to these fields over time. MR fields are not static, however, they vary rapidly.

Concerns for varying magnetic fields include cardiac fibrillation and peripheral nerve stimulation. The threshold for cardiac stimulation appears to be well above typical MR field strengths used, except if very long pulse durations occur, but the trend in clinical MRI is towards shorter and shorter gradient pulses, in order to reduce imaging time. Uncomfortable nerve stimulation is thought to be controllable by maintaining the "maximum exposure level for time-varying magnetic fields... to a *dB/dt* of 80% of the median perception threshold for normal operation, and 100% of the median for controlled

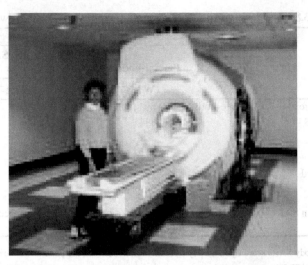

Figure 13.15 MRI scanner. (From www.ngdc.com.)

operation"[10] (*dB/dt* is the time rate of change of the magnetic field). No effects on pregnant subjects or their unborn children have been documented, but concerns about this sensitive population cause the decision thresholds for treatment to be generally higher for these subjects.

Safety Procedures

In addition, SARs from the RF fields around these devices is of concern. SAR levels were defined in the ICINRP document on MR Procedures.[10] See Table 13.8.

The intense magnetic fields around MR scanners create other hazards. Perhaps the mildest of them is that all of one's electronic banking and credit cards

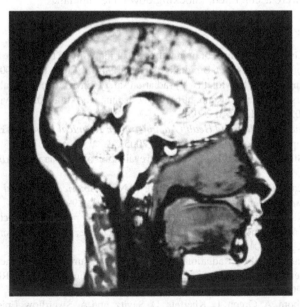

Figure 13.16 MR brain image. (From http://www.mind-disorders.com/Kau-Nu/Magnetic-resonance-imaging.html.)

Table 13.8 SAR levels valid at environmental temperatures below 24°C.

	Whole body SAR (W kg^{-1})	Partial-body SAR		Local SAR (averaged over 10 g tissue)		
		(W kg^{-1})		(W kg^{-1})		
Body region→ Operating mode↓	Whole-body	Any except head	Head	Head	Trunk	Extremities
Normal	2	2–10^a	3	10^b	10	20
Controlled	4	4–10^a	3	10^b	10	20
Restricted	>4	>(4–10)^a	>3	10^b	>10	>20
Short term SAR	The SAR limit over any 10 s period should not exceed 3 times the corresponding average SAR limit.					

[a] Partial-body SARs scale dynamically with the ratio r between the patient mass exposed and the total patient mass:

−normal operating mode: $SAR = (10 - 8 \cdot r)$ W kg^{-1}

−controlled operating mode: $SAR = (10 - 6 \cdot r)$ W kg^{-1}

The exposed patient mass and the actual SAR levels are calculated by the SAR monitor implemented in the MR system for each sequence and compared to the SAR limits.

[b] In cases where the eye is in the field of a small local coil used for RF transmission, care should be taken to ensure that the temperature rise is limited to 1°C.

can be instantly erased! More seriously, though, ferromagnetic objects can be accelerated in the intense fields near the magnet creating minor to major hazards for patients and personnel. A six-year-old boy undergoing an MRI scan was killed when an oxygen tank was pulled towards the MR scanner during the exam, patients with implanted cardiac pacemakers have died during MR exams, intravenous (IV) bottle support poles have been set in motion, cutting patients, scissors have become airborne and hurt people, and the steel tines of a forklift were accelerated, knocking down a technician.[11]

Endnotes

1. W. B. Grant, An estimate of premature cancer mortality in the US due to inadequate doses of solar ultraviolet-B radiation, *Cancer* **94** (6), 1867–1875 (2002).

2. J. Myung Chul, Ultraviolet (UV) radiation safety. Environmental health and safety, (University of Nevada Reno, http://www.ehs.unr.edu/Portals/0/Ultraviolet radiation safety 050707.pdf, April 2005).

3. *Occupational Safety and Health Administration Technical Manual*, TED 01-00-015 (US Dept. of Labor, Washington, DC (2005).

4. D. H. Sliney and W. A. Palmisano, The evaluation of laser hazards. *AIHA J.* **20**, 425 (1968).

5. American National Standards Institute (ANSI), *American National Standard for Safe Use of Lasers* (ANSI Z136.1-1993, New York, 1993).

6. S. E. Rademacher, Base level management of laser radiation protection program. (Occupational and Environmental Health Directorate, Brooks Air Force Base, TX, AL-TR-1991-0012, 1991).

7. National Council on Radiation Protection and Measurements, A practical guide to the determination of human exposure to radiofrequency fields. NCRP Report No. 119, (NCRP, Bethesda, MD, 1993).

8. A. Ahlbom, A. Green, L. Kheifets, D. Savitz, and A. Swerdlow, Epidemiology of health effects of radiofrequency exposure. ICNIRP (International Commission for

Non-Ionizing Radiation Protection) Standing Committee on Epidemiology. *Environmental Health Perspectives*, **112** (17),1741–1754 (2004).

9. A. Ahlbom, E. Cardis, A. Green, M. Linet, D. Savitz, and A. Swerdlow,. Review of the epidemiologic literature on EMF and health. ICNIRP (International Commission for Non-Ionizing Radiation Protection) Standing Committee on Epidemiology. *Environmental Health Perspectives*, **109** (Supplement 6), 911–933 (2001).

10. R. Matthes, Medical magnetic resonance (MR) procedures: Protection of patients. The International Commission on Non-Ionizing Radiation Protection. *Health Phys.* **87** (2),197–216 (2004).

11. www.fda.gov/cdrh.

Index